Basic Concepts of Plant Biotechnology (with MCQs)

By

Vijay Prakash
(Ph.D. Molecular Biology and Biotechnology)
Jawaharlal Nehru Krishi Vishwavidhyala,
Jabalpur (Madhya Pradesh)

Niraj Tripathi
(Ph.D. Biotechnology)
Awadhesh Pratap Singh University
Rewa (M.P.)

An imprint of Scientific Publishers (India)

Published by:
Competition Tutor
5-B, New Pali Road, P.O. Box 91
Jodhpur 342 001 (India)

E-mail: competitiontutor@gmail.com
Website: www.competitiontutor.co.in

An imprint of Scientific Publishers (India)

ISBN: 978-93-86652-14-0

Printed in India

Dr. V.S. Tomar
Vice Chanceller
Jawaharlal Nehru Krishi
Vishwa Vidhyalaya, Jabalpur

Foreword

Biotechnology is now beginning to be viewed as an informational science based on molecular mechanism of cell. In a simplistic sense, genetic information of a cell transmitting from parents to progeny through different molecular processes viz., replication, transcription and translation collectively called central dogma. This book evolved from a course in molecular biology that is a part of biotechnology and life sciences. This book regarding molecular biology and crop improvement is a large knowledge pack-capsule. This text will provide a quick tour of this subject and also test the grasping power of the reader and student which are preparing for many competition exams, CSIR-NET, DBT-JRF, ARS-NET, ICMR etc. as each chapter is following by objective question and answers. Hope this collection will help to move forward in the subject under target.

Place: Jabalpur **V.S. Tomar**

Preface

The aim of the book entitled **"Basic Concepts of Plant Biotechnology (with MCQs)"** is to cover a wide area of Molecular Biology especially useful for the all graduate and post-graduate students of Agricultural Biotechnology and life sciences to quickly grasp the facts comprehensively. This book contains almost 3000 multiple choice Questions. This book will be proved very useful and increase the knowledge about molecular biology to all students of Life Sciences and Biotechnology who involve in the preparation of all competitive exams like ARS, ICAR-NET, CSIR-NET, DBT-JRF, ICMR-JRF, PSC etc. As with this first edition, we have been careful to include only the information that we believe is essential for good student understanding of the subject and for rapid revision when exams appear on the horizon. Do use the book not only to get to grips with the subject but also as aready source of elusive information. We hope and believe that you will find it will be very useful for students.

We are deeply obligated to Dr. Sharad Tiwari, Professor and Director, Biotechnology Centre for this passionate and unreserved support, guidance, encouragement and patience which contributed to the success of this book. We are extremely thankful to Dr. Navinder Saini, Senior Scientist, Division of Genetics, IARI, New Delhi for his valuable and timely suggestions and your blessing during the preparation of book. We are extremely thankful to Dr. Anita Babbar, Professor, Department of Plant Breeding and Genetics, Dr. Iti Gontia-Mishra, Post-doctoral DBT Visiting Fellow, Biotechnology Centre, JNKVV, Jabalpur to share valuable experiences and support during this book. We thank to friends and colleagues, Dr. Swapnil Sapre (Ph.D. Scholar) and juniors Dr. Vishwa Vijay Thakur (Ph.D. Scholar), Dr. Satish Kachare (Ph.D. Scholar), Mr. Jagat Kumar (SRF, IARI, New Delhi), Dr. Arpita Shrivastava (Ph.D. Scholar), Akanksha Tiwari (Ph.D. Scholar), Ritu Sharma (Ph.D. Scholar), Sumana Sikadar (Ph.D. Scholar), Ruchika Mishra (Ph.D. Scholar), Nishtha Singh, Monika Mishra, for their utmost cooperation rendered at the times of need during the preparation of this book. Finally, we owe entire of this achievement to our parents who have tremendously encouraged us throughout our academic career as well as whole life.

– Vijay Prakash

– Niraj Tripathi

– Vinay Prakash

– Niraj Tripathi

Contents

Synopsis

The first edition of the book entitled **"Basic Concepts of Plant Biotechnology (with MCQs)"** has been publishing when the recombinant DNA and sequencing of human and many plant genomes have been completed. It prepared on the demand of students that have poor knowledge of molecular biology, especially central dogma, is process of transmission of genetic information in biological system. Therefore, we prepared and compiled many questions on central dogma inspired from different biotechnological and biochemical resource material. We have made an effort to include each simple and difficult question on central dogma as could be possible. This book contains almost 3000 multiple choice questions as well as fill in the blanks with answers covering all aspects of Molecular Biology systems of Prokaryotes and Eukaryotes. Inwriting the first edition, the aim is to provide all simple and difficult questions for weak students in plant Molecular Biology that have no more knowledge and have more problems in solving the questions. Therefore, in this book we included questions belongs to all basic concept of Molecular Biology which will provide strong knowledge to students preparing for competitive exams of life science like CSIR-NET, DBT-JRF, ICMR-JRF, ICAR-NET, ARS, PSC, graduate and post-graduate exams. In the first chapter, objective question on different biomolecules like carbohydrates, lipids, and proteins have been included to understand basic concepts about them. Second chapter includes fundamental questions for nucleic acid structure and functions have been presented that gives basic information about nucleic acid and its role in biological systems. Some questions are based on figures that illustrated key approaches for identifying possible structure, function and metabolism of nucleotides (purine and pyrimidine) and proteins. In third chapter, we summarized structure and functions of genes and chromosomes including their numbers and genome size in different organisms. Forth chapter describes DNA replication and its role in central dogma of molecular biology. Many questions on structures and functions of various enzymes, different factors and their active sites and role in replication provided in this chapter. In fifth chapter, we included causes of DNA damage and mutations, its processes and different DNA repair systems like base excision repair, SOS response etc. Sixth section provides a lot of questions on transcription and RNA processing including different transcription factors, their binding sites and role, RNA editing and intron splicing in prokaryotic and eukaryotic systems. Seventh chapter is based on process of protein synthesis, genetic codes, tRNA structure, amino acyl tRNA, translation in prokaryotes, translation in eukaryotes, protein targeting and protein glycosylation. Eighth chapter provides information about different gene expression systems, operon and tryptophan regulation, promoter, enhancers, open reading frames (ORFs), gene controlling, oncogenes and disease causing factors. We provided two special ninth and tenth chapters in last 'Techniques and tools in molecular biology' and 'Biotechnology in Crop Improvement" for agricultural biotechnology students in which all questions are included regarding to molecular and biotechnological techniques, methods and processes like ribozymes, molecular farming, MAS, QTL, RNAi, antisense RNA technology, recombinant DNA technology, microarray, sequencing, molecular markers etc., help in the improvement of traits, control of abiotic and biotic stress factors that negatively affect crop growth and development.

1

Biomolecules: Structures and Functions

All living organisms originated on the earth and are made up of thousands of chemical elements and compunds. These have also vast diversity in their structures and functions of chemical composition. The small chemical elements and compounds form different types of molecules which compose of all forms of life. Carbon, hydrogen and oxygen found abundantly in living system. Biomolecules are organic molecules especially macromolecules including proteins, carbohydrates, lipids and nucleic acids. Biomolecules consist mainly of carbon and hydrogen with nitrogen, oxygen, sulphur, and phosphorus.

There are four major classes of biomolecules:

- Carbohydrates • Lipids • Proteins • Nucleic acids

Carbohydrates

Carbohydrates are considered as good source of energy and also called as sugars. Carbohydrates are organic compounds that contain hydrogen and oxygen in the ratio 2:1. Carbohydrates are considered as hydrates of carbon. Carbohydrates are polyhydroxy aldehydes or ketones, which are substances with hydroxy (OH) and aldehyde (CHO) or ketone (C=O) functional groups. Carbohydrates are classified into three categories, monosachharides, disachharides and polysachharides. Monosaccharides are simple sugars that are composed of 3-7 carbon atoms containing a free aldehyde or ketone group. These groups act as reducing agents, therefore these are known as reducing sugars. Disaccharides are made of two monosaccharide units which linked to each other with glycosidic bonds. Monosaccharides and disaccharides are sweet, crystalline and water soluble sugars. Polysaccharides are polymers of monosaccharides. They are sweetless and complex carbohydrates that are insoluble in water and not found in crystalline form.

Classification of Carbohydrates. Cellulose in polysachhardies	
Simple Carbohydrates	
Monosaccharides or simple sugars (1 sugar unit)	Fructose, galactose, glucose, mannose, tagatose, xylose.
Disaccharides (2 sugar units)	Isomaltose, isomaltulose, lactose, maltose, sucrose, trehalose, trehalulose.
Sugar alcohols (polyols)	Erythritol, glycerol, hydrogenated starch hydrolysates, isomalt, lactitol, maltitol, mannitol, sorbitol, xylitol.
Complex Carbohydrates	
Oligosaccharides (3-10 sugar units)	Fructooligosaccharides (FOS), galactooligosaccharides (GOS), human milk oligosaccharides (HMO), isomalto-oligosaccharides (IMO), maltotriose, mannan oligosaccharides (MOS), raffinose, stachyose and verbascose.
Polysaccharides (>10 sugar units)	Chondroitin sulfate, dextrins, glycosaminoglycans, starch, cellulose.

Dietary fiber	Soluble fiber: *acacia* (Arabic gum), agar-agar, algin, arabynoxylan, beta-glucan, carrageenan.

Enzymes involved in carbohydrate degradation and metabolism

Enzyme	Function.
Salivary alpha-amylase	Splits cooked starch and glycogen (at alpha 1-4 bonds) to dextrins, isomaltose and maltose.
Pancreatic alpha-amylase	Splits cooked and uncooked starch and glycogen (at alpha 1-4 bonds) to dextrins, isomaltose and maltose.
Glucoamylase	Splits starch (at alpha 1-6 bonds) to smaller carbohydrates and cleaves glucose molecules, one by one, from starch (at alpha 1-4 bonds).
Sucrase	Splits sucrose to glucose and fructose.
Lactase	Splits lactose to glucose and galactose.
Maltase (alpha-glucosidase)	Splits maltose to two glucose molecules.
Isomaltase (dextrinase)	Splits isomaltose to two glucose molecules; splits isomaltulose and trehalulose to glucose and fructose; splits starch (at alpha 1-6 bonds) to dextrins.
Trehalase	Splits trehalose to two glucose molecules.

List of major disachharides

Disaccharide	Monosachharide units	Bond
Sucrose	Glucose + Fructose	$\alpha(1\rightarrow2)\beta$
Lactose (*milk sugar*)	Galactose + Glucose	$\beta(1\rightarrow4)$
Maltose (*malt sugar*)	Glucose + Glucose	$\alpha(1\rightarrow4)$
Trehalose	Glucose + Glucose	$\alpha(1\rightarrow1)\alpha$
Cellobiose	Glucose + Glucose	$\beta(1\rightarrow4)$

List of major polysachharides

Polysaccharide	Composition	Functions
Starch	Amylose (straight chain of glucose) and amylopectin (branched chain of glucose)	Storage of reserve food
Glycogen	Polymer of glucose	Storage of reserve food
Callose	Polymer of glucose	Formed often as a response to wounds
Insulin	Polymer of fructose	Storage of reserve food
Cellulose	Polymer of glucose	Cell wall matrix
Pectin	Polymer of galactose and its derivatives	Cell wall matrix
Hemi cellulose	Polymer of pentoses and sugar acids	Cell wall matrix
Lignin	Polymer of glucose	Cell wall matrix
Chitin	Polymer of acetylglucosamine (Derivative of glucose	Polymer of acetylglucosamine (a derivative of glucose)
Murein	Hetero-polymer of linear glycan strands of alternating β1,4-linked N-acetylglucosamine and N-acetylmuramic acid residues	Structural protection
Hyaluronic acid	Polymer of sugar acids	Ground substance, protection

Heparin	Closely related to chrondroitin	Anticoagulant
Gums and mucilages	Polymers of sugars and sugar acids	Retain water in dry seasons

Lipids

Lipids are the main class of biomolecules which are unable to dissolve in water. Lipids have hydrophobic, nonpolar nature and made up mostly of hydrocarbon chains. Lipids are generally esters of fatty acids and building blocks of biological membranes. Fatty acids can be unsaturated and saturated fatty acids. Lipids occur in plasma membrane that has three classes based on the type of hydrophilic head (i) Glycolipids, whose head are formed by 1-15 oligosaccharide units (ii) Phospholipids contain a positively charged head which are linked to the negatively charged phosphate groups (iii) Sterols, whose head contain a steroid ring.

List of fatty acids and lipids and their function in cell

Class	Primary location	Function
	Classification of Fatty acids	
Saturated fatty acids (carbon number: double bonds)		
Palmitic acid (16:0). Stearic acid (18:0). Arachidic acid (20:0)	Component of complex lipids in plasma membrane or in lipid droplets when esterified	Energy source: β oxidation in mitochondria
Unsaturated fatty acids (carbon number:double bonds)		
Oleic acid (18:1), Arachidonic acid (20:4), Eicosapentaenoic acid (20:5)	Component of complex lipids in plasma membrane	Energy source: β oxidation in mitochondria
Eicosanoids		
Prostaglandins Leukotrienes Thromboxanes	Intracellular	Signal transmission
	Classification of Lipids	
Glycerolipids		
Triglycerides	Lipid droplets	Energy storage
Phospholipids		
Phosphatidylcholine Phosphatidylserine Phosphatidylethanolamine Phosphatidylinositol Phosphatidic acid	Lipid bilayers	Structural Signal transmission
Phosphosphingolipids		
Sphingomyelin	Cell membranes	Structural Signal transmission
Glycosphingolipids		
Sterol lipids		
Cholesterol	Cell membranes (containing cholesterol esters)	Structural building block for other sterol lipid

Proteins

Proteins are heteropolymers of stings of amino acids. Amino acids are joined together by the peptide bond which is formed between the carboxyl group and amino group of successive amino acids. Proteins are formed are 20 different amino acids, depending on the number of amino acids and the sequence of amino acids. There are four levels of protein structure:

- **Primary structure** – It is composed of a long chain of amino acids arranged in a specific sequence and is non-functional protein.
- **Secondary structure** – The long chain of small peptide chains are folded and arranged in a helix shape, where the amino acids interact by the formation of hydrogen bonds. This structure is divided into *alpha*-helix and *beta*-pleated sheet. Silk fibres have secondary structure. Functional parts of secondary structure are turns, loops, and motifs.
- **Tertiary structure** – Long polypeptide chains have more stability due to folding and coiling by the formation of ionic or hydrophobic bonds or disulphide bridges. Functional part of tertiary structure is domains.
- **Quaternary structure** – It is an assembly of more than one polypeptide or subunits, Example: Haemoglobin, insulin.

Standard amino acid abbreviations and their properties

Sl	Amino acid	3-letter symbol	1-letter symbol	Side chain	Side chain polarity	Side chain charge
1.	Alanine	Ala	A	Aliphatic	nonpolar	neutral
2.	Arginine	Arg	R	Basic	basic polar	positive
3.	Asparagine	Asn	N	Amide	polar	neutral
4.	Aspartic acid	Asp	D	Acid	acidic polar	negative
5.	Cysteine	Cys	C	Sulfur containing	nonpolar	neutral
6.	Glutamic acid	Glu	E	Acid	acidic polar	negative
7.	Glutamine	Gln	Q	Amide	polar	neutral
8.	Glycine	Gly	G	Aliphatic	nonpolar	neutral
9.	Histidine	His	H	basic aromatic	basic polar	positive
10.	Isoleucine	Ile	I	Aliphatic	nonpolar	neutral
11.	Leucine	Leu	L	Aliphatic	nonpolar	neutral
12.	Lysine	Lys	K	Basic	basic polar	positive
13.	Methionine	Met	M	Sulfur containing	nonpolar	neutral
14.	Phenylalanine	Phe	F	Aromatic	nonpolar	neutral
15.	Proline	Pro	P	Cyclic	nonpolar	neutral
16.	Serine	Ser	S	Hydroxyl containing	polar	neutral
17.	Threonine	Thr	T	Hydroxyl containing	polar	neutral
18.	Tryptophan	Trp	W	Aromatic	nonpolar	neutral
19.	Tyrosine	Tyr	Y	Aromatic	polar	neutral
20.	Valine	Val	V	Aliphatic	nonpolar	neutral
21.	Selenocysteine	Sec	U	Selenium containing	nonpolar	neutral

22.	Pyrrolysine	Pyl	O	Pyrroline side-chain	nonpolar	positive
Last two amino acids (Selenocysteine, Pyrrolysine) are found in some species and coded by stop codons						

Limiting amino acids in some plant proteins

Crop	Limiting amino acids
Cereal grains and millets	Lysine, Threonine
Rice and soybeans	Methionine
Legumes	Methionine, Tryptophan
Groundnuts	Methionine, Lysine, Threonine
Sunflower seeds	Lysine
Green leafy vegetables	Methionine

Important proteins and their functions

Protein	Function
Insulin	Enzyme for sugar metabolism
Cytochrome c	Enzyme for cell respiration
Growth Hormone	Used in anti-aging treatment
Hemoglobin	Oxygen transport in blood
Hexokinase	Enzyme for glycolysis
Gamma Globulin	Part of immune system in blood
Myosin	Muscle action

Key Points

- Carbohydrate contains carbon (C), hydrogen (H) and oxygen (O) atoms.
- Carbohydrates are hydrates of carbon.
- Structurally, carbohydrates are polyhydroxy aldehydes and ketones.
- Generally, hydrogen–oxygen atom ratio is 2:1 in carbohydrates.
- Common empirical formula is $C_x(H_2O)_y$.
- Exceptions = deoxyribose ($C_5H_{10}O_4$).
- Grape sugr - glucose, cane sugar - sucrose and milk sugar - lactose.
- Polysaccharides serve for the storage of energy (starch in plants and glycogen in animal).
- Structural components in plant cell wall are cellulose.
- The smallest monosaccharides contain three carbons that can be dihydroxyacetone and D- and L-glyceraldehydes.
- Monosachharides generally have two configurations that are called anomers, α and β anomers of glucose.
- Monosachharides have asymmetric carbon atoms which connected four different groups or atoms. Therefore, these have more than one isomer. For example, glucose contains four assymetric carbon atoms, possible isomers are 2^4=16.
- The D or L forms are made according to the orientation of the asymmetric carbon.

- Ring structure with five and six atoms are called furanose and pyranose forms, respectively.
- Lyxose is a component of lyxoflavin found in the human heart.
- Ribulose and xylulose occur in the pentose phosphate pathway.
- Galactose is found in galactolipids in plant cell membranes.
- Mannose occurs in human metabolism, especially in the glycosylation.
- Fructose is also called as fruit sugar.
- Trehalose is a major sugar of insects and made by two glucose molecules.
- The formula of unmodified disaccharides is $C_{12}H_{22}O_{11}$.
- Sucrose is the most abundant disaccharide and composed of α-D-glucose molecule and α-D-fructose molecule connected by α-1→2-glycosidic bond.
- Lactose is a disaccharide composed of one β-D-galactose molecule and one β-D-glucose molecule linked with β-1→4 glycosidic bond.
- Maltose made up two α-glucose units by linking α-1,4 glycosidic bond.
- A reducing sugar has a free aldehyde or ketone group, eg. galactose, glucose and fructose, lactose and maltose.
- Non-reducing sugars: sucrose and trehalose.
- Carbohydrate consumed in food yields 3.87 calories of energy per gram for simple sugars and 3.57 to 4.12 calories per gram for complex carbohydrate.
- Glucemic index (GI) and glycemic load concepts characterize food behavior during human digestion.
- GI is a measure of how quickly food glucose is absorbed, while glycemic load is a measure of the total absorbable glucose in foods.
- Plants synthesize carbohydrates from carbon dioxide and water during photosynthesis.
- Oxidation of one gram of carbohydrate yields approximately 4 kcal of energy, while the oxidation of one gram of lipids yields about 9 kcal.
- Glycolysis and citric acid cycle are catabolic processes.
- Lipids are a group of naturally occurring molecules including fats, olis, waxes, sterols, fat-soluble vitamins (such as vitamins A, D, E, and K).
- Fats are known as triglycerides which are esters of three fatty acid chains and alcohol glycerol.
- Oil normally refers to a fat containing short or unsaturated fatty acid chains that is liquid at room temperature.
- Fats are solid at room temperature.
- Fats are generally hydrophobic and soluble in organic solvents and insoluble in water.
- Essential fatty acids cannot be synthesized in the body from simpler constituents.
- There are two essential fatty acids (EFAs) in human nutrition: *alpha*-linolenic acid (an *omega*-3 fatty acid) and linoleic acid (an *omega*-6 fatty acid).
- Fats and other lipids are broken down by enzymes called lipases in the pancreas.
- Saturated fats have no double bonds between the carbons in the aliphatic chain.
- Unsaturated fats have one or more double bonded carbons in the aliphatic chain.

- Nomenclature is based on the non-acid (non-carbonyl) end of the chain. This end is called the *omega* end or the n-end.
- *Alpha*-linolenic acid is called an omega-3 fatty acid because the 3rd carbon from that end is the first double bonded carbon in the chain.
- Unsaturated fats divided into cis fats which are the most common in nature and trans fats which are rare in nature.
- Liquid *cis*-unsaturated fats *viz.,* vegetable oils are hydrogenated to produce saturated fats.
- Polyunsaturated oils go rancid when they react with oxygen in the air.
- Trans fats are generated during hydrogenation as contaminants created by an unwanted side reaction on the catalyst during partial hydrogenation.
- Saturated fats are typically solid at room temperature.
- Olive and linseed oils are unsaturated and liquid.
- Fats serve as energy sources for the body and as stores for energy in excess of what the body needs immediately.
- One gram of fat releases about 37 kJ (8.8 kcal).
- Vitamins A, D, E, and K are fat-soluble.
- Fatty acids = A chain of an acetyl-CoA primer with malonyl-CoA ormethylmalonyl-CoA groups.
- Glycerolipids = mono-, di-, and tri-substituted glycerols called triglycerides.
- Glycerophospholipids usually called as phospholipids which found in biological membranes. eg.
- Major phospholipids are phosphatidylcholine (lecithin), phosphatidylethanolamine and phosphatidylserine.
- Sphingolipids are synthesized *de novo* from amino acid serine and a long-chain fatty acyl CoA, then converted into ceramides, phosphosphingolipids, glycosphingolipids and other compounds.
- Predominant sterol in fungal cell membranes is ergosterol.
- Polyketides are synthesized by polymerization of acetyl and propionyl subunits.
- Beta oxidation is the metabolic process by which fatty acids are broken down in the mitochondria and/or in peroxisomes to generate acetyl-CoA.
- Protein was first described by Dutch chemist Gerardus Johannes Mulder and termed by Swedish chemist Jöns Jacob Berzelius in 1838.
- Protein is derived from the Greek word *proteios* meaning "primary".
- James B. Sumner showed that the enzyme urease was a protein in nature.
- Armour Hot Dog Co. purified 1 kg of pure bovine pancreatic ribonuclease A.
- Linus Pauling described regular protein secondary structures by hydrogen bonds.
- Insulin is the first protein which sequenced by Frederick Sanger in 1953 who won Nobel Prize for this achievement in 1958.
- Hemoglobin and myoglobin are the first protein structures which solved by Max Perutz and Sir John Cowdery Kendrew, respectively in 1958.
- Protein Data Bank has over 115,000 atomic-resolution structures of proteins.

- The sequence of amino acid residues in a protein is defined by the sequence of a gene, which is encoded in the genetic code.
- Genetic code specifies 20 standard amino acids.
- Two new amino acids (selenocysteine and pyrrolysine) were identified in certain archaea. These synthesized during post-translational modification.
- Protein purification techniques: ultracentrifugation, precipitation, electrophoresis, and chromatography.
- Techniques to study protein structure and function: immunohistochemistry, site-directed mutagenesis, X-ray crystallography, nuclear magnetic resonance (NMR) and mass spectrometry.

Primary structure of protein	amino acid sequence	Peptide bonds
Secondary structure	α-helix, β-sheet and turns	Hydrogen bonds
Tertiary structure	Three-dimensional shape of an entire protein	Salt bridges, hydrogen bonds, disulfide bonds
Quaternary structure	Polypeptide chains	Hydrogen-bonding, disulfide bridges and salt bridges

- Proteins divided into three main classes: globular proteins, fibrous proteins, and membrane proteins.
- X-ray crystallography and NMR spectroscopy used to study tertiary or quaternary structure of a protein.
- Circular dichroism is a technique for determining internal β-sheet/α-helical composition of proteins.
- Cryoelectron microscopy is used to produce lower-resolution structural information about very large protein complexes of viruses.
- Electron crystallography can produce high-resolution information especially for two-dimensional crystals of membrane proteins.
- Protein Data Bank (PDB) is a freely available resource for structural data of proteins.
- The molecules bound and acted upon by enzymes are called substrates.
- The region of the enzyme that binds the substrate and contains the catalytic residues is known as the active site.
- Enzyme aspartokinase which catalyses the first step in the synthesis of lysine, methionine, and threonine from aspartate.
- Antibodies are protein components of an adaptive immune system whose main function is to bind antigens or foreign substances in the body and target them for destruction.
- Antibodies can be secreted into the extracellular environment or anchored in the membranes of specialized B cells known as plasma cells.
- Receptors and hormones are highly specific binding proteins.
- *In silico* is computational method to study proteins.
- Proteome is the total complement of proteins of a cell or cell type.

Multiple Choice Questions

(i) Carbohydrates

1. Which of the following are the most abundant biomolecules on Earth?
(A) Proteins (B) Lipids
(C) Carbohydrates (D) Nucleic acids

2. Which of the following process convert CO_2 and H_2O into cellulose and other plant products?
(A) Respiration (B) Photosynthesis
(C) Transpiration (D) Vernalization

3. Enantiomers are designated by
(A) A and B isomer (B) D and L isomer
(C) R and S isomers (D) D and L apimers

4. Which process is the central energy-yielding pathway in most nonphotosynthetic cells?
(A) Oxidation of carbohydrates
(B) Oxidation of Proteins
(C) Oxidation of Lipids
(D) Both oxidation and reduction of carbohydrates

5. .... serve as structural and protective elements in the cell walls of bacteria and plants and in the connective tissues of animals.
(A) Water-soluble carbohydrate polymers
(B) Organic solvent-soluble carbohydrate polymers
(C) Insoluble carbohydrate polymers
(D) Insoluble protein polymers

6. What is called the hybrid molecules that covalently made up of more complex carbohydrate polymers and proteins or lipids that act as signals that determine the intracellular location?
(A) Glycosides
(B) Glycoconjugates
(C) Glycolysis
(D) Conjugate acid and base

7. What are polyhydroxy aldehydes or ketones or substances that yield on hydrolysis?
(A) Proteins
(B) Lipids
(C) Nucleic acids
(D) Carbohydrates

8. What are carbohydrates?
(A) Polyhydroxy aldehydes or ketones
(B) Polyhydroxy cynaides
(C) Polyhydroxy amides
(D) Polyhydroxy glycerol

9. The starch content of pulses is about
(A) 50% (B) 60%
(C) 70% (D) 80%

10. The starch content of wheat is about
(A) 50% (B) 60%
(C) 70% (D) 80%

11. Simplest sugars are known as ....
(A) Monosaccharides
(B) Oligosaccharides
(C) Polysaccharides
(D) Disaccharides

12. Monosaccharide is
(A) A single polyhydroxy base unit
(B) A single polyhydroxy acid unit
(C) A single polyhydroxy aldehyde or ketone unit
(D) A single polyhydroxy amide or ketone unit

13. Which of the most abundant monosaccharide in nature sometimes referred to as?
(A) Dextrin (B) Dextrose
(C) Acetal (D) Cellulose

14. Minimum how many carbon atoms should be present in monosaccharides to have cyclic structures?
(A) Four (B) Five
(C) Six (D) Three

15. .... made up of short chains of monosaccharide units or residues.
(A) Monosaccharides
(B) Oligosaccharides
(C) Polysaccharides
(D) Homopolysaccharides

16. Which type of bond present in short chains of monosaccharide units of oligosaccharides?
(A) Ionic bond (B) Hydrogen bond
(C) Covalent bond (D) Glycosidic bond

17. Monosaccharide units in maltose are ....
(A) Two β-D-glucose units
(B) β –D-glucose and α-D-glucose units
(C) Two α-D-glucose units
(D) α-D-glucose and β-D-fructose

18. How many monosaccharide units present in disaccharides?
(A) Four (B) Five
(C) Two (D) Three

19. Which of the following is not a disaccharide sugar?
(A) Sucrose (B) Trehlose
(C) Xylose (D) Maltose

20. Which of the following is known as cane sugar?
(A) Sucrose (B) Glucose
(C) Mannose (D) Maltose

21. Monosaccharide units in sucrose are ....
(A) β-D-galactose and β-D-glucose
(B) β –D-glucose and α-D-fructose
(C) α-D-glucose and α-D-glucose
(D) α-D-glucose and β-D-fructose

22. In cells, most oligosaccharides do not occur as free entities but are joined to nonsugar molecules in glycoconjugates. These nonsugar molecules may be.......
(A) Vitamins (B) Lipids and proteins
(C) Enzymes (D) Phosphate groups

23. How many monosaccharide units found in oligosaccharides?
(A) 10 (B) 50-100
(C) 2-50 (D) 2-10

24. What is called a carbon atom that attach to four different atoms or groups?
(A) Epimer
(B) Chiral centre
(C) Stereoisomer carbon
(D) Paralogous carbon

25. .... is crystalline solids that are freely soluble in water but insoluble in nonpolar solvents.
(A) Glucose (B) Cellulose
(C) Pectin (D) Lignin

26. Monosaccharide units in lactose are...........................
(A) β-D-galactose and β-D-glucose
(B) β –D-glucose and α-D-fructose
(C) α-D-glucose and α-D-glucose
(D) α-D-glucose and β-D-fructose

27. Which does not contain any asymmetric carbon atoms?
(A) Glucose
(B) Dihydroxyacetone
(C) Glyceraldehyde
(D) Lactose

28. Which has only one chiral carbon with two different optical isomers?
(A) Glucose
(B) Dihydroxyacetone
(C) Glyceraldehyde
(D) Lactose

29. What are enantiomers?
(A) Different optical apimers
(B) Different crystal structures
(C) Different in solubility
(D) Different optical isomers

30. Agarose is a polysaccharide which is formed by repeating unit of agarobiose disaccharide made up of ....
(A) L-galactose and 3,6-anhydro-L-galactopyranose
(B) D-galactose and 3,6-anhydro-D-galactopyranose
(C) D-galactose and 3,6-anhydro-L-galactopyranose
(D) L-galactose and 3,6-anhydro-D-galactopyranose

31. Which technique is used to represent the absolute configurations of biomolecules?
(A) X-ray crystallography
(B) SAGE
(C) Microarray
(D) Autoradiography

32. Which is used to represent three-dimensional sugar structures on paper?
(A) Michal-mentan equation
(B) Watson-crick model
(C) Fischer projection formulas
(D) Ramchandran plot

33. Which of the following group related to pentose sugar?
(A) Deoxyribose and ribose
(B) Glucose and galactose
(C) Sucrose and maltose
(D) Raffinose and Starch

34. Which of the following group related to trisachharide sugar?
(A) Ribose (B) Starch
(C) Sucrose (D) Raffinose

35. Plant cell walls contain large amounts, of what polysaccharide?
(A) Carbon (B) Water
(C) Chitin (D) Cellulose

36. Which sugar present as insoluble state in cell?
(A) Galactose and mannose
(B) Dextrin and inulin
(C) Starch and cellulose
(D) Glucose and fructose

37. Which sugar does not reduce Fehling's solution and does not form an osazone?
(A) Sucrose (B) Lactose
(C) Maltose (D) Glucose

38. Which polysachharide is widely distributed in higher plants as the component of cell wall?
(A) Chitin (B) Pectin
(C) Starch (D) Cellulose

39. Starch is converted by the enzyme .... into dextrin and maltose.
(A) Maltase (B) Zymase
(C) Invertase (D) Diastase

40. Which of the following is not a class of monosachharide?
(A) Tetroses (B) Pentoses
(C) Cellulose (D) Hexoses

41. Which of the following functional group is generally not present in sugars?
(A) Aldehyde (B) Alcohol
(C) Keton (D) Carboxylic

42. Which of the following sugar is insoluble in aceton and many other organic solvents?
(A) Starch (B) Glucose
(C) Cellulose (D) Glycogen

43. Which of the following group is responsible for reducing action in sugars?
(A) Aldehyde and alcohol
(B) Alcohol and ketone
(C) Aldehyde and Ketone
(D) Carboxylic and amino

44. What is reaction take place during reducing action of sugars?
(A) Copper salts in hot alkaline solution by forming cuprous oxide
(B) Copper salts in hot alkaline solution by forming cupric oxide
(C) Iron salts in hot acidic solution by forming ferrus oxide
(D) Zinc salts in hot alkaline solution by forming zinc oxide

45. Which of the following pentose sugar is found in Gum Arabic and Cherry Gum?
(A) Ribose (B) Arabinose
(C) Xylose (D) Lyxose

46. Which sugar widely distributed in woody plants?
(A) Ribose (B) Arabinose
(C) Xylose (D) Lyxose

47. Monosaccharide units in trehalose are ....
(A) Two β-D-glucose units
(B) β –D-glucose and α-D-glucose units
(C) Two α-D-glucose units
(D) α-D-glucose and β-D-fructose

48. Which of the following is not used for testing of glucose?
(A) Moore's test
(B) Fehling test
(C) Osazone test
(D) *alpha*-nephthol reaction

49. Agar-agar is mucilage obtained from Red Seaweeds related to ....
(A) Rhodophyceae (B) Chlorophyceae
(C) Xanthophyceae (D) Pteridophyta

50. Trommer's test is used to detect ....
(A) Sugar (B) Amino acid
(C) Nucleic acid (D) Lipids

51. Which of the following is reducing sugar?
(A) Glucose (B) Sucrose
(C) Cellulose (D) Glycogen

52. Seliwanoff's test is used to detect ....
(A) Lipids (B) Amino acid
(C) Nucleic acid (D) Sugar

53. Sucrose is hydrolyzed by the enzyme .... into glucose and fructose.
(A) Maltase (B) Zymase
(C) Invertase (D) Glucanase

54. Maltose is hydrolyzed by the enzyme .... into two molecules of glucose.
(A) Maltase (B) Zymase
(C) Invertase (D) Diastase

55. Which substance occurs as a soluble "reserve material" in the cell-sap of the underground stems, roots and also leaves of Dahlia *(Dahlia variabilis)* and Chicory *(Cichorium intybus)*?
(A) Glycogen (B) Gums
(C) Dextrin (D) Inulin

56. "Vegetable ivory" which is the endosperm of the Palm *(Phytelephas macrocarpa)* contains considerable quantities of a mannan and is used as a source of .... ?
(A) Glucose (B) Fructose
(C) Galactose (D) Mannose

57. Which are translucent and amorphous in the solid state and soluble in water giving sticky colloidal solutions?
(A) Glycogen (B) Gums
(C) Dextrin (D) Inulin

58. Which does not produce gums?
(A) *Acacia senegal*
(B) *Azadirachta indica*
(C) *Prunus cerasus*
(D) *Astragalus gummifer*

59. Which colour is produced after reaction between cellulose and guaiacol?
(A) Violet (B) Red-violet
(C) Cherry-red (D) Yellow-green

60. .... products are found in the walls of corky and cuticularized tissues.
(A) Adipo-cellulose
(B) Hemi-cellulose
(C) Pecto-cellulose
(D) Ligno-cellulose

61. Match between the percentages of pure cellulose and various lignified tissues.

i.	Cotton fibre	p.	54%
ii.	Flax and Hemp fibre	q.	35-38 %
iii.	Jute	r.	88.3%
iv.	Beech and Oak wood	s.	72-73 %

(A) i—p, ii—q, iii—r, iv—s
(B) i—s, ii—r, iii—q, iv—p
(C) i—r, ii—s, iii—p, iv—q
(D) i—q, ii—p, iii—r, iv—s

62. Furfural phloroglucide method used to detect and estimate ....
(A) Cellulose (B) Dextrin
(C) Mucin (D) Pectin

63. Which substance occurs as a soluble "reserve material" in the cell-sap of the underground stems, roots and also leaves of a number of plants of compositae family?
(A) Inulin (B) Dextrin
(C) Mucin (D) Pectin

64. Agarose obtained from see weed (red algae) ....
(A) *Acrochaetium efflorescens*
(B) *Gelidium amansii*
(C) *Atractophora hypnoides*
(D) *Schmitzia hiscockiana*

65. Which of the following plant enzyme does not hydrolyze carbohydrates?
(A) Diastase (B) Invertase
(C) Zymase (D) Maltase

66. The number of molecules of ATP produced by the total oxidation of acetyl CoA in TCA cycle is
(A) 6 (B) 8
(C) 10 (D) 12

67. One molecule of glucose gives .... molecules of CO_2 in EM-TCA cycle.
(A) 6 (B) 3
(C) 1 (D) 2

68. One molecule of glucose gives molecules of CO_2 in one round of HMP shunt.
(A) 6 (B) 1
(C) 2 (D) 3

69. When O_2 supply is inadequate, pyruvate is converted to
(A) Phosphopyruvate (B) Acetyl CoA
(C) Lactate (D) Alanine

70. For glycogenesis, Glucose should be converted to
(A) Glucuronic acid (B) Pyruvic acid
(C) UDP glucose (D) Sorbitol

71. Before pyruvic acid enters the TCA cycle it must be converted to
(A) Acetyl CoA (B) Lactate
(C) α-ketoglutarate (D) Citrate

72. Acetyl CoA is not used for the synthesis of
(A) Fatty acid (B) Cholesterol
(C) Pyruvic acid (D) Citric acid

73. Pyruvate kinase requires .... ions for maximum activity.
(A) Na^+ (B) K^+
(C) Ca^{2+} (D) Mg^{2+}

74. Synthesis of Glucose from amino acids is termed as
(A) Glycolysis (B) Gluconeogenesis
(C) Glycogenesis (D) Lipogenesis

75. How many ATP molecules will be required for conversion of 2-molecules of Lactic acid to Glucose?
(A) 2 (B) 4
(C) 8 (D) 6

76. The number of isomers of glucose is
(A) 2 (B) 4
(C) 8 (D) 16

77. Two sugars which differ from one another only in configuration around a single carbon atom are termed
(A) Epimers (B) Anomers
(C) Optical isomers (D) Stereoisomers

78. Isomers differing as a result of variations in configuration of the —OH and —H on carbon atoms 2, 3 and 4 of glucose are known as
(A) Epimers (B) Anomers
(C) Optical isomers (D) Steroisomers

79. The most important epimer of glucose is
(A) Galactose (B) Fructose
(C) Arabinose (D) Xylose

80. α-D-glucose and β -D-glucose are
(A) Stereoisomers (B) Epimers
(C) Anomers (D) Keto-aldo pairs

81. Compounds having the same structural formula but differing in spatial configuration are known as
(A) Stereoisomers (B) Anomers
(C) Optical isomers (D) Epimers

82. Fluoride inhibits .... and arrests glycolysis.
(A) Glyceraldehyde-3-phosphate dehydrogenase
(B) Aconitase
(C) Enolase
(D) Succinate dehydrogenase

83. The distinguishing test between monosaccharides and dissaccharide is
(A) Bial's test (B) Seliwanoff's test
(C) Barfoed's test (D) Hydrolysis test

84. Out of 24 moles of ATP formed in TCA cycle, 2 molecules of ATP can be formed at "substrate level" by which of the following reaction?
(A) Citric acid → Isocitric acid
(B) Isocitrate → Oxaloacetate
(C) Succinic acid → Fumarate
(D) Succinyl-CoA → Succinyl acid

85. An allosteric enzyme responsible for controlling the rate of T.C.A cycle is
(A) Malate dehydrogenase
(B) Isocitrate dehydrogenase
(C) Fumarase
(D) Aconitase

86. The ratio that approximates the number of net molecule of ATP formed per mole of glucose oxidized in presence of O_2 to the net number formed in absence of O_2 is
(A) 4 : 1 (B) 10 : 2
(C) 12 : 1 (D) 18 : 1

87. A specific inhibitor for succinate dehydrogenase is
(A) Arsinite (B) Malonate
(C) Citrate (D) Cyanide

88. Whcih of the following features are common to monosaccharides?
(A) Contain asymmetric centres
(B) 2 types – aldoses and ketoses
(C) Exist as ring structures in solution
(D) Include glucose, galactose and raffinose

89. Glucose enters the cells by
(A) Insulin independent transport
(B) Insulin dependent transport
(C) Enzyme mediated transport
(D) Both (A) and (B)

90. Tissues form lactic acid from glucose. This phenomenon is termed as
(A) Aerobic glycolysis
(B) Oxidation
(C) Oxidative phosphorylation
(D) Anaerobic glycolysis

91. Substrate level phosphorylation in TCA cycle is in step:
(A) Isocitrate dehydrogenase
(B) Malate dehydrogenase
(C) Aconitase
(D) Succinate thiokinase

92. Two examples of substrate level phosphorylation in EM pathway of glucose metabolism are in the reactions of
(A) 1,3 bisphosphoglycerate and phosphoenol pyruvate
(B) Glucose-6 phosphate and Fructo-6-phosphate
(C) 3 phosphoglyceraldehyde and phosphoenolpyruvate
(D) 1,3 diphosphoglycerate and 2-phosphoglycerate

93. Under anaerobic conditions the glycolysis of one mole of glucose yields .... moles of ATP.
(A) One (B) Two
(C) Eight (D) Thirty

94. During glycolysis, Fructose 1, 6 diphosphate is decomposed by the enzyme:
(A) Enolase a (B) Fructokinase
(C) Aldolase
(D) Diphosphofructophosphatose

95. Oxidative decarboxylation of pyruvate requires
(A) NADP+ (B) Cytichromes
(C) Pyridoxal phosphate
(D) CoA-SH

96. Under anaerobic conditions the glycolysis one mole of glucose yields .... moles of ATP.
(A) One (B) Two
(C) Eight (D) Thirty

97. The purple ring of Molisch reaction is due to
(A) Furfural
(B) Furfural + α Napthol
(C) Napthol
(D) Furfurol + H_2SO_4 + α -Naphthol

98. Which of the following metabolite integrates glucose and fatty acid metabolism?
(A) Acetyl CoA (B) Pyruvate
(C) Citrate (D) Lactate

99. How many steps and enzymes are involved in glycolysis, respectively?
(A) 9 and 10 (B) 9 and 9
(C) 10 and 10 (D) 10 and 11

100. An amphibolic pathway among the following is
(A) HMP shunt (B) Glycolysis
(C) Citirc acid cycle (D) Gluconeogenesis

101. .... acts in the conversion of hexoses into pentoses and in NADPH regeneration.
(A) Glycolysis
(B) Kreb's cycle
(C) Pentose phosphate pathway
(D) Glycogenesis

102. .... is an essential antioxidant in cells which prevents oxidative damage.
(A) FAD (B) NADPH
(C) Glucose (D) Pyrophosphate

103. What is called the conversion of excess glucose into glycogen which prevents excessive osmotic pressure buildup inside the cell?
(A) Glycolysis
(B) Kreb's cycle
(C) Pentose phosphate pathway
(D) Glycogenesis

104. What is called the breakdown of glycogen into glucose which provides a glucose supply for glucose-dependent tissues?
(A) Glycolysis
(B) Glycogenolysis
(C) Pentose phosphate pathway
(D) Glycogenesis

105. .... *de novo* synthesis of glucose molecules from simple organic compounds.
(A) Gluconeogenesis
(B) Glycolysis
(C) Glycogenolysis
(D) Pentose phosphate pathway

(ii) Fats and Lipids

106. What is fat?
(A) An ester or glyceride of a fatty acid
(B) Polymers of amino acids
(C) Polymers of monosachharides
(D) Sequences of only saturated fally acids

107. What is glycerol?
(A) An aromatic amino acid
(B) Unsaturated fatty acid
(C) Trihydric alcohol
(D) Oil

108. Choose correct match

	Crop		**Oil %**
i.	Maize (*Zea Mays*)	p.	24
ii.	Coconut (*Cocos nucifera*)	q.	51
iii.	Castor-oil (*Ricinus communis*)	r.	65
iv.	Cotton (*Gossypium herbaceum*)	s.	4

(A) i—p, ii—q, iii—r, iv—s
(B) i—q, ii—p, iii—s, iv—r
(C) i—r, ii—s, iii—p, iv—q
(D) i—s, ii—r, iii—q, iv—p

109. Which of the following statement about fat and lipids in plants is false?
(A) They float upon water due to insolubility in water
(B) They are soluble in ether, benzene, chloroform, carbon tetrachloride
(C) Fat is formed by the reaction of the trihydric alcohol and a fatty acid
(D) Fats are an esters or glyceride of a fatty acid

110. Potassium salt, potassium stearate of the fatty acid, stearic acid, is termed
(A) Soap (B) Oil
(C) Wax (D) Gum

111. A fat is heated with an alkaline hydroxide called
(A) Saponification (B) Hydrolysis
(C) Acidification (D) Intensification

112. What best describes phospholipid molecules?
(A) Their chemical formula is $(CH_2O)n$
(B) They form cell membranes
(C) They are proteins
(D) They are ionic

113. A group of naturally occurring molecules like fats, oils, waxes, sterols, fat-soluble vitamins (A, D, E and K) are considered as ....
(A) Nucleic acids (B) Carbohydrates
(C) Proteins (D) Lipds

114. Lipids may be broadly defined as
(A) Hydrophobic or proteiolytic small molecules
(B) Hydrophilic or amphiphilic small molecules
(C) Hydrophobic or amphiphilic small molecules
(D) Hydrophilic or amphoteric small molecules

115. Isoprene group is found in ....
(A) Sterol lipids and prenol lipids
(B) Fatty acids and phospholipids
(C) Glycerolipids and polyketides
(D) Sphingolipids and saccharolipids

116. .... synthesized by chain-elongation of an acetyl-CoA primer with malonyl-CoA ormethylmalonyl-CoA groups.
(A) Nucleic acids (B) Carbohydrates
(C) Proteins (D) Fatty acids

117. Fatty acids are made of a hydrocarbon chain that contains the molecule with
(A) Only a polar, hydrophilic end
(B) Only a nonpolar, hydrophobic end
(C) Both a polar, hydrophilic end and a nonpolar, hydrophobic end
(D) N terminus and C-terminus in their hydrocarbon chain

118. What is commonly used as a building-block of more structurally complex lipids?
(A) Nucleotides (B) Fatty acids
(C) Monosachharides (D) Amino acids

119. A fatty acid may be found in either a cis or trans geometric isomerism due to

(A) Double bonds in hydrocarbon chain
(B) Triple bonds in hydrocarbon chain
(C) Single bonds in hydrocarbon chain
(D) Both double and triple bonds in hydrocarbon chain

120. Match between unsaturated fatty acids and number of double bonds

i.	Linolenic acid	p.	Two
ii.	Oleic acid	q.	Four
iii.	Linoleic acid	r.	Three
iv.	Arachidonic acid	s.	One

(A) i—p, ii—q, iii—r, iv—s
(B) i—q, ii—p, iii—s, iv—r
(C) i—r, ii—s, iii—p, iv—q
(D) i—s, ii—r, iii—q, iv—p

121. Which of the following is not a function of lipids?
(A) Storage material of metabolic fuel
(B) Structural components of membranes
(C) Transport form of metabolic fuel
(D) Involved in protein synthesis

122. Correct match between biological lipids and functions

i.	Carotene	p.	Hormones
ii.	Vitamin D derivatives	q.	Pigments
iii.	Eicosanoids	r.	Cofactors
iv.	Vitamin K	s.	Signaling molecules

(A) i—p, ii—q, iii—r, iv—s
(B) i—q, ii—p, iii—s, iv—r
(C) i—r, ii—s, iii—p, iv—q
(D) i—s, ii—r, iii—q, iv—p

123. Which of the following biological lipid serve as Signaling molecule?
(A) Carotene
(B) Vitamin D derivatives
(C) Phosphatidylinositol derivatives
(D) Vitamin K

124. How many carbons present in long chain of hydrocarbon in fatty acids?
(A) 10-20 (B) 4-36
(C) 2-50 (D) 5-25

125. Correct match between biological fatty acids and number of carbon

i.	Palmitic acid	p.	12
ii.	Linoleic acid	q.	14
iii.	Myristic acid	r.	18
iv.	Lauric acid	s.	16

(A) i—p, ii—q, iii—r, iv—s
(B) i—q, ii—p, iii—s, iv—r
(C) i—r, ii—s, iii—p, iv—q
(D) i—s, ii—r, iii—q, iv—p

126. Correct match between Common name and systematic name of fatty acids

	Common Name		**Systematic Name**
i.	Lauric acid	p.	Cis-Δ^9-Hexadecenoic
ii.	Palmitic acid	q.	Dodecanoic
iii.	Stearic acid	r.	Hexadecanoic
iv.	Oleic acid	s.	Octadecanoic

(A) i—q, ii—s, iii—p, iv—r
(B) i—p, ii—s, iii—r, iv—q
(C) i—q, ii—r, iii—s, iv—p
(D) i—s, ii—q, iii—p, iv—r

127. *Cis* and *trans* configuration is found in ....
(A) All saturated fatty acids
(B) All unsaturated fatty acids
(C) All essential fatty acids
(D) All nonessential fatty acids

128. What is the property of saturated fatty acids?
(A) No single bond
(B) No double bond
(C) One or more than one double bond
(D) Only one double bond in all fatty acids

129. All nonessential fatty acids synthesized from ....
(A) Acetyl coenzyme A
(B) Tryptophan
(C) Pyruvic acid
(D) All are vitamin derivatives

130. From which Acetyl coenzymeA derived that used for the synthesis of nonessential fatty acids?
(A) Glucose oxidation
(B) Amino acid oxidation
(C) Pyruvate degradation
(D) Vitamin K

131. Why fatty acids used as detergents?
(A) Due to presence of double bonds
(B) Due to amphipathic nature
(C) Due to Insoluble in water
(D) Due to Soluble in organic solvents

132. What are the reason of polarity and nonpoarity in fatty acids?
(A) Presence of double bonds
(B) Polarity due to -CH3 and Nonpolarity due to –COOH
(C) Polarity due to -COOH and Nonpolarity due to -CH3
(D) Due to *cis* and *trans* configuration

133. Melting point of fatty acids depends on ....
....
(A) Only double bonds
(B) Length of hydrocarbon chain and degree of unsatuation
(C) Amphipathic nature and cis and trans configuration
(D) Only degree of unsaturation

134. Which of the following statement is true about fatty acids?
i. Unsaturated fatty acids have lower melting point than saturated fatty acids
ii. Melting point changed due to length of chain and degree of unsaturation
iii. Nonessential fatty acids can be synthesized from acetyl coenzyme A
iv. Unsaturated fatty acids have double bonds with cis and trans configuration

(A) Only i
(B) i and iii
(C) i, ii and iv
(D) i, ii, iii and iv

135. Triacylglycerols are ....
(A) Triesters of unsaturated fatty acids
(B) Amides of fatty acids and glycerol
(C) Triesters of saturated fatty acids
(D) Triesters of fatty acids and glycerol

136. What are the properties of triglycerols?
i. Nonpolar, hydrophobic in nature
ii. All are unsaturated
iii. They have no charge
iv. All are sarurated

(A) Only i
(B) i and iii
(C) i, ii and iv
(D) i, ii, iii and iv

137. Fats are hydrolyzed with ...., produced soap.
(A) NaOH or KOH
(B) $Ca(OH)_2$ and HCl
(C) Iodine
(D) Chloroform and benzene

138. What is saponification?
(A) Acetylation of fats with acetic acid
(B) Hydroxylation of fats with alkalis
(C) Reaction of fats with iodine
(D) Unsatuartion of fats

139. What is called the number of KOH required to saponify one gram of fat ?
(A) Iodine number
(B) Unsaturation number
(C) Saponification number
(D) Respiration quotient

140. Iodine number is used to determine ....
....
(A) Degree of saturation
(B) Degree of unsaturation
(C) Nature of polarity and nonpolarity
(D) Melting point

141. How many quantity of iodine is called iodine number that added to 100g sample of fat or oil?
(A) 100 gram (B) One gram
(C) 10 gram (D) 0.1 gram

142. What is wax?
(A) Esters of only unsaturated fatty acids
(B) Esters of fatty acids and glycerol
(C) Esters of only saturated fatty acids and alcohol
(D) Esters of fatty acids and alcohol

143. What is the major component of bee wax known as animal wax?
(A) Triacontanoyllaurate
(B) Triacontanoylpalmitate
(C) Triacontanoylstearate
(D) Triacontanoyllinolate

144. What is triacontanoylpalmitate?
(A) Esters of palmitic acid with glycerol
(B) Esters of steric acid with alcohol triacontanol
(C) Esters of only saturated fatty acids with trietahnol
(D) Esters of palmitic acid with alcohol triacontanol

145. Which of the following most numerous phospholipid molecules found in cell membranes?
(A) Shingophospholipid
(B) Phosphoglycerides
(C) Ceramide
(D) Eicosanoid

146. Which of the following is the precursor for all phosphoglyceride molecules?
(A) Phosphatidic acid
(B) Lecithin
(C) Cephalin
(D) Cardiolipin

147. Simplest phospholipid is ....
(A) Phosphatidylcholine
(B) Phosphotidic acid
(C) Phosphatylglycerol
(D) Lecithin

148. Find the correct match

i.	Phosphatidylcholine	p.	Cardiolipin
ii.	Phosphotidylethano-lamine	q.	Lecithin
iii.	Diphosphotidylglycerol	r.	Ceramide
iv.	Sphingophospholipid	s.	Cephalin

(A) i—q, ii—s, iii—p, iv—r
(B) i—p, ii—s, iii—r, iv—q
(C) i—q, ii—r, iii—s, iv—p
(D) i—s, ii—q, iii—p, iv—r

149. Correct match phospholipids and net charge

Common Name	Net charge
i. Lecithin	Zero
ii. Cephalin	Zero
iii. Cardiolipin	-2
iv. Phosphotidylserine	-1

(A) i, ii, and iii (B) iii and iv
(C) Only iv (D) i, ii, iii and iv

150. What is plasmalogens?
(A) One hydrocarbon group linked to C1 position of glycerol as vinyl ether
(B) One hydrocarbon group linked to C2 position of glycerol as vinyl ether
(C) One hydrocarbon group linked to C3 position of glycerol as vinyl ether
(D) One hydrocarbon group linked to C1 position of glycerol as amide

151. What is sphingophospholipid?
(A) A long chain of ether alcohol
(B) A long chain of thaol
(C) A long chain of fatty acid and glycerol
(D) A long chain of amino alcohol

152. Product of acylation of amino group of sphingosine with fatty acid?
(A) Globoside
(B) Cephalin
(C) Ceremide
(D) Cerebroside

153. Sphingomyeline contains
(A) Phosphatidylserine
(B) Phosphatidylcholine
(C) Phosphatidylglycerol
(D) Phosphatidylinositol

154. Which of the following lipid contain sugar?
(A) Phospholipid
(B) Glycolipid
(C) Steroid
(D) Sphingolipid

155. Find the correct match between lipid molecules and their types.

i.	Ceremide	p.	Oligopolysachharides with N-acetylnuraminic acid
ii.	Cerebroside	q.	Type of sphingolipid
iii.	Globosides	r.	A single sugar linked to ceremide
iv.	Gangliosides	s.	Two or more sugar linked to ceremide

(A) i—q, ii—s, iii—p, iv—r
(B) i—p, ii—s, iii—r, iv—q
(C) i—q, ii—r, iii—s, iv—p
(D) i—s, ii—q, iii—p, iv—r

156. Match between disease and enzymes deficiency?

i. Tay-Sachs p. β-Hexosaminidase A
ii. Gaucher's q. α-Galactosidase A
iii. Krabbe's r. Galactocerebrosidase
iv. Fabry's s. β-Glucosidase

(A) i—q, ii—s, iii—p, iv—r
(B) i—p, ii—s, iii—r, iv—q
(C) i—q, ii—r, iii—s, iv—p
(D) i—s, ii—q, iii—p, iv—r

157. Match between disease and accumulating sphingolipids?

i. Tay-Sachs p. Deacylated galactocerebroside
ii. Gaucher's q. Glucocerebroside
iii. Krabbe's r. Trihexosylceramide
iv. Fabry's s. Gangilioside GM2

(A) i—q, ii—s, iii—p, iv—r
(B) i—p, ii—s, iii—r, iv—q
(C) i—q, ii—r, iii—s, iv—p
(D) i—s, ii—q, iii—p, iv—r

158. Match between disease and enzymes?

i. Tay-Sachs p. Loss of myelin
ii. Gaucher's q. Blindness and muscle weekness
iii. Krabbe's r. Skin rash and kidney failure
iv. Fabry's s. Liver and spleen enlargement

(A) i—q, ii—s, iii—p, iv—r
(B) i—p, ii—s, iii—r, iv—q
(C) i—q, ii—r, iii—s, iv—p
(D) i—s, ii—q, iii—p, iv—r

159. What are steroids?
(A) Complex derivatives of leukotrienes
(B) Complex derivatives of alkloids
(C) Complex derivatives of triterpenes
(D) Complex derivatives of vitamin K

160. Which of the following essential sterol for animal cell membranes?
(A) Ergestrol (B) Eicosanoid
(C) Cholesterol (D) Sphingosine

161. Which of the following essential sterol for cell membranes of fungi?
(A) Ergesterol (B) Eicosanoid
(C) Cholesterol (D) Sphingosine

162. Which of the following sterol is not found in plants and fungi?
(A) Stigmasterol (B) Ergesterol
(C) Cholesterol (D) Campesterol

163. Which of the following group included in eicosanoids?
(A) Cholecterol and ergesterol
(B) Stigmasterol and sitosterol
(C) Sphingomyelin and ceramide
(D) Prostanoids and leukotrienes

164. Which of the following pathways included in the production of eicosanoids?
i. Cyclooxygenase pathways
ii. Lipooxygenase pathway
iii. Xanthine oxidage pathway
iv. Carboxypeptidase pathway
(A) i and ii (B) ii and iii
(C) iii and iv (D) i and iv

165. Cyclooxygenase and lipooxygenase pathways involve in .... synthesis, respectively
(A) Sphingomyeilin and fatty acids
(B) Cholecterol and ergesterol
(C) Prostaglandins and leukotrienes
(D) Waxes and oils

166. Which of the following is first lipid identified in human semen?
(A) Sphingomyeilin (B) Prostaglandins
(C) Cholecterol (D) Leukotrienes

167. Which of the following is most prevalent prostaglandin precursor in human?
(A) Arachidonic acid (B) Lauric acid
(C) Palmitic acid (D) Stearic acid

168. Which of the following inhibit the synthesis of prostaglandin?
(A) Streptomycin (B) Aspirin
(C) Prostacyclins (D) Penicillin

169. What is major problem in the transport and digestion of lipids?
(A) Presence of unsaturated fatty acids
(B) Insolubility in water
(C) Solubility in organic solvents
(D) Amphipathic nature

170. In humans, a dietary essential fatty acid is
(A) Palmitic acid (B) Stearic acid
(C) Oleic acid (D) Linoleic acid

171. If the fatty acid is esterified with an alcohol of high molecular weight instead of glycerol, the resulting compound is
(A) Lipositol (B) Plasmalogen
(C) Wax (D) Cephalin

172. The importance of phospholipids as constituent of cell membrane is because they possess
(A) Fatty acids
(B) Both polar and nonpolar groups
(C) Glycerol
(D) Phosphoric acid

173. Phospholipid acting as surfactant is
(A) Cephalin
(B) Phosphatidyl inositol
(C) Lecithin
(D) Phosphatidylserine

174. Oil which contains cyclic fatty acids and once used in the treatment of leprosy is
(A) Elaidic oil (B) Rapeseed oil
(C) Lanoline (D) Chaulmoogric oil

175. Salkowski test is performed to detect
(A) Glycerol (B) Cholesterol
(C) Fatty acids (D) Vitamin D

176. The number of ml of N/10 KOH required to neutralize the fatty acids in the distillate from 5 gm of fat is called
(A) Reichert-Meissel number
(B) Polenske number
(C) Acetyl number
(D) Non volatile fatty acid number

177. The number of milliliters of 0.1 N KOH required to neutralize the insoluble fatty acids from 5 gms of fat is called
(A) Acid number (B) Acetyl number
(C) Halogenation (D) Polenske number

178. Deterioration of food (rancidity) is due to presence of
(A) Cholesterol
(B) Vitamin E
(C) Peroxidation of lipids
(D) Phenolic compounds

179. 'Drying oil', oxidized spontaneously by atmospheric oxygen at ordinary temperature and forms a hard water proof material is
(A) Coconut oil (B) Peanut oil
(C) Rape seed oil (D) Linseed oil

180. Long chain fatty acids are first activated to acetyl-CoA in
(A) Cytosol (B) Microsomes
(C) Nucleus (D) Mitochondria

181. The enzyme acyl-CoA synthase catalyses the conversion of a fatty acid of an active fatty acid in the presence of
(A) AMP (B) ADP
(C) ATP (D) GTP

182. The enzymes of β-oxidation are found in
(A) Mitochondria (B) Cytosol
(C) Golgi apparatus (D) Nucleus

183. Alkaptonuria occurs due to deficiency of the enzyme:
(A) Maleylacetoacetate isomerase
(B) Homogentisate oxidase
(C) p-Hydroxyphenylpyruvate hydroxylase
(D) Fumarylacetoacetate hydrolase

184. β-Oxidation of odd-carbon fatty acid chain produces
(A) Succinyl CoA (B) Propionyl CoA
(C) Acetyl CoA (D) Malonyl CoA

185. The percentage of polyunsaturated fatty acids in soyabean oil is
(A) 62 (B) 10
(C) 3 (D) 2

186. The energy yield from complete oxidation of products generated by second reaction cycle of β-oxidation of palmitoyl CoA will be
(A) 5 ATP (B) 12 ATP
(C) 17 ATP (D) 34 ATP

187. Dietary fibre denotes
(A) Undigested proteins
(B) Plant cell components that cannot be digested by own enzymes
(C) All plant cell wall components
(D) All non digestible water insoluble polysaccharide

188. Dietary fibres are rich in
(A) Cellulose (B) Glycogen
(C) Starch (D) Proteoglycans

189. Minimum dietary fibre is found in
(A) Dried apricot (B) Peas
(C) Bran (D) Cornflakes

190. The acetyl CoA formed on β-oxidation of all long chain fatty acids is metabolized under normal circumstances to
(A) CO_2 and water (B) Cholesterol
(C) Fatty acids (D) Ketone bodies

191. The major storage form of lipids is
(A) Esterified cholesterol
(B) Glycerophospholipids
(C) Triglycerides
(D) Sphingolipids

192. The nitrogenous base in lecithin is
(A) Ethanolamine (B) Choline
(C) Serine (D) Betaine

193. β-Oxidation of fatty acids requires all the following coenzymes except
(A) CoA (B) FAD
(C) NAD (D) NADP

194. Which of the following can be oxidized by β-oxidation pathway?
(A) Saturated fatty acids
(B) Monosaturated fatty acids
(C) Polyunsaturated fatty acids
(D) All of these

195. alpha-Oxidation of fatty acids occurs mainly in
(A) Liver (B) Brain
(C) Muscles (D) Adipose tissue

196. Fat content of pulses is about
(A) 5% (B) 10%
(C) 15% (D) 20%

197. Find the group of unsaturated fatty acids.
(A) Stearic acid, Palmitoleic acid, Oleic acid
(B) Lauric acid, Myristic acid, Palmitic acid
(C) Oleic acid, Linoleic acid, Linolenic acid
(D) Arachidonic acid, Lauric acid, Linoleic acid

198. At room temperature (25°C) an unsaturated fatty acid will have ....
(A) Solid consistency
(B) Liquid consistency
(C) Waxy consistency
(D) Cannot be predicted

199. 1 molecule of palmitic acid on total oxidation to CO_2 will yield molecules of ATP (as high energy bonds)
(A) 83 (B) 154
(C) 25 (D) 129

200. The number of OH groups in fatty acids can be expressed as ....
(A) Acetyl number
(B) Polenske number
(C) Reichert-Meissl number
(D) Iodine number

201. Which is a lipid database?
(A) SwissProt (B) PDBTM
(C) LMSD (D) SWISS-MODEL

202. Who elucidated the steps in beta-oxidation in 1904?
(A) Alberts Bruce (B) Franz Knoop
(C) Johns Hopkins (D) Jonathan Ling

203. BCKA decarboxylase enzyme is essential for the synthesis of ....
(A) Saturated fally acids
(B) Branched-chain fatty acids
(C) Single-chain fatty acids
(D) Unbranched fatty acids

204. The solubility of fatty acids in water ....
(A) Increase with increase in chain length and fewer the double bonds
(B) Increase with increase in chain length and increase in double bonds
(C) Decrease with chain length and decrease with number of double bonds
(D) Decrease with chain length and increase with double bonds

205. In naturally occurring unsaturated fatty acids, the double bonds are in ....
(A) Cis conformation
(B) Trans conformation
(C) A mixture of cis and trans conformation
(D) Cis and trans conformation alternatively

205. Which of the following effect produce by inhibiting of enzyme cyclooxyganase by aspirin?
(A) Blocks the synthesis of prostaglandins
(B) Blocks the synthesis of Phospholipids
(C) Blocks the synthesis of sphingomyelins
(D) Blocks the synthesis of steoids

207. Which of the following enzymes involve in the synthesis of thromoxanes derived from prostaglandins in blood platelets?
(A) Prostacyclin synthase
(B) Cyclooxygenase
(C) Thromboxane synthase
(D) β-Glucosidase

208. Match between eicosanoids and their biological effect?

i.	Prostaglandins	p.	Thromboxane antagonists
ii.	Thromboxanes	q.	Platelet agrregation
iii.	Prostacyclins	r.	Bronchoconstriction and leukotaxis
iv.	Leukotrienes	s.	Regulation of nerve transmission

(A) i—q, ii—s, iii—p, iv—r
(B) i—p, ii—s, iii—r, iv—q
(C) i—q, ii—r, iii—s, iv—p
(D) i—s, ii—q, iii—p, iv—r

209. What is the major function of lipid protein complexes in blood plasma?
(A) Form lipid transport system
(B) Act as precursors for different biological pathways
(C) Stimulate the formation RBCs
(D) Stimulate the formation RBCs, WBCs and platelets

210. Chylomicrons are .... that found in blood plasm(A)
(A) Glycolipid (B) Lipoprotein
(C) Phospholipid (D) Steroid

(iii) Amino acids and Proteins

211. Proteins are linear polymers of monomer units called
(A) Nucleoties
(B) Monosachharides
(C) Amino acids
(D) Glycerides

212. What is insulin?
(A) Water soluble lipid
(B) Polysachharide
(C) Nucleic acid derivative
(D) Protein hormone

213. What is the function of insulin?
(A) To provide energy to body organelles
(B) To maintain blood sugar at appropriate levels
(C) To maintain urine secretion from body
(D) To maintain sugar level in urine

214. What is the basic feature of primary structure of protein?
(A) A specific sequence of amino acids
(B) well-defined structures of amino acids in a single protein molecule
(C) Presence of α and β sheets
(D) 3-D structure of amino acids

215. How many insulin molecules assembled in its quaternary structure?
(A) Two (B) Five
(C) Six (D) Ten

216. How many standard amino acids built protein structure?
(A) 100 (B) 50
(C) 10 (D) 20

217. What are the building blocks of proteins?
(A) Amino acids (B) Nucleotides
(C) Monosachharides (D) Lipids

218. Which type of carbon present as central carbon in all amino acids?
(A) β-carbon (B) α-carbon
(C) γ-carbon (D) σ-carbon

219. Central carbon atom in amino acid linked to ....
(A) An amino group
(B) A carboxylic acid group
(C) A hydrogen atom and a distinctive R group
(D) All of the above

220. Amino acid differ to each other due to ....
(A) An amino group
(B) A carboxylic acid group
(C) A hydrogen atom
(D) A distinctive R side chain

221. What is the configuration of central carbon atom in amino acids?
(A) Hexagonal (B) Tetrahedral
(C) Spherical (D) Coiled

222. What is *chiral* carbon atom?
(A) Carbon attaches four different groups
(B) Carbon linked to amino and carboxylic groups
(C) Carbon has four mirror-image forms
(D) Carbon has tetrahedral configuration

223. What are called the mirror-image forms of chiral carbon containing compounds?
(A) Isoschizomers (B) Apimers
(C) Isomers (D) Pallindromes

224. What are the mirror-image forms of chiral carbon containing compounds?
(A) R and F isomers (B) P and Q isomers
(C) L and D apimers (D) L and D isomers

225. The four different substituents of an asymmetric carbon atom is called as
(A) Chiral carbon
(B) Pallindrome carbon
(C) Polar carbon
(D) Charged carbon

226. What are the stereoisomers of the four different substituents containing asymmetric carbon atom?
(A) L (right) and D (left)
(B) T (True) and F (False)
(C) P and Q arms
(D) S (*sinis-ter*) and R (*rectus*)

227. Amino acids in solution at neutral pH exist predominantly as ....
(A) Positive ions (B) Dipolar ions
(C) Negative ions (D) Partially positive

228. What is called the dipolar ion form of amino acid in solution at neutral pH?
(A) Cation (B) Anaion
(C) Neutron (D) Zwitterions

229. Which statement is true about dopolar form of amino acids?
(A) Amino group is uncharged and carboxlic group is deprotonated
(B) Amino group is negative and carboxlic group is positive
(C) Amino group is protonated and carboxlic group is deprotonated
(D) Amino group is deprotonated and carboxlic group is protonated

230. Ionization state of amino acids is varied with ....
(A) Temperature
(B) pH
(C) Side chain
(D) Concentration of molecules

231. In acid solution, amino acids contain ....
(A) $-NH_3^+$ + and –COOH
(B) $-NH_3^+$ and $-COO^-$
(C) $-NH_3$ and –COO+
(D) $-NH_3$ and –COOH

232. In basic solution, amino acids contain ....
(A) $-NH_3^+$ + and –COOH
(B) $-NH_3^+$ + and $-COO^-$
(C) $-NH_3$ and –COOH
(D) $-NH_3$ and $-COO^-$

233. What happen, as amino acids is dissolved in water and the pH is raised?
(A) Carboxylic acid is the first group to add a proton
(B) Amino is the first group to remove a proton
(C) Carboxylic acid is the first group to remove a proton
(D) No structural change

234. Amino acids are vary in .... due to different side chains?
(A) Size and shape
(B) Charge and chemical reactivity
(C) Hydrogen-bonding capacity and hydrophobic character
(D) All of the above

235. Which is the simplest amino acid?
(A) Methionine (B) Proline
(C) Glycine (D) Trptophan

236. Why glycine is considered as simplest amino acid?
(A) It has a hydrogen atom as its side chain
(B) It contains imino group instead of amino group
(C) It has aromatic side chain
(D) It has low molecular weight

237. Which feature about glycine is not true?
(A) Glycine is simplest amino acid
(B) It shows optical activity
(C) It is unique amino acid having achiral carbon
(D) It contains a hydrogen atom as its side chain

238. Which does amino acid contain aliphatic side chain and thioether group?
(A) Glycine (B) Tryptophan
(C) Phenylalanine (D) Methionine

239. Which does amino acid contain additional chiral center and hydrophobic aliphatic side chains?
(A) Leucine (B) Isoleucine
(C) Arginine (D) Methionine

240. .... contains an imidazole group?
(A) Histidine (B) Arginine
(C) Lysine (D) Tryptophan

241. Name of amino acids that contain acidic side chain?
(A) Aspargine and glutamine
(B) Arginine and lysine
(C) Aspartate and glutamate
(D) Tyrosine and cysteine

242. Match single letter code for following amino acids.

i.	Aspartate	p.	K
ii.	Tryptophan	q.	Q
iii.	Lysine	r.	D
iv.	Glutamine	s.	W

	i	ii	iii	iv
(A)	p	q	r	s
(B)	r	s	p	q
(C)	q	r	p	s
(D)	s	q	r	p

243. Proteins are linear polymers formed by linking the .... of one amino acid to the .... of another amino acid with a peptide bon(D)
(A) α-arboxyl group and α-amino group
(B) α-hydrogen and side chain
(C) α-carboxyl group and hydrogen atom
(D) α-amino group and R-group

244. Peptide bond is a type of ....
(A) Phosphodiester bond
(B) Glycosidic bond
(C) Amide bond
(D) Electrostatic bond

245. Which of the following molecule is removed after forming a peptide bond?
(A) NH_3 (B) H_2O
(C) OH^- (D) CO_2

246. Which of the following statement is not true?
(A) A series of amino acids joined by peptide bonds form a polypeptide chain.
(B) Each amino acid unit in a polypeptide is called a residue
(C) A polypeptide chain has polarity
(D) The ends of a polypeptide chain contain same an α-amino group

247. The ends of a polypeptide chain contain ....
(A) α-carboxyl group at both ends
(B) α-amino group at both ends
(C) α-amino group at one end and α-carboxyl group at the other
(D) α-amino group at one end and R-group at the other

248. Polypeptide chain is always read in ....
(A) C → N Termini (B) N→ C termini
(C) 5' → 3' direction (D) 3' → 5' direction

249. Write in single letter code of polypeptide chain Tyr-Gly-Gly-Phe-Leu
(A) GLFGY (B) YLFGY
(C) YGGFL (D) LFGGY

250. What is the mean molecular weight of an amino acid residue?
(A) 100 daltons (B) 110 daltons
(C) 20 daltons (D) 198 daltons

251. One *dalton* is equal to
(A) One atomic mass unit
(B) One ppm
(C) One kg
(D) One atm

252. Who develop atomic theory of matter?
(A) Thomson (B) Rhaderford
(C) John Dalton (D) Neils bore

253. What is disulfide bond?
(A) A cross-link formed by the oxidation of a pair of cysteine residues
(B) Linkage between two nucleotides
(C) Linkage between any two amino acids
(D) Connection between two sugar and base

254. A unit of mass equal to 1000 daltons
(A) 1 km (B) 1 kg
(C) 1 kd (D) 1 kilo

255. Proteins have unique amino acid sequences that are specified by ....
(A) DNA (B) Genes
(C) Nucleotides (D) RNA

256. Who determined the amino acid sequence of insulin as a protein hormone?
(A) Frederick Sanger (1953)
(B) Watson and Crick (1953)
(C) Harshey and Chase (1952)
(D) H.G. Khorana (1972)

257. The amino acid sequence of a protein hormone is determined by Frederick Sanger in 1953.
(A) Heamoglobin (B) Niacin
(C) Insulin (D) ABA

258. The amino acid sequence of a protein is referred to as its
(A) Primary structure
(B) Secondary structure
(C) Tertiary structure
(D) Quaternary structure

259. What is DNA?
(A) Mutation unit
(B) Specify the sequence of amino acids in a protein
(C) Expression unit
(D) The molecule of heredity

260. The peptide bond is essentially
(A) Planar
(B) Triangular
(C) V-shaped
(D) Zigzag

261. How many atoms present in the same plane of a peptide bond?
(A) Ten (B) Two
(C) Six (D) Five

262. Which of the following is not present in same plan of a peptide bond?
(A) α-carbon atoms from the linked two amino acid
(B) NH group and from the second amino acid
(C) NH group and from the first amino acid
(D) CO group from the first amino acid

263. Which of the following characteristic is not of peptide bond?
(A) Peptide bond has double-bond character
(B) Almost all peptide bonds in proteins has *trans* configuration
(C) CN distance in a peptide bond is typically 1.32 Å
(D) It is charged bond

264. Which of the following conformation are possible for a planar peptide bond?
(A) *Trans* and *cis* configuration
(B) R and S configuration
(C) D and L configuration
(D) P and Q configuration

265. Which of the following bond angle specify the rotation of peptide bonds
(A) *phi* (Φ) and *psi* (ψ)
(B) *delta* (λ) and *gamma* (γ)
(C) *alpha* (α) and *beta* (β)
(D) *epsilon* (ε) and *omega* (ω)

266. The angles of rotation *phi* (Φ) and *psi* (ψ) determine ....
(A) 3-D structure of proteins
(B) The path of the polypeptide chain
(C) Solubility of protein in water
(D) Amount of bond energy

267. What is called the dihedral angles of rotation *phi* (Φ) and *psi* (ψ) in proteins?
(A) Torsion angles
(B) Hybridization angles
(C) Direction angles
(D) Twisting angles

268. Who given the two dimensional plot of protein showing torsion angles?
(A) Christian Anfinsen
(B) G. N. Ramachandran
(C) Robert Corey
(D) Linus Pauling

269. What is called the two dimensional plot of protein showing torsion angles?
(A) Linus Pauling
(B) Robert Corey
(C) Ramchandran plot
(D) Anfinsen plot

270. In which structure polypeptide chains can fold into regular structures such as the alpha helix, the beta sheet, and turns and loops?
(A) Primary structure
(B) Secondary structure
(C) Tertiary structure
(D) Quaternary structure

271. Who proposed the two periodic structures are *α-helix* (alpha helix) and the *β-pleated* sheet (*beta* pleated sheet) in secondary structure of proteins?
(A) Christian Anfinsen and Holly
(B) Linus Pauling and Robert Corey
(C) G.N. Ramchandran and Christian Anfinsen
(D) Maxam Gilbert and Sanger

272. *alpha* helix, *beta* pleated sheet, β turn and *omega* (ω) loop are the characteristic structures of ....
(A) Primary structure
(B) Secondary structure
(C) Tertiary structure
(D) Quaternary structure

273. Which of the following technique is used to reveal three-dimensional structures of thousands of proteins?
(A) Electrophoresis and HPLC
(B) SAGE and protein microarray
(C) TEM and SEM
(D) X-ray crystallographic and nuclear magnetic resonance

274. Which of the following statement is not true about myoglobin protein?
(A) Myoglobin is the carbon dioxide carrier in muscle
(B) It is a single polypeptide chain of 153 amino acids
(C) The capacity of myoglobin to bind oxygen depends on the presence of heme
(D) Two histidine residues play critical roles in binding iron and oxygen

275. Which of the following protein found in the outer membranes of many bacteria that function in hydrophobic environments?
(A) Cyclin (B) Myoglobin
(C) Porin (D) Flagellin

276. Protein domains are the characteristics of ......
(A) Primary structure
(B) Secondary structure
(C) Tertiary structure
(D) Quaternary structure

277. Which of the following is the oxygen-carrying protein in muscles?
(A) Hemoglobin (B) Myoglobin
(C) Porin (D) Keratin

278. Which statement is not true regarding to four levels of protein architecture?
(A) Primary structure is the amino acid sequence
(B) Secondary structure is the sequence and the pattern of disulfide bonds
(C) Tertiary structure contains domains of proteins
(D) Quaternary structure contains a dimer consisting of two identical subunits.

279. Which of the following is the oxygen-carrying protein in blood?
(A) Hemoglobin (B) Myoglobin
(C) Porin (D) Keratin

280. Which of the statement is not true about human hemoglobin?
(A) It contains four heme groups containing polypeptide chains
(B) Hemoglobin molecule exists as an α2β2 tetramer
(C) Protein leghemoglobin has same characteristics to human hemoglobin
(D) It carries oxygen from the lungs to tissues

281. Who revealed the relation between the amino acid sequence of a protein and its conformation working on enzyme ribonuclease?
(A) Christian Anfinsen
(B) G.N. Ramachandran
(C) Robert Corey
(D) Linus Pauling

282. Which of the following enzyme was used to reveale the relation between the amino acid sequence of a protein and its conformation by Christian Anfinsen?
(A) DNase (B) RNase
(C) Barnase (D) DNA polymerase

283. How many amino acids consisted in a single polypeptide chain of ribonuclease?
(A) 150 amino acid residues
(B) 100 amino acid residues
(C) 70 amino acid residues
(D) 124 amino acid residues

284. How many disulfide bonds cross-link a single polypeptide chain consisting of 124 amino acid residues in ribonuclease?
(A) Two (B) Six
(C) Four (D) Only one

285. Which of the following chemical agents effectively disrupt the noncovalent bonds in polypeptide chain?
(A) Urea
(B) Guanidinium chloride
(C) β-mercaptoethanol
(D) Both a and b

286. Which of the following chemical agents effectively disrupt the disulfide bonds in polypeptide chain?
(A) Acrylamide
(B) Guanidinium chloride
(C) β-mercaptoethanol
(D) Ethanol

287. A mutation has changed an isoleucine residue of a protein to Glutamic acid, which statement best describes its location in a hydrophilic exterior
(A) On the surface since it is hydrophilic in nature
(B) In side the core of the protein since it is hydrophobic in nature
(C) Any where inside or outside
(D) Inside the core of protein since it has a polar but uncharged side chain

288. Choose the incorrect statement.
(A) Only L amino acids are found in the biological system
(B) Glycine is optical inactive
(C) Tyrosine is a modified amino acid
(D) Selenocysteine is 21st amino acid

289. Which of the following amino acid carriy a net positive charge at the physiological pH?
(A) Valine (B) Leucine
(C) Isoleucine (D) Lysine

290. Which of the following amino acids is a precursor for a mediator of allergies and inflammation?
(A) Histidine (B) Tyrosine
(C) Phenyl Alanine (D) Tryptophan

291. Mother of a mal-nourished child has been instructed to include a complete protein in diet for her child, which out of the followings proteins should be recommended?
(A) Pulses (B) Wheat
(C) Soy Protein (D) Milk

292. Amino acid can participate in hydrogen bonding except one
(A) Serine (B) Cysteine
(C) Threonine (D) Valine

293. Which out of the following amino acid is a precursor of niacin (Vitamin)?
(A) Tyrosine (B) Threonine
(C) Tryptophan (D) Phenylalanine

294. The greatest buffering capacity at physiological pH would be provided by a protein rich in which of the following amino acids?
(A) Serine (B) Cysteine
(C) Alanine (D) Histidine

295. Which of the following amino acid is a limiting amino acid in pulses?
(A) Leucine (B) Lysine
(C) Methionine (D) Glutamine

296. All the below mentioned proteins are metalloproteins except
(A) Carbonic anhydrase
(B) Xanthine oxidase
(C) Lactate dehydrogenase
(D) Superoxide dismutase

297. Amino acids is most compatible with an α-helical structure?
(A) Tryptophan (B) Alanine
(C) Leucine (D) Proline

298. In scurvy, which amino acid that is normally part of collagen is not synthesized?
(A) Hydroxy Tryptophan
(B) Hydroxy Tyrosine
(C) Hydroxy Alanine
(D) Hydroxy Proline

299. Which of the amino acids is the uncharged derivative of an acidic amino acid?
(A) Cystine (B) Tyrosine
(C) Glutamine (D) Serine

300. Which of the following has an imino (>NH) group instead of amino group (-NH_2)?
(A) Isoleucine (B) Proline
(C) Tyrosine (D) Serine

Answer-Key

Q	Ans	Q	Ans	Q	Ans	Q	Ans	Q	Ans	Q	Ans	Q	Ans	Q	Ans
1.	(C)	41.	(D)	81	(A)	121.	(D)	161.	(A)	201.	(C)	241.	(C)	281.	(A)
2.	(B)	42.	(B)	82.	(C)	122.	(B)	162.	(C)	202.	(B)	242.	(B)	282.	(B)
3.	(B)	43.	(C)	83.	(C)	123.	(C)	163.	(D)	203.	(B)	243.	(A)	283.	(D)
4.	(A)	44.	(A)	84.	(D)	124.	(B)	164.	(A)	204.	(C)	244.	(C)	284.	(C)
5.	(C)	45.	(B)	85.	(B)	125.	(D)	165.	(C)	205.	(A)	245.	(B)	285.	(D)
6.	(B)	46.	(C)	86.	(B)	126.	(C)	166.	(B)	206.	(A)	246.	(D)	286.	(C)
7.	(D)	47.	(C)	87.	(B)	127.	(B)	167.	(A)	207.	(C)	247.	(C)	287.	(A)
8.	(A)	48.	(D)	88.	(C)	128.	(C)	168.	(B)	208.	(D)	248.	(B)	288.	(C)
9.	(B)	49.	(A)	89.	(D)	129.	(A)	169.	(B)	209.	(A)	249.	(C)	289.	(D)
10.	(C)	50.	(A)	90.	(D)	130.	(A)	170.	(D)	210.	(B)	250.	(B)	290.	(A)
11.	(A)	51.	(A)	91.	(D)	131.	(B)	171.	(C)	211.	(C)	251.	(A)	291.	(D)
12.	(C)	52.	(D)	92.	(A)	132.	(C)	172.	(B)	212.	(D)	252.	(C)	292.	(D)
13.	(B)	53.	(C)	93.	(B)	133.	(B)	173.	(C)	213.	(B)	253.	(A)	293.	(C)
14.	(A)	54.	(A)	94.	(C)	134.	(D)	174.	(D)	214.	(A)	254.	(C)	294.	(D)
15.	(B)	55.	(C)	95.	(D)	135.	(D)	175.	(B)	215.	(C)	255.	(B)	295.	(C)
16.	(D)	56.	(D)	96.	(B)	136.	(B)	176.	(A)	216.	(D)	256.	(A)	296.	(C)
17.	(C)	57.	(B)	97.	(B)	137.	(A)	177.	(D)	217.	(A)	257.	(C)	297.	(B)
18.	(C)	58.	(B)	98.	(A)	138.	(B)	178.	(C)	218.	(B)	258.	(A)	298.	(D)
19.	(C)	59.	(D)	99.	(C)	139.	(C)	179.	(D)	219.	(D)	259.	(D)	299.	(C)
20.	(A)	60.	(A)	100.	(C)	140.	(B)	180.	(A)	220.	(D)	260.	(A)	300.	(B)
21.	(D)	61.	(C)	101.	(C)	141.	(B)	181.	(C)	221.	(B)	261.	(C)		
22.	(B)	62.	(D)	102.	(B)	142.	(D)	182.	(A)	222.	(A)	262.	(C).		
23.	(C)	63.	(A)	103.	(D)	143.	(B)	183.	(B)	223.	(C)	263.	(A)		
24.	(B)	64.	(B)	104.	(B)	144.	(D)	184.	(D)	224.	(D)	264.	(A)		
25.	(A)	65.	(C)	105.	(A)	145.	(B)	185.	(A)	225.	(A)	265.	(B)		
26.	(A)	66.	(D)	106.	(A)	146.	(A)	186.	(D)	226.	(D)	266.	(A)		
27.	(B)	67.	(A)	107.	(C)	147.	(B)	187.	(B)	227.	(B)	267.	(A)		
28.	(C)	68.	(B)	108.	(D)	148.	(A)	188.	(A)	228.	(D)	268.	(B)		
29.	(D)	69.	(C)	109.	(C)	149.	(D)	189.	(B)	229.	(C)	269.	(C)		
30.	(C)	70.	(C)	110.	(A)	150.	(A)	190.	(A)	230.	(B)	270.	(B)		
31	(A)	71.	(A)	111.	(A)	151.	(D)	191.	(C)	231.	(A)	271.	(B)		
32.	(C)	72.	(C)	112.	(B)	152.	(C)	192.	(B)	232.	(D)	272.	(B)		
33.	(A)	73.	(B)	113.	(D)	153.	(B)	193.	(D)	233.	(C)	273.	(D)		
34.	(D)	74.	(B)	114.	(C)	154.	(B)	194.	(D)	234.	(D)	274.	(A)		
35.	(D)	75.	(D)	115.	(A)	155.	(C)	195.	(B)	235.	(C)	275.	(C)		
36.	(C)	76.	(D)	116.	(D)	156.	(B)	196.	(A)	236.	(A)	276.	(B)		
37.	(A)	77.	(A)	117.	(C)	157.	(D)	197.	(C)	237.	(B)	277.	(B)		
38.	(D)	78.	(A)	118.	(B)	158.	(A)	198.	(B)	238.	(D)	278.	(B)		
39.	(D)	79.	(A)	119.	(A)	159.	(C)	199.	(D)	239.	(B)	279.	(A)		
40.	(C)	80.	(C)	120.	(C)	160.	(C)	200.	(A)	240.	(A)	280.	(C)		

2

Structure and Functions of Nucleic Acids

Johann Friedrich Miescher (1869) isolated a substance from nuclei of human white blood cells and called as "nuclein" which presently known as nucleic acids. Griffith also proved that nuclein is acidic and phophorus containing substance. Before discovery of nucleic acids, most of all the scientists considered that protein must be the genetic material responsible for inheritance of traits because proeins were quite complex molecules, whereas DNA was a relatively simple molecule in comparison to proteins.

In 1928, F. Griffith performed a significant experiment on transformation in bacterium *Streptoccus pneumoniae* (previously known as *Diplococcus pneumonia*). He concluded from the experiment that there was a genetic substance called as "transforming principle" in the heat treated smooth type III bacteria (SIII) which transformed the live mutant rough type II (RII) bacteria contains capsule polysaccharide. But he could not exactly prove that what was the "transforming priciple" protein or DNA, or what? In 1944, Avery, McCarty and McCleod provided strong evidence that the "transforming agent" was in fact DNA (and not protein) which was not degraded by protease enzyme (trypsin, protein degrading enzyme) and RNase (RNA degrading enzyme). They finally proved that DNA is only genetic material. More strong evidence for DNA as genetic material obtained from the experiment of Hershey and Chase worked on bateriophage T2 (a virus) which attacks the bacteria *E. coli*. They used ^{32}P labeled ATP in DNA and ^{35}S labeled methionine in protein. They concluded that Phage T2 transferred genetic material (mainly DNA) to the bacteria during infection and proved that DNA was the "stuff" of genetic inheritance.

Nucleic acids formulated as alternating phosphoric acid (P) and nucleosides (N):

$$\sim P - N - P - N' - P - N'' - P - N''' - P - N \sim$$

Information is stored or encoded in the DNA polymer represented by four nucleotides, guanosine (G), cytidine (C), adenosine (A) and thymidine (T). Total content of DNA of a cell is called genome and total content of RNA is called transcriptome. Three types of RNA found in a cell. Ribosomal RNA (rRNA) is the major component of ribosomes, messenger RNA (mRNA) and transfer RNA (tRNA). Watson and Crick proposed the right-handed double helix structure of DNA called also B-DNA. They showed that DNA has 10.4 base pairs per turn, and rises 3.4 Å in each turn. Two alternating grooves found, a wide and deep major groove (22Å wide), and a shallow and narrow minor groove (12Å wide). Other helical structures of DNA have also been observed, and are designated by A and Z.

Nomenclature of bases, nucleosides, and nucleotides

RNA		
Base	Ribonucleoside	Ribonucleotide (5′-monophosphate)
Adenine (A)	Adenosine	Adenylate (AMP)
Guanine (G)	Guanosine	Guanylate (GMP)

Uracil (U)	Uridine	Uridylate (UMP)
Cytosine (C)	Cytidine	Cytidylate (CMP)
DNA		
Base	Deoxyribonucleoside	Deoxyribonucleotide (5′-monophosphate)
Adenine (A)	Deoxyadenosine	Deoxyadenylate (dAMP)
Guanine (G)	Deoxyguanosine	Deoxyguanylate (dGMP)
Thymine (T)	Thymidine	Thymidylate (TMP)
Cytosine (C)	Deoxycytidine	Deoxycytidylate (dCMP)

Role of nucleotides in biological processes

1. Nucleotides are the activated precursors of nucleic acids and essential for the replication and transcription of the genome.
2. Adenine triphosphate (ATP) is an universal currency of energy. Guanine triphosphate (GTP) also serves as an energy source in many biological processes.
3. Nucleotide derivatives such as UDP-glucose participate used in biosynthetic processes such as the formation of glycogen.
4. Nucleotides are essential components of signal-transduction pathways. cyclic AMP and cyclic GMP are second messengers that transmit signals both within and between cells. ATP acts as the donor of phosphoryl groups transferred by protein kinases.

Nucleotide biosynthesis

Two important biosynthesis pathways of nucleotides: de novo pathways and salvage pathways. In de novo pathways, the nucleotide bases are composed from simpler compounds. A pyrimidine base is assembled first and then attached to ribose. A purine base is synthesized in segments directly onto a ribose-based structure. In salvage pathways, bases are recovered and reconnected to a ribose unit. Both de novo and salvage pathways lead to the synthesis of ribonucleotides. Nucleotide biosynthesis is regulated by feedback inhibition. NAD^+, FAD, and Coenzyme A are formed from ATP.

Salvage pathway: recovery and recycling of nucleotides obtained in the diet

Activated ribosome (PRPP) + base → Nucleotide

De Novo pathway: nucleotides are constructed from simple precursors

Activated ribose (PRPP) + amino acids + ATP + CO_2 → Nucleotide

Disorders of nucleotide metabolisms

Lesch-Nyhan Syndrome: It is described by William Nyhan and Michael Lesch in 1964. It caused due to hereditary deficiency of hypoxanthine-guanine phosphoribosyltransferase in mostly males. Major symptoms of the disease are mental retardation; spasticity; bizarre tendency to self-mutilate.

Orotic aciduria: It causes megaloblastic anemia, mental retardation, and stunted growth due to hereditary deficiency of the enzyme orotate phosphoribosyl transferase (a bifunctional enzyme) that catalyzes two reactions in the pyrimidine biosynthesis pathway: conjugation of orotic acid to ribose 5'-phosphate and decarboxylation of orotidine 5-phosphate to uridine 5-phosphate.

Key points

- Nucleic acid discovered by Friedrich Miescher in 1869 and called nuclein.
- Nucleic acid term coined by Richard Altmann in 1889.
- Nucleic acid consists of nucleotides that are linked together by phosphodiester bonds.
- Nucleotides are made up of 3 components: Nitrogenous base, Sugar and Phosphate group.
- Nitrogenous bases are

  (i) Adenine (ii) Guanine, (iii) Cytosine, (iv) Thymine (present only in DNA) and (v) Uracil (present only in RNA).
- Pentose sugar found in DNA called deoxyribose and in RNA it is called ribose.
- Adenine and guanine are purines and form a glycosidic bond between their 9th nitrogen and the 1' -OH group of the nucleic acids.
- Cytosine, thymine and uracil are Pyrimidines, glycosidic bond forms between 1st nitrogen and the 1' -OH of the nucleic acids.
- Purines consist of one six membered and one five membered ring containing nitrogen. Pyrimidines have only six membered nitrogen containing ring.
- A purine always pairs with a pyrimidine by hydrogen bonds. guanosine (G) pairs with cytosine (C) with three hydrogen bonds and adenine (A) pairs with thymine (T) or uracil (U) with two hydrogen bonds.
- The polarity in DNA and RNA caused by the presence of oxygen and nitrogen atoms in the backbone.
- DNA's secondary structure is predominantly determined by base-pairing of the two polynucleotide strands wrapped around each other to form a double helix.
- The four simple components in the secondary structure of RNA are helices, loops, bulges, and junctions.
- Stem-loop or hairpin loop found mostly in secondary structure of RNA.
- Stem-loop is formed when the RNA chains fold back on themselves to form a double helical tract called the stem, the unpaired nucleotides forms single stranded region called loop.
- Bulges and inner loops are designed by leaving of the double helical tract on either one strand or on both strands by unpaired nucleotides.
- Pseudoknot is a RNA secondary structure first recognized in turnip YMV.
- DotKnot-PW method is used for comparative pseudoknots prediction.
- B-DNA is the most common form of DNA *in vivo* and is a narrower, elongated helix than A-DNA.
- B-DNA base pairs are almost vertical to the helix axis. The sugar pucker which determines the shape of the α-helix, whether the helix exist in the A-form or in the B-form, occurs at the C2'-endo.
- RNA-DNA duplexes are mostly A-form.
- Base pairs in A-DNA sloped comparatively to the helix axis, and are moved from the axis. The sugar pucker occurs at the C3'-endo and in RNA 2'-OH inhibits C2'-endo conformation.
- Z-DNA is a left-handed double-helix *in vivo* but its function is unclear.
- Z-DNA's major groove is not really a groove, and it has a narrow minor groove.

- Z-DNA bases form helix axis with zigzag backbone.
- Supercoiling is characterized by the linking number, twist and writhe. The linking number (Lk) for circular DNA is defined as the number of times one strand would have to pass through the other strand to completely separate the two strands.
- Linking number for circular DNA only changed by breaking of a covalent bond in one of the two strands. Always an integer, the linking number of DNA is the sum of two components: twists (Tw) and writhes (Wr). Lk = Tw + Wr
- Twists are the number of times the two strands of DNA are twisted around each other and writhes are number of times the DNA helix crosses over itself.
- DNA is negatively supercoiled and has the tendency to unwind. The separation of strands is easier in negatively supercoiled DNA than in relaxed DNA.
- The two components of supercoiled DNA are solenoid and plectonemic nature. The plectonemic supercoil is found in prokaryotes, while the solenoidal supercoiling in eukaryotes.
- Quaternary structure of nucleic acids is similar to protein quaternary structure.

Multiple Choice Questions

1. Which of the following is mostly used as the energy currency in metabolic processes?
   (A) GTP (B) ATP
   (C) UTP (D) CTP
2. What is the function of nucleotides?
   (A) Essential chemical links in the response of cells to hormones
   (B) Structural components of an array of enzyme cofactors
   (C) Constituents of nucleic acids (DNA and RNA)
   (D) All of the above
3. Which are the molecular repositories of genetic information?
   (A) Proteins (B) Enzymes
   (C) Carbohydrates (D) Nucleic Acids
4. Which has the ability to store and transmit genetic information from one generation to the next?
   (A) Nucleic acids (B) Proteins
   (C) Carbohydrates (D) Nucleosides
5. The amino acid sequences in the structure of protein is determined by
   (A) The nucleotide sequences of RNA
   (B) The nucleotide sequences of DNA
   (C) The monosaccharide sequences of carbohydrates
   (D) Fatty acid sequences of Lipids
6. What are the building blocks of nucleic acids?
   (A) Nucleosides (B) Fatty acids
   (C) Amino acids (D) Nucleotides
7. The nucleotide sequences of every RNA is specified by
   (A) Nucleotide sequences in the DNA
   (B) Nucleoside sequences in the DNA
   (C) Amino acid sequence in the Protein
   (D) Nucleoside sequences in the RNA
8. Which is contains the information required for the synthesis of a functional biological product (protein or RNA)?
   (A) A segment of a fatty acid molecule
   (B) A segment of a sugar molecule
   (C) A segment of a DNA molecule
   (D) A segment of a amino acid molecule
9. What is called a segment of a DNA molecule that contains the information required for the synthesis of a functional biological product (protein or RNA)?
   (A) Gene (B) Codon
   (C) Chromosome (D) Genome
10. Which types of major RNAs are found in cells?
    (A) Ribosomal RNAs (rRNAs)
    (B) Messenger RNAs (mRNAs)
    (C) Transfer RNAs (tRNAs)
    (D) All of the above
11. Which types of RNAs are the components of ribosome that carry out the synthesis of proteins?
    (A) Ribosomal RNAs (rRNAs)
    (B) Messenger RNAs (mRNAs)
    (C) Transfer RNAs (tRNAs)
    (D) Heterogenous nuclear RNA (hnRNA)
12. Which types of RNAs are intermediaries that carry genetic information from one or a few genes to a ribosome?
    (A) Ribosomal RNAs (rRNAs)
    (B) Messenger RNAs (mRNAs)
    (C) Transfer RNAs (tRNAs)
    (D) Heterogeneous nuclear RNA (hnRNA)
13. Which types of RNAs work as adapter molecules that faithfully translate the information in mRNA into a specific sequence of amino acids?
    (A) Ribosomal RNAs (rRNAs)
    (B) Messenger RNAs (mRNAs)
    (C) Transfer RNAs (tRNAs)
    (D) Heterogeneous nuclear RNA (hnRNA)
14. What are the characteristic components of nucleotides?
    (A) A nitrogenous base
    (B) A pentose sugar
    (C) A phosphate
    (D) All of the above
15. Which is the nitrogen-containing component of the nucleotide?
    (A) A base (B) A pentose sugar
    (C) A phosphate (D) Ribose and deoxyribose

16. The molecule that has a nitrogenous base and a pentose sugar but not has the phosphate group is called ....
(A) Nucleoside (B) Nucleotide
(C) Purine (D) Pyrimidine

17. Which are the major types of the nitrogenous bases?
(A) Pyrimidines and Purines
(B) Proteins and Purines
(C) DNA and RNA
(D) Amino acids and Nucleotides

18. Which are the heterocyclic compounds of the common nucleotides?
(A) Nitrogenous base and phosphate group
(B) A pentose sugar and phosphate group
(C) Nitrogenous base and sugar
(D) Nitrogenous base, phosphate group and sugar

19. In the pentoses of nucleotides and nucleosides, the carbon numbers are given a .... designation to distinguish them from the numbered atoms of the nitrogenous bases.
(A) Percentage (%) (B) Comma (,)
(C) Star (*) (D) Prime (')

20. By which bond is joined the base of a nucleotide covalently to the 1' carbon of the pentose, and the phosphate is linked to the 5' carbon, respectively?
(A) *N*-β-glycosyl bond and phosphodiester bond
(B) Phosphodiester bond and N-β-glycosyl bond
(C) Hydrogen bond and phosphodiester bond
(D) *N*-β-glycosyl bond and hydrogen bond

21. Which nitrogen atoms participate in joining the base of purines and pyrimidines covalently bind to the 1' carbon of the pentose?
(A) N-1 of pyrimidines and N-7 of purines
(B) N-1 of pyrimidines and N-9 of purines
(C) N-3 of pyrimidines and N-3 of purines
(D) N-9 of pyrimidines and N-1 of purines

22. The *N*-β-glycosyl bond is formed between base and sugar by removal of ....
(A) CO_2 (B) H_2O
(C) NH_3 (D) H_2O and NH_3

23. Which are the major purine bases found in both DNA and RNA?
(A) Adenine (A) and guanine (G)
(B) Adenine (A) and cytosine (C)
(C) Thymine (T) and guanine (G)
(D) Uracil (U) and cytosine (C)

24. Which is one of the major pyrimidine base found in both DNA and RNA?
(A) Cytosine (B) Adenine
(C) Thymine (D) Uracil

25. Which is one of the major pyrimidine base found only in RNA but not in DNA?
(A) Cytosine (B) Adenine
(C) Thymine (D) Uracil

26. Which is the one of the major pyrimidine base found only in DNA but not in RNA?
(A) Cytosine (B) Adenine
(C) Thymine (D) Uracil

27. How many kinds of pentose sugars found in nucleic acids?
(A) Two (B) Five
(C) Four (D) Three

28. Which pentose sugars are contained in DNA and RNA, respectively?
(A) 2'-deoxy-D-ribose and D-ribose
(B) D-ribose and 2'-deoxy-D-ribose
(C) 3'-deoxy-D-ribose and 2'-D-ribose
(D) 2'-D-ribose and 3'-deoxy-D-ribose

29. In which form present both pentose sugars (2'-deoxy-D-ribose and D-ribose) in nucleotides?
(A) β-furanose form (closed five-membered ring)
(B) β-furanose form (closed six-membered ring)
(C) β-pyranose form (closed five-membered ring)
(D) β-pyranose form (closed six-membered ring)

30. The pentose ring is not planar in nucleotides but occurs in one of a variety of conformations generally described as ....
(A) Compacted
(B) Puckered
(C) Cruciform
(D) Hairpin

31. Which are the structural units of DNAs and RNAs?

(A) Ribonucleoside 5'-monophosphates and deoxyribonucleoside 5'-monophosphates

(B) Deoxyribonucleoside 5'-monophosphates and ribonucleoside 5'-monophosphates

(C) Deoxyribonucleoside 3'-monophosphates and ribonucleoside 3'-monophosphates

(D) Ribonucleoside 3'-monophosphates and deoxyribonucleoside 3'-monophosphates

32. Which nitrogenous base was first isolated from guano (bird manure)

(A) Thymine (B) Cytosine

(C) Guanine (D) Adenine

33. Which nitrogenous base was first isolated from thymus tissues.

(A) Thymine (B) Cytosine

(C) Guanine (D) Adenine

34. DNA and RNA have deoxyribose and ribose sugar, respectively, which are differ in ....

(A) SH group on the 2' carbon in ribose replaced with OH

(B) OH group on the 5' carbon in deoxyribose replaced with H

(C) H group on the 2' carbon in ribose replaced with OH

(D) H group on the 3' carbon in deoxyribose replaced with OH

35. Specific long sequences of .... in DNA are the repository of genetic information.

(A) A, U, G, and C

(B) Ser-Try-Tyr-Phn

(C) Glucose-Fructose-Sucrose-Mannose

(D) A, T, G, and C

36. Which Structure of thymine and uracil, respectively?

(i)

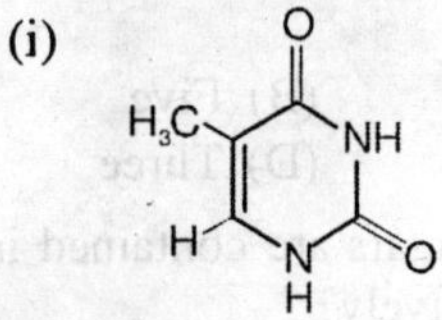

(ii) (iii) (iv)

(A) i & iv (B) ii & iii (C) iii & iv (D) iii & ii

37. Which is the common structure of nucleotide?

(A)

(B)

(C)

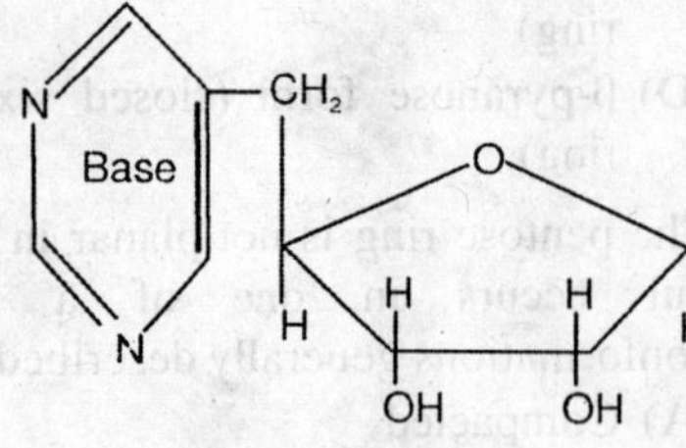

(D)

38. Which are common structures of ribose and deoxyribose sugars?

(i)

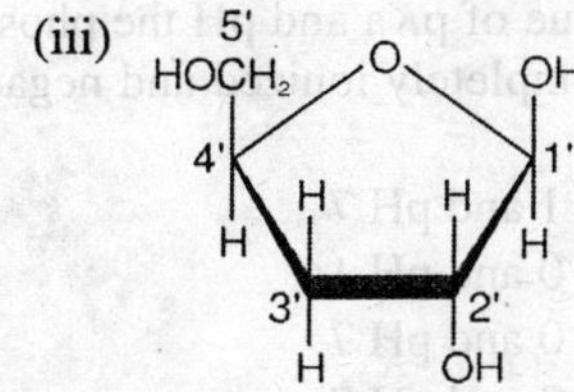

(ii)

(iii)

(iv)

(A) i & iii (B) ii & iii (C) i & ii (D) ii & iv

39. Ribofuranose rings in ribonucleotides can exist in .... different puckered conformations.

(A) Three (B) Two
(C) One (D) Four

40. In DNA, phosphate group and bases are linked to sugar at .... and ...., respectively.

(A) 3' carbon & 4' carbon
(B) 5' carbon & 3' carbon
(C) 3' carbon & 2' carbon
(D) 2' carbon & 5' carbon

41. What is chemical name of thymine and adenine?

(A) 2, 4-dioxy-uracil and 6'-amino purine
(B) 5'-methyl uracil and 6'-amino purine
(C) 2, 6'-dioxypyrimidine and 2'-amino-6'-oxypurine
(D) 2, 4'-dioxy-5'methyl pyrimidine and 6'-amino purine

42. Which nucleotides are formed by enzymatic and alkaline hydrolysis of RNA?

(A) Adenosine 2'-monophosphate
(B) Guanosine 3',5'-cyclic monophosphate
(C) Ribonucleotides 2',3'-cyclic monophosphate
(D) All of the above

43. In which form are minor bases found in DNA and RNA?

(A) Methylated forms of the major bases
(B) Sulphonated forms of the major bases
(C) Ethylated forms of the major bases
(D) Methylsulfonated forms of the major bases

44. What are the roles of altered or unusual bases (methylated) in DNA molecules?

(A) Regulation or protection of the genetic information
(B) Polymerization of the genetic information
(C) Splicing of the genetic information
(D) Mutation of the genetic information

45. Which bond covalently linkes successive nucleotides in both DNA and RNA?

(A) Phosphodiester linkage
(B) N-glycosidic bond
(C) Hydrogen bond
(D) Wander wall bond

46. The successive nucleotides in both DNA and RNA covalently linked through phosphate-group "bridges," formed between ....

(A) 3'-phosphate group of one nucleotide and 5'-hydroxyl group of next nucleotide
(B) 5'-phosphate group of one nucleotide and 3'-hydroxyl group of next nucleotide
(C) 5'-phosphate group of one nucleotide and 2'-hydroxyl group of next nucleotide
(D) 5'-phosphate group of one nucleotide and 1'-hydroxyl group of next nucleotide

47. Which residues are formed the covalent backbones of nucleic acids
(A) Alternating nitrogenous bases and pentose residues
(B) Alternating phosphate, pentose residues and the nitrogenous bases
(C) Alternating phosphate and nitrogenous bases
(D) Alternating phosphate and pentose residues

48. In DNA and RNA, the covalent backbones consist of alternating phosphate and pentose residues but nitrogenous bases may be regarded as .... joined to the backbone at regular intervals.
(A) Major group (B) Partial group
(C) Side groups (D) Alkyl group

49. The backbones of both DNA and RNA are
(A) Basic and Hydrophobic
(B) Basic and Hydrophilic
(C) Acidic and Hydrophobic
(D) Acidic and Hydrophilic

50. What is the nature of the hydroxyl groups of the sugar form hydrogen bonds with water?
(A) Hydrophobic (B) Hydrophilic
(C) Amphiteric (D) Isotonic

51. Which is the following structure of Adenine?
(A)

(B)

(C)

(D)

52. On which value of p*K*a and pH the phosphate groups are completely ionized and negatively charged
(A) p*K*a near 1 and pH 7
(B) p*K*a near 0 and pH 14
(C) p*K*a near 0 and pH 7
(D) p*K*a near 7 and pH 7

53. Positive charges of which ions and compounds neutralize the negative charges of phosphate groups in DNA.
(A) Proteins, metal ions, and polyamines
(B) Carbohydrates, amines and metal ions
(C) Divalent ions, lipids and proteins
(D) Inorganic compounds, metal ions and acids

54. All the phosphodiester linkages have the same orientation along the chain in each linear nucleic acid strand with a specific polarity of distinct ....
(A) 3' and 4' ends (B) 2' and 5' ends
(C) 5' and 4' ends (D) 5' and 3' ends

55. RNA is hydrolyzed rapidly under alkaline conditions, but DNA is not in the test tube, because RNA has ...., but not in DNA
(A) 2'-hydroxyl groups
(B) 3'-hydroxyl groups
(C) 5'-hydroxyl groups
(D) 2 and 3'-hydroxyl groups

56. .... is the first product of the action of alkali on RNA and are rapidly hydrolyzed further to yield a mixture of 2'- and 3'-nucleoside monophosphates.
(A) Cyclic 3'-monophosphate nucleotides
(B) Cyclic 2'-monophosphate nucleotides
(C) Cyclic 2', 3'-monophosphate nucleotides
(D) 2', 3'-monophosphate nucleotides

57. By convention, the structure of a single strand of nucleic acid is always written with the .... at the left and the .... at the right direction.

(A) 3’ end, 5’ end and 3’→5’
(B) 5’ end, 3’ end and 5’→3’
(C) 5’ end, 2’ end and 5’→2’
(D) 2’ end, 5’ end and 2’→5’

58. Which is not suitable structure for simpler representations of pentadeoxyribonucleotide?
(A) 5’ pA-C-G-T-AOH3’
(B) 5’ pApCpGpTpA 3’
(C) 5’ pACGTA 3’
(D) 5’ ACGTAp 3’

59. An oligonucleotide polymers contain .... nucleotides.
(A) 100 (B) ≥50
(C) ≤50 (D) 1000

60. What is called a longer nucleic acid that has more than 50 nucleotides in a linear chain?
(A) Polynucleotide (B) Oligonucleotide
(C) Linker (D) Adapter

61. Tautomeric forms of purines are depending on the
(A) Temperature (B) Humidity
(C) Light (D) pH

62. Which of the following is not a tautomeric form of uracil?

(A) Lactum (B) Lactim

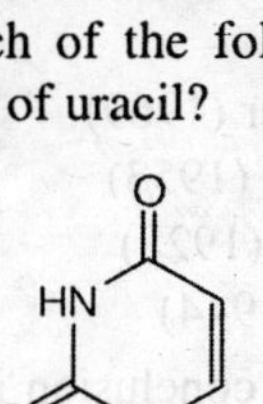

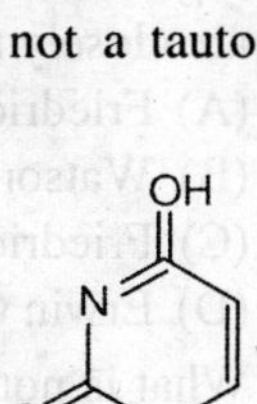

(C) Double Lactim (D) Lactom

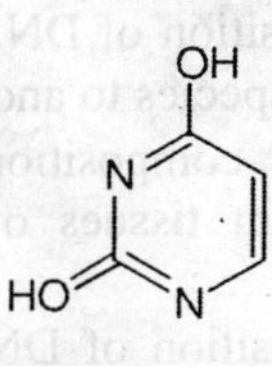

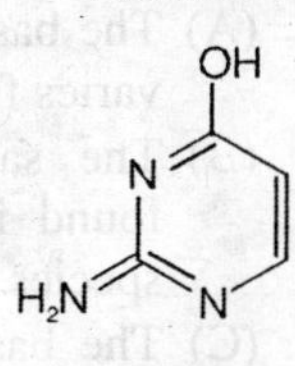

63. What are the names of tautomeric forms of uracil?
(A) Lactam, lactim, and double lactim forms
(B) Lactic, lactim, and double lactim forms
(C) Leucine, Lactic, and lactim
(D) Lysine, Leucine, and lactim

64. All nucleotide bases absorb UV light, and nucleic acids are characterized by a strong absorption at wavelengths near ....
(A) 230 nm (B) 260 nm
(C) 300 nm (D) 280 nm

65. Bases become charged and their solubility in water .... at acidic or alkaline pH
(A) Increase
(B) Decrease
(C) No changeable
(D) First increase then decrease

66. .... interactions are positioned two or more bases with the planes of their rings parallel (like a stack of coins) are one of two important modes of interaction between bases in nucleic acids.
(A) Van der waal bond
(B) Hydrophilic stacking
(C) Hydrophobic stacking
(D) Both Hydrophobic and Hydrophilic stacking

67. Which type of interactions and bonds between the bases are found which help to minimize contact of the bases with water and very important in stabilizing the three-dimensional structure of nucleic acids.
i. Vander Waals
ii. Dipole-dipole interactions
iii. Hydrogen bonds
iv. Hydrophobic stacking
(A) i and iii (B) I, ii, iii, iv
(C) I, iii, iv (D) I, ii, iv

68. Which are the most important functional groups of pyrimidines and purines?
(A) Ring nitrogens, carbonyl groups, and endocyclic amino groups
(B) Ring nitrogens, hydroxyl groups, and exocyclic amino groups
(C) Ring nitrogens, carbonyl groups, and exocyclic amino groups
(D) Ring nitrogens, aldehyde groups, and exocyclic amino groups

69. Which bond between bases permit a complementary association of two strands of nucleic acid?
(A) Vander Waals
(B) Dipole-dipole interactions
(C) Hydrogen bonds
(D) Hydrophobic stacking

70. Who are defined the most important hydrogen-bonding patterns for complimentary in association of two strands of nucleic acid?
(A) James D. Watson and Francis Crick (1953)
(B) Erwin Chargaff (1940)
(C) Rosalind Franklin and Maurice Wilkins (1962)
(D) Measolson and Stahal (1957)

71. Hydrogen bond formed between which bases in two strands of nucleic acid?
(A) A to T (or C) and G to U
(B) A to T (or G) and C to U
(C) A to T (or U) and G to C
(D) A to C (or U) and G to T

72. Specific pairing of bases in DNA that permits the .... of genetic information.
(A) Translation (B) Duplication
(C) Amination (D) Transformation

73. Hierarchical levels of complexity on nucleic acids based on the storage and utilization of a cell's genetic information, RNA structure and protein structure, is representative by ....
....
(A) Primary, secondary, tertiary, quaternary
(B) Primary, tertiary
(C) Primary, secondary, tertiary
(D) Secondary, tertiary, quaternary

74. The .... of a nucleic acid is its covalent structure and nucleotide sequence.
(A) Primary structure
(B) Secondary structure
(C) Tertiary structure
(D) quartly structure

75. Stable structure taken up by some or all of the nucleotides in a nucleic acid can be referred to as ....
(A) Secondary structure
(B) Primary structure
(C) Tertiary structure
(D) quarterly structure

76. Who discovered the structure of DNA?
(A) Watson and Crick (1953)
(B) Sutton and Bovery (1954)
(C) Harshey and Chase (1952)
(D) Avery, McCleod and McCarty (1944)

77. The complex folding of large chromosomes within eukaryotic chromatin and bacterial nucleoid is generally considered ....
(A) Secondary structure
(B) Primary structure
(C) Tertiary structure
(D) quarterly structure

78. Who began the biochemical investigation of DNA?
(A) Friedrich Miescher (1969)
(B) Altman (1970)
(C) Friedrich Griffith (1928)
(D) Watson and Crick (1953)

79. Who gave the most important clues to determine the distinctive base compositions in DNA?
(A) Friedrich Miescher (1969)
(B) Erwin Chargaff (1940s)
(C) Friedrich Griffith (1928)
(D) Watson and Crick (1953)

80. Who proposed the different ratios of four nucleotide bases in the DNAs of different organisms and the amounts of certain bases are closely related?
(A) Friedrich Miescher (1969)
(B) Watson and Crick (1953)
(C) Friedrich Griffith (1928)
(D) Erwin Chargaff (1944)

81. What is not found the conclusion in his work on base composition of DNAs of different organisms by E. Chargaff?
(A) The base composition of DNA generally varies from one species to another
(B) The same base composition in DNA found in different tissues of the same species
(C) The base composition of DNA changes with an organism's age or environment
(D) In all cellular DNAs (A = T), (G = C)

82. According to Chargaff'rules, DNA contains .... in all cellular DNAs.
i. (A = T) and (G = C)
ii. (A = G) and (T = C)
iii. (A + T)/ (G + C) = 1
iv. (A + G)/ (T + C) = 1
(A) i & iii (B) i & iv
(C) ii & iii (D) ii & iv

83. Who concluded that the sum of the purine residues equals the sum of the pyrimidine residues?
(A) Friedrich Miescher (1969)
(B) Watson and Crick (1953)
(C) Erwin Chargaff (1944)
(D) Friedrich Griffith (1928)

84. What is called the quantitative relationships of bases in DNA
(A) Altman's rules (B) Franklin's rules
(C) Chargaff's rules (D) Wilkins's rules

85. What is the importance of the quantitative relationships of bases in DNA according to Chargaff's rules?
(A) In the study of three dimensional structure of DNA
(B) Find the pathway of genetic information encoded in DNA (Central dogma)
(C) how genetic information transfer from one generation to next generation
(D) All of the above

86. If the DNA of an organism contained 30% adenine, what is the amount of cytosine in DNA?
(A) 30% (B) 40%
(C) 20% (D) 10%

87. In samples of DNA isolated from two unidentified species of bacteria, X and Y, adenine makes up 32% and 17%, respectively, of the total bases. What are the relative proportions of adenine, guanine, thymine, and cytosine, respectively?
(A) X = 32, 18, 32, 18 and Y = 17, 33, 17, 33
(B) X = 32, 36, 32, 36 and Y = 17, 33, 17, 33
(C) X = 32, 18, 17, 33 and Y = 17, 33, 32, 18
(D) X = 17, 33, 17, 33 and Y = 32, 18, 32, 18

88. In which experiment used isotopically labeled bacteriophage T2?
(A) Hershey-Chase experiment
(B) Friedrich Miescher experiment
(C) Avery, McLeod and McCarty experiment
(D) Friedrich Griffith experiment

89. Which isotopes were used to prove DNA as genetic material in Hershey-Chase experiment?
(A) ^{35}P in DNA and ^{35}S in protein coats
(B) ^{35}P in DNA and ^{32}S in protein coats
(C) ^{32}P in DNA and ^{32}S in protein coats
(D) ^{32}P in DNA and ^{35}S in protein coats

90. Why labeled 32P in DNA and 35S in protein coats in Hershey-Chase experiment, not reverse labeled?
(A) Because DNA contains no sulfur and viral protein contains no phosphorus
(B) Because viral protein contains no sulfur and DNA contains no phosphorus
(C) Because DNA contains both sulfur and phosphorus
(D) Because viral protein contains both sulfur and phosphorus

91. What was called the empty viral coats obtained from each suspension of phage-infected cells agitated in a blender to shear the viral capsids from the bacteria?
(A) Viroid (B) Ghosts
(C) Prions (D) Virion

92. What found the results when the cells infected with the ^{32}P-labeled phage DNA?
(A) Labeled viral DNA entered the cells
(B) Viral ghosts contain radioactivity
(C) Viral DNA contained no radioactivity
(D) Genetic message introduced by viral protein, not by viral DNA

93. What found the results when the cells infected with 35S-labeled phage protein?
(A) Labeled viral proteins entered the cells
(B) Viral ghosts contained radioactivity
(C) Viral DNA contains radioactivity
(D) Genetic message introduced by viral protein, not by viral DNA

94. What conclusion was found for replication from Hershey-Chase experiment?
(A) Genetic message introduced by viral protein, not by viral DNA
(B) Genetic message introduced by viral DNA, not by viral protein
(C) Genetic message introduced by both viral DNA and viral protein
(D) Genetic message not introduced by both viral DNA and viral protein

95. DNA is a .... according to Watson and Crick.
(A) Triple helix structure
(B) Single helix structure

(C) Basic nature
(D) Double helix structure

96. Who used the powerful method of X-ray diffraction to analyze structure of DNA fibers?
(A) Rosalind Franklin and Maurice Wilkins
(B) Friedrich Miescher
(C) Watson and Crick
(D) Erwin Chargaff

97. What were the conclusions found through x-ray diffraction pattern of DNA?
(A) DNA molecules are helical
(B) Three-dimensional model of the DNA molecule
(C) Distance between two base pairs (3.4 Å) and Helix length of 34 Å
(D) All of the above

98. Which is the active form of vitamin B_{12} that participates in intramolecular group transfers between adjacent carbons?
(A) NAD^+
(B) Riboflavin
(C) Adenosine is 5'-deoxyadenosylcobalamin
(D) FAD

99. Who postulated a three dimensional model of DNA structure?
(A) Friedrich Miescher (1969)
(B) Watson and Crick (1953)
(C) Erwin Chargaff (1940s)
(D) Rosalind Franklin and Maurice Wilkins (1950s)

100. Which form of DNA proposed by Watson and Crick in 1953?
(A) A-form (B) Z-form
(C) H-form (D) B-form

101. What are found the characteristics in B-form of DNA?
(A) Major groove and minor groove
(B) Right handed double helix
(C) Hydrophilic backbones of alternating sugar and phosphate groups
(D) All of the above

102. What found the conformation in the furanose ring of each deoxyribose?
(A) C-4' endo conformation
(B) C-2' endo conformation
(C) C-3' endo conformation
(D) C-5' endo conformation

103. Which are the purine and pyrimidine bases of both strands in DNA?
(A) Purines = Adenine and Guanine, Pyrimidines = Cytosine and Thymine
(B) Purines = Cytosine and Thymine, Pyrimidines = Adenine and Guanine
(C) Purines = Adenine and Thymine, Pyrimidines = Cytosine and Guanine
(D) Purines = Adenine and Guanine, Pyrimidines = Cytosine and Uracil

104. If duplex DNAs isolated from human cells and from mouse cells are completely denatured by heating, then mixed and kept at 65°C for many hours. Which is formed after annealing in this experiment?
(A) RNA-DNA hybrid
(B) Transposable DNA
(C) Hybrid DNA duplexes
(D) Mutated DNA duplex

105. A major groove and minor groove on the surface of the duplex create due to
(A) The offset pairing of the two strands
(B) Equality of bases in both strands
(C) Hydrophilic backbones of alternating sugar and phosphate groups
(D) Angle and Pitch length of DNA helix

106. G with C and A with T are paired through .... and .... hydrogen bonds, respectively in DNA double helix.
(A) Two and Three (B) Three and Two
(C) One and Two (D) Four and Three

107. Hydrogen bonds found between bases in nucleic acids based on ....
(A) Watson's rules
(B) Franklin's rules
(C) Chargaff's rules
(D) Wilkins's rules

108. How symbolized three hydrogen bonds between G and C but only two between A and T?
(A) GUC and AUT
(B) GqC and AUT
(C) GqC and AqT
(D) GUC and AUT

109. The Separation of paired DNA strands is more difficult the higher the ratio of .... to .... base pairs.

(A) G=C and A≡T (B) G≡C and A≡T
(C) G≡C and A=T (D) G=C and A=T

110. What is believable orientation in both DNA double strands to each other that was confirmed by x-ray analysis?

(A) Antiparallel orientation
(B) Parallel orientation
(C) 3'-5' direction (D) All of the above

111. How confirmed the antiparallel orientation in DNA strands?

(A) DNA polymerases experiment
(B) X-ray diffraction analysis
(C) Chargaff's rules (D) All of the above

112. Which is the correct answer for following matches regarding to B-DNA?

Sl.	Column 1	Sl.	Column 2
i.	Helix length	a	20A°
ii.	Distance between two stacked bases	b	C-2' endo
iii.	No. of base pairs per helical turn	c	3.4A°
iv.	Diameter of helix	d	6°
v.	Sugar pucker conformation	e	34A°
vi.	Base tilt normal to the helix axis	f	10.5

(A)		(B)		(C)		(D)	
i.	a	i.	e	i.	f	i.	d
ii.	b	ii.	c	ii.	c	ii.	a
iii.	c	iii.	f	iii.	b	iii.	e
iv.	d	iv.	a	iv.	e	iv.	f
v.	e	v.	b	v.	a	v.	b
vi.	f	vi.	d	vi.	d	vi.	c

113. Which is not correct statement about B-DNA?

(A) Two antiparallel polynucleotide chains
(B) Two polynucleotide chains are not identical in either base sequence
(C) Two polynucleotide chains are complementary to each other
(D) Adenine bonds with guanine and Cytosine with thymine

114. Which forces are held both strands of DNA together?

(A) Hydrogen bonding
(B) Base-stacking interactions
(C) Hydrophobic bonding
(D) (A) and (B)

115. The segments of a mouse DNA strand form base-paired regions with segments of a human DNA strand reflects .... under appropriate condition at 65°C for many hours?

(A) Evolutionary Heritage
(B) Differences between both DNAs
(C) Mutation effect
(D) chemical properties of DNA

116. Which is the most popular form of DNA which suggests a mechanism for the transmission of genetic information?

(A) A-DNA (B) H-DNA
(C) B-DNA (D) Z-DNA

117. Which is the form of DNA that contains left handed helical sense and 12 base pairs per helical turn?

(A) A-DNA (B) H-DNA
(C) B-DNA (D) Z-DNA

118. Which is the incorrect answer in following matches?

Options	Form of DNA	Helix length (A°)	Diameter (A°)	No. of base pairs per helical turn (bp)
(A)	A-DNA	26	26	11
(B)	B-DNA	34	20	10.5
(C)	Z-DNA	37	18	12
(D)	H-DNA	40	25	16

119. Which is the essential feature of the DNA duplex for replication?
(A) Complementary base pairing
(B) 10.5 base pairs per helical turn
(C) Helix length and diameter
(D) Right handed helical sense

120. Which is essential requirement for DNA replication?
(A) Base-pairing rules
(B) Separation of the two strands
(C) Synthesizing a complementary strand for each strand
(D) All of the above

121. Different organisms generally have some proteins and RNAs with similar functions and, often, similar structures. This is called....................
(A) Hypochromic effect
(B) Hyperchomic effect
(C) Evolutionary Heritage
(D) Genetic diversity

122. In which technique formed the spots as a cross in the center denote a helical structure?
(A) X-ray diffraction pattern of DNA
(B) NMR Technique
(C) Autoradiography of DNA
(D) FISH

123. Which of the following the pathways of the synthesis of nucleotides are used in cells?
i. de novo pathways
ii. S_N2 pathway
iii. A simulated folding pathway
iv. Salvage pathways
(A) i & ii (B) ii & iii
(C) i & iii (D) i & iv

124. What measurements were revealed in the original model of B-DNA that proposed by Watson and Crick in 1953?
(A) 10 base pairs per turn of the helix
(B) 34 Å (3.4 nm) the helix length
(C) Right handed and antiparallel pattern of duplex
(D) All of the above

125. Which model was proposed for studying the structure of DNA by Watson and Crick in 1953?
(A) Space-filling DNA model
(B) X-ray diffraction pattern of DNA
(C) Autoradiography of DNA
(D) Nucleosome model

126. Which are the characteristics of DNA molecule?
(A) Flexibility
(B) Acidic nature
(C) 5'→ 3' directional
(D) All of the above

127. Which molecules or groups are involved in backbone of nucleic acids?
(A) Purin bases and ribose sugar
(B) Phosphodeoxyribose
(C) Pyrimidine bases and ribose sugar
(D) Purine and Pyrimidine bases

128. How can produce bending, stretching, and unpairing of the DNA strands.
(A) Thermal fluctuation
(B) pH Changing
(C) Salt concentration changing
(D) Cold treatment

129. In which case the more extensive DNAs between two species will hybridize?
(A) More genetic diverse relationship
(B) Far away evolutionary relationship
(C) More mutated DNAs
(D) Closer evolutionary relationship

130. Why purines in purine nucleotides are restricted to stable conformations with respect to deoxyribose?
(A) Cause of steric constraints
(B) Stability of DNA
(C) Thermal fluctuation
(D) DNA flexibility

131. Which are called the two possible stable conformations with respect to deoxyribose in purines nucleotides due to steric constraints?
(A) syn and anti (B) uni and di
(C) exo and endo (D) cis and trans

132. Pyrimidines in DNA are generally restricted to the anti conformation because of steric interference between ....
....
(A) The sugar and the carbonyl oxygen at C-2 of the pyrimidine

(B) The sugar and the carbonyl oxygen at C-4 of the pyrimidine
(C) The Phosphate and the carbonyl oxygen at C-2 of the pyrimidine
(D) The Phosphate and the carbonyl oxygen at C-4 of the pyrimidine

133. Which form of DNA was proposed by Watson-Crick
(A) A-DNA (B) Z-DNA
(C) B-DNA (D) H-DNA

134. Which form of DNA is the most stable structure for a random-sequence DNA molecule under physiological conditions?
(A) A-DNA (B) Z-DNA
(C) H-DNA (D) B-DNA

135. Which is the left-handed DNA form?
(A) A-DNA (B) Z-DNA
(C) B-DNA (D) H-DNA

136. Which of the following characteristic is not of A-DNA?
(A) The number of base pairs per helical = 11
(B) Base tilt to the helix axis about = 20°
(C) Helix Diameter = 26 A°
(D) Left-handed DNA form

137. Which form of DNA is a more radical departure from the B structure
(A) A-DNA (B) Z-DNA
(C) C-DNA (D) H-DNA

138. Which of the following characteristic is not of Z-DNA?
(A) The number of base pairs per helical = 12
(B) Base tilt to the helix axis about = 7°
(C) Helix Diameter = 18 A°
(D) Right-handed DNA form

139. Which of DNA has zigzag appearance in its backbone?
(A) A-DNA (B) Z-DNA
(C) B-DNA (D) H-DNA

140. Which DNA form has specific sequences of pyrimidines that alternate with purines, especially alternating C and G or 5-methyl-C and G residues?
(A) A-DNA (B) Z-DNA
(C) B-DNA (D) H-DNA

141. What is the requirement to form the left-handed helix in Z-DNA?
(A) Purine in syn and pyrimidines in anti conformation
(B) Pyrimidine in syn and purines in anti conformation
(C) Suagr in syn and phosphate in anti conformation
(D) Phosphate in syn and sugar in anti conformation

142. Evidences for which DNA form found as short stretches in both prokaryotes and eukaryotes?
(A) A-DNA (B) Z-DNA
(C) B-DNA (D) H-DNA

143. What may be the role of Z-DNA tracts in both prokaryotes and eukaryotes?
(A) Regulation the expression of some genes and genetic recombination
(B) Cell signaling
(C) Chromosomal linkage and crossing over
(D) Energy transfer

144. Which atoms of purines participate in the hydrogen bonding of triplex DNA (H-DNA) that is referred to as Hoogsteen positions?
(A) N-6, O6, and N6 (B) N-7, O6, and N6
(C) N-7, O7, and N6 (D) N-7, O6, and N7

145. What is called the non-Watson-Crick pairing in triplex DNA called H-DNA?
(A) Chargaff's pairing
(B) Double and triple H-bond pairing
(C) Hoogsteen pairing
(D) Specific base pairing

146. Who first recognized H-DNA in 1963?
(A) Watson-Crick (B) Karst Hoogsteen
(C) F. Sanger (D) Wilkins and Chargaff

147. Hoogsteen pairing allows the formation of ....
(A) Single stranded DNA
(B) Triplex DNAs (C) Duplex DNAs
(D) Quadruplex DNA

148. What is the requirement for the C≡G • C triplet to be most stable at low pH?
(A) A protonated cytosine
(B) A protonated adinosine
(C) A methylated cytosine
(D) A Methylated adenosine

149. What is the value of the p*K*a of proronated cytosine in the triplex that is altered from its normal value of 4.2?
(A) >5.5 (B) >6.5
(C) >7.5 (D) >10.5

150. Why human DNA hybridizes much more extensively with mouse DNA than with DNA from yeast?
(A) More genetic diverse relationship
(B) Far away evolutionary relationship
(C) More mutated DNAs
(D) Closer evolutionary relationship

151. Which base residues in DNA sequences occurs readily with a very high proportion in four DNA strands to form a tetraplex (quadruplex)?
(A) Guanosine residues
(B) Adenosine residues
(C) Cytosine residues
(D) Thymine residues

152. What are the potential applications of synthetic DNA strands designed to pair with polypyrimidine or polypurine sequences to form triplex DNA could disrupt gene expression that control cellular metabolism at commercial scale?
(A) Medicinal
(B) Agricultural
(C) Study of molecular mechanisms
(D) All of the above

153. Which sequences of double stranded nucleic acids with two folds symmetry, for superimpose one repeat on the other, it must be rotated 180° about the horizontal axis then 180° about the vertical axis?
(A) Palindromes sequences
(B) Mirror repeat sequences
(C) SINEs
(D) LINEs

154. Which has a symmetric sequence within each strand with superimposing one repeat on the other requires only a single 180° rotation about the vertical axis.
(A) Palindromes sequences
(B) Mirror repeat sequences
(C) Promoter sequences
(D) Cozak sequences

155. In which DNA form can be considered a sequence of alternating T and C residues as a mirror repeat centered about a central T or C and purine strand with alternating A and G residues which produces a sharp bend in the DNA.
(A) A-DNA (B) Z-DNA
(C) B-DNA (D) H-DNA

156. Messenger RNAs Code for ....
(A) Polypeptide Chains
(B) Glucose chain
(C) Lipid chain
(D) Nucleotide chain

157. Which acts as an intermediary by using the information encoded in DNA to specify the amino acid sequence of a functional protein In gene expression?
(A) Glucose (B) Tryptophan
(C) RNA (D) cDNA

158. Where does protein synthesis occur on ribosomes whereas the DNA of eukaryotes largely present in the nucleus?
(A) Cytoplasm (B) Nucleolus
(C) Nucleus (D) Nucleoplasm

159. What is the way of genetic material that carried by DNA?
(A) From the cytoplasm to the nucleus
(B) From the nucleus to the cytoplasm
(C) From the cytoplasm to nucloide
(D) From the mitochondria to the nucleus

160. Where RNA found in cell?
(A) In nucleus (B) In cytoplasm
(C) In both the nucleus and the cytoplasm
(D) In all cell organelles

161. Which is the factory of ribosome synthesis in cell?
(A) Endoplasmic Reticulum
(B) Nucleolus
(C) Nucleus
(D) Golgi complex

162. Which is the factory of protein synthesis in cell?
(A) Endoplasmic Reticulum
(B) Nucleolus
(C) Nucleus
(D) Ribosome

163. How can increase protein synthesis?
(A) By an increase in the amount of cytoplasmic DNA
(B) By an increase in the amount of cytoplasmic cDNA
(C) By an increase in the amount of cytoplasmic RNA
(D) By an increase in the amount of cytoplasmic ER

164. What are the applications of DNA hybridization techniques?
(A) Identification of an individual in crime scene
(B) Prediction of the onset of a disease before symptoms appear
(C) Finding the genetic similarities between two different species
(D) All of the above

165. Who proposed the name "messenger RNA" (mRNA) for that portion of the total cellular RNA carrying the genetic information from DNA to the ribosomes?
(A) Jacob and Monod (1961)
(B) Watson and Crick (1953)
(C) Holy (1970)
(D) R. Franklin and M. Wilkins (1950s)

166. Who provide the templates that specify amino acid sequences in polypeptide chains?
(A) tRNA (B) rRNA
(C) SnRNA (D) mRNA

167. The size of mRNAs can vary greatly in length that is depend on ....
(A) A particular gene
(B) All mRNA have same size
(C) DNA template
(D) The number of amino acids

168. What is called the process of forming mRNA on a DNA template?
(A) Transcription (B) Replication
(C) Translation (D) Transformation

169. In prokaryotes, a single mRNA molecule may code for ....
(A) One polypeptide chains
(B) Several polypeptide chains
(C) One or several polypeptide chains
(D) One or several DNA chains

170. What is called a single mRNA molecule that carries the code for only one polypeptide in prokaryotes?
(A) Monocistronic (B) Polycistronic
(C) Unicistronic (D) Repeat cistronic

171. What is called a single mRNA molecule that carries the code for two or more different polypeptides in prokaryotes?
(A) Monocistronic (B) Polycistronic
(C) Unicistronic (D) Repeat cistronic

172. In eukaryotes, most mRNAs are ....
....
(A) Monocistronic (B) Polycistronic
(C) Unicistronic (D) Repeat cistronic

173. What is called the functional unit of gene according to benzer?
(A) Muton (B) Cistron
(C) Recon (D) Transposon

174. By which is set the minimum length of an mRNA in eukaryotes?
(A) The length of the polypeptide chain
(B) The length of the DNA
(C) RNA polymerase activity
(D) Promoter sequence

175. How many nucleotides requires of RNA coding sequence for a polypeptide chain of 100 amino acid residues
(A) 200 (B) 150
(C) 100 (D) 300

176. What is codon?
(A) Triplet of nucleotide
(B) Duplet of nucleotide
(C) Quadrate of nucleotide
(D) Triplet of amino acid

177. How many nucleotides in one codon encode for each amino acid?
(A) One (B) Two
(C) Three (D) Four

178. Which of the following metabolic precursor does not participate in the synthesis of nucleotides in de novo pathway in cells?
(A) Tryptophan
(B) Ribose 5'-phosphate
(C) Formic acid
(D) CO_2 and NH_3

179. What is shown in the following figure?

Gene1 Gene2 Gene3

(A) Monocistronic mRNA
(B) Polycistronic mRNA
(C) Unicistronic mRNA
(D) Repeat cistronic mRNA

180. What is called the white regions in following figure of polycistronic mRNA?

Gene1 Gene2 Gene3

(A) Coding RNA (B) Repeated RNA
(C) Noncoding RNA (D) Exons

181. Transfer RNAs serve as .... molecules in protein synthesis.
(A) Adapter (B) Template
(C) Messenger (D) Transporter

182. What is the component of ribosomes?
(A) Messenger RNAs
(B) Transfer RNAs
(C) Ribosomal RNAs
(D) Heterogenous RNAs

183. What are ribozymes?
(A) RNAs with enzymatic activity
(B) Ribosomes with enzymatic activity
(C) Enzymes for RNAs
(D) Enzymes for Ribosomes

184. What is the product of transcription of DNA?
(A) Always single-stranded RNA
(B) Double-stranded RNA and
(C) Single-stranded RNA
(D) Antisense RNA

185. What is assumed the confirmation of the product of transcription of DNA?
(A) Zig-zag structure
(B) Left-handed helical
(C) Right-handed helical
(D) Triplex structure

186. What is the reason to assume the right-handed helical confirmation of the transcript of DNA?
(A) Base stacking interactions
(B) Hydrogen bonding
(C) Phosphodiester bonding
(D) Covalent and non-covalent bonding

187. Base stacking interactions is the main reason to assume the right-handed helical confirmation of the product of transcription of DNA which are stronger between
(A) Two purines or two pyrimidines
(B) A purine and a pyrimidine
(C) Two purines and a pyrimidine
(D) Sugar and bases

188. Which is the following base pairing found in RNA.
(A) A pairs with C and G pairs with U
(B) G pairs with C and A pairs with T
(C) G pairs with C and A pairs with U
(D) C pairs with U and A pairs with G

189. Thymine is found in which RNA structure?
(A) rRNA (B) tRNA
(C) mRNA (D) snRNA

190. Which is the characteristic of the formation of DNA-RNA duplex?
(A) Watson-Crick base pairing
(B) Hoogsteen base pairing
(C) 2'-OH group in both RNA and DNA
(D) Uracil base in both DNA and RNA

191. Which does play a major role in stabilizing RNA structures as DNA structure?
(A) Weak interactions
(B) Hydrogen bonding
(C) Negative charge
(D) Concentration of AT bases

192. Why does interchange the conformation of A-DNA, B-DNA and Z-DNA?
(A) Due to salt or temperature conditions
(B) Due to UV absorption
(C) Due to change in negative charge
(D) Due to base pairing

193. Breaks in the regular A-form helix of RNA caused by mismatched or unmatched bases in one or both strands that form ....
(A) Hoogsteen loops
(B) Hairpin loops
(C) RNA-DNA loops
(D) Bulges or Internal loops

194. Hairpin loops form between………..in single stranded RNA by mismatched or unmatched bases.

(A) Self-complementary sequences away from each other
(B) Pallindrome sequences
(C) Self-complementary sequences located at ends of strand
(D) Nearby self-complementary sequences

195. Hairpin loops are the most common type of .... in RNA.
(A) Primary structure
(B) Secondary structure
(C) Tertiary structure
(D) Quarterly structure

196. Which specific short base sequences are often found at the ends of RNA hairpins and are known to form particularly tight and stable loops.
(A) UUCG (B) AATT
(C) AUTC (D) ATUU

197. What is the function of specific short base sequences are found at the ends of RNA hairpins?
(A) Act as starting points for the folding of a 3D-RNA molecule
(B) Act as starting points for the folding of a 2D-RNA molecule
(C) Act as termination points for the folding of a 3D-RNA molecule
(D) Act as termination points for the folding of a 3D-RNA molecule

198. Which is found structural contributions by hydrogen bonds that are not part of standard Watson-Crick base pairs at the ends of RNA hairpins?
(A) 1'-hydroxyl group of ribose paired with other groups
(B) 3'-hydroxyl group of ribose paired with other groups
(C) 2'-hydroxyl group of ribose paired with other groups
(D) 5'-hydroxyl group of ribose paired with other groups

199. Hairpin structural properties are evident in the structure of the phenylalanine transfer RNA of yeast that responsible for inserting Phe residues into polypeptides and in two RNA enzymes or ribozymes whose functions depend on their ....

(A) Three-dimensional structures
(B) Two-dimensional structures
(C) One-dimensional structures
(D) None of these

200. What are X, Y and Z regions in secondary structure of RNA in following figure?

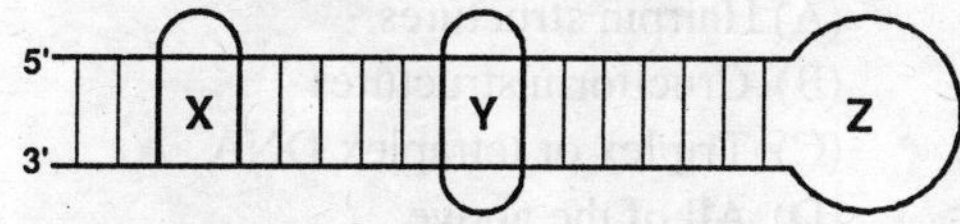

(A) X=Internal loop, Y=Bulge and Z=Hairpin
(B) X=Bulge, Y=Internal loop and Z=Hairpin
(C) X=Bulge, Y=hairpin and Z= Internal loop
(D) X=Hairpin, Y=Internal loop and Z=Bulge

201. Which enzyme has hairpin loops and non-Watson-Crick G=U base pairs in the secondary structure of the M1 RNA component of *E.Coli.*
(A) RNase A (B) RNase H
(C) RNase One (D) RNase P

202. What is the function of a segment of mRNA called intron act as a ribozyme in the ciliated protozoan *Tetrahymena thermophila* in mRNA strand?
(A) Catalyzes the translation of mRNA into proteins
(B) Catalyzes the formation of 3'-poly A-tail
(C) Catalyzes the synthesis of 5'-cap
(D) Catalyzes its own excision from between exons

203. Which process is performed for showing DNA as genetic material isolated from one bacterial strain and enter into the cells of another strain in Avery-MacLeod-McCarty experiment?
(A) Transcription (B) Transformation
(C) Conjugation (D) Transduction

204. Who proved that DNA of a bacteriophage but not protein coat carries genetic message for replication in host cell?
(A) Hershey-Chase
(B) Avery-MacLeod-McCarty
(C) F. Meischer (D) F. Griffith

205. Which structural form of DNA does exist in triple strand form?
(A) B-DNA (B) A-DNA
(C) Z-DNA (D) H-DNA

206. Which structure(s) can form DNA strands with appropriate sequences?
(A) Hairpin structures
(B) Cruciform structures
(C) Triplex or tetraplex DNA
(D) All of the above

207. Which does transfer genetic information from DNA to ribosomes for protein synthesis?
(A) Ribosomal RNA (B) Transfer RNA
(C) Messenger RNA (D) hn RNA

208. Transfer RNA is act as .... in protein synthesis.
(A) Template (B) Adapter
(C) Enzyme (D) Linker

209. Which structural form can be folded into hairpins, double-stranded regions, or complex loops?
(A) Double stranded RNA
(B) Single RNA strands
(C) DNA-RNA duplexes
(D) All of the above

210. Which property is responsible in the role of DNA as a repository of genetic information?
(A) Inherent stability (B) Negative charge
(C) Phosphodiester bonds
(D) Cot1/2 value

211. Which of the following can form the hairpin or cruciform structures?
(A) DNA (B) RNA
(C) Proteins (D) Both a and b

212. Which processes may be intimately linked to slowly accumulating, irreversible alterations of DNA?
(A) Carcinogenesis (B) Aging
(C) Respiration (D) Both a and b

213. Solutions of carefully isolated, native DNA are highly viscous at pH 7.0 and room temperature (25°C). What will be happened if the solution is subjected to extremes of pH or to temperatures above 80°C?
(A) Its viscosity decreases sharply without a physical change
(B) Its viscosity decreases sharply with structural and physical changes
(C) Its viscosity decreases sharply with physical change
(D) Its viscosity increases sharply with physical change

214. What is called the disruption of the hydrogen bonds between paired bases and base stacking causes unwinding of the DNA double helix to form two single strands?
(A) Renaturation
(B) Denaturation
(C) Break phosphodiester bonds
(D) Annealing

215. Which physical factors are involved in denaturation or melting of double-helical DNA by disruption of the hydrogen bonds between paired bases and of base stacking causes unwinding of the double helix to form two single strands?
(A) pH and Temperature
(B) Pressure and pH
(C) Salt concentration and UV light
(D) pH, Temperature, Salt concentration and UV light

216. Which types of bonds are involved in denaturation or melting of double-helical DNA?
(A) Hydrogen bonds
(B) Covalent bonds
(C) Wander val bonds
(D) Phosphodieter bonds

217. What happened if the temperature or pH is returned to the range in which most organisms live, the unwound segments of the two strands spontaneously rewind to yield the intact duplex?
(A) Saturation
(B) Denaturation
(C) Break phosphodiester bonds
(D) Annealing

218. Renaturation process carried out in two steps, first is slow and second much faster. Which is second step?
(A) Finding the two strands each other by random collisions and form a short segment of complementary double helix

(B) Remaining unpaired bases successively as base pairs
(C) Two strands "zipper" themselves together to form the double helix
(D) Both (B) and (C)

219. What is called the decreasing of UV absorption when two complementary nucleic acids strands are paired?
(A) Hypochromic effect
(B) Cancer-causing effect
(C) Therapeutic effects
(D) Hyperchromic effect

220. Which of the following amino acid work as precursor for the synthesis of pyrimidines and purines, respectively, in de novo pathway in cells?
(A) Glycine for purines and aspartate for pyrimidines
(B) Tryptophan for purines and histidine for pyrimidines
(C) Proline for purines and glycine for pyrimidines
(D) Arginine for purines and Tyrosine for pyrimidines

221. How many ATP molecules are consumed in de novo adenylate (AMP) biosynthesis?
(A) 2 (B) 4
(C) 6 (D) 5

222. Renaturation of a double-stranded nucleic acid produces .... in which UV absorption decreases in two complementary paired strands?
(A) Hypochromic effect
(B) Cancer-causing effect
(C) Hyperchromic effects
(D) Antagonistic effect

223. Denaturation of a double-stranded nucleic acid produces in which an increase in absorption into two complementary strands.
(A) Hypochromic effect
(B) Cancer-causing effect
(C) Hyperchromic effects
(D) Antagonistic effect

224. Denatured form of DNA can be detected by monitoring .... in which is done the transition from double-stranded DNA to the single-stranded.
(A) Absorption of infrared-rays
(B) Absorption of β-rays
(C) Absorption of UV rays
(D) Absorption of X-rays light

225. Maximum UV absorption is shown by
(A) ssRNA
(B) ds DNA
(C) Mononucleiotides
(D) ssDNA

226. When will be more melting temperature (Tm) for the transition from double-stranded DNA to the single-stranded?
(A) Higher contents of AqC base pairs
(B) Higher contents of AqT base pairs
(C) Higher contents of GqC base pairs
(D) Higher contents of GqT base pairs

227. Melting temperature (Tm) is related to
(A) Association of two ssDNA strands
(B) Separation of dsDNA strand
(C) Joining of two nucleotides
(D) restriction digestion of dsDNA

228. The overall rate of de novo purine and pyrimidine nucleotide biosynthesis is regulated by
(A) Feedback inhibition
(B) Forward inhibition
(C) Backward inhibition
(D) Opinion inhibition

229. Careful determination of the melting point of a DNA specimen is carried out under fixed conditions of ....
(A) pH and ionic strength
(B) Pressure and pH
(C) Salt concentration and UV light
(D) pH, Temperature, Salt concentration and UV light

230. If denaturation conditions are carefully controlled and regions that are rich in AT base pairs will specifically denature while most of the DNA remains double-stranded. Such denatured regions called bubbles can be visualized with
(A) Electron microscopy
(B) Simple microscopy
(C) Compound microscopy
(D) Scanning microscopy

231. In which process must occur strand separation of DNA in vivo?
(A) DNA replication
(B) DNA transcription
(C) Translation
(D) Both (A) and (B)

232. The DNA sites where these strand separation processes are initiated which are often rich in ....
(A) CT base pairs
(B) GC base pairs
(C) AG base pairs
(D) AT base pairs

233. At neutral pH, denaturation of a double helical RNA often requires temperatures .... or higher than denaturation of a DNA molecule with a comparable sequence.
(A) 100°C (B) 50°C
(C) 10°C (D) 20°C

234. The stability of an RNA-DNA hybrid is generally.....................
(A) Intermediate between RNA and DNA
(B) More than DNA
(C) Less than RNA
(D) No change in stability

235. The melting temperature depends on
(A) pH and Ionic strength
(B) Base composition of the DNA
(C) The size of DNA (D) All of the above

236. What detect by using the ability of two complementary DNA strands to pair with one another can be used?
(A) Similar RNA sequences in two different species or within genome of single specie
(B) Similar DNA sequences in two different species or within single species genome
(C) Similar RNA sequences in two different species
(D) Similar amino acid sequences in two different species

237. If duplex DNAs isolated from human cells and from mouse cells are completely denatured by heating, then mixed and kept at 65°C for many hours. What will not happen?
(A) Formation of Mouse duplex DNA
(B) Formation of Human duplex DNA
(C) Formation of Mouse and human single stranded DNA
(D) Formation of Mouse and human duplex DNA

238. What is the angle between bases and the long axis in B-DNA?
(A) 60° (B) 20°
(C) 90° (D) 34°

239. What is the role of Base-stacking interactions in DNA duplex?
(A) Stability of the double helix
(B) Form antiparallel pattern
(C) Responsible for negative charge on DNA
(D) Determine 5'-3' direction

240. Each preexisting DNA strand functions as a .... to lead the synthesis of one complementary strand.
(A) Repeat sequences
(B) Nonparental DNA
(C) Template
(D) Transcript

241. Structural variation in DNA reflects ....
(A) The different possible conformations of the deoxyribose
(B) Rotation in the phosphodeoxyribose backbone
(C) Free rotation about the C-1'–N-glycosyl bond
(D) All of the above

242. What are the characteristics of triplex DNA strands regarding to bases?
(A) Only pyrimidines or only purines in a strand
(B) Two pyrimidine strands and one purine strand
(C) Two purine strands and one pyrimidine strand
(D) All of the above

243. Which carries genetic information from DNA to protein biosynthetic machinery of ribosome?
(A) cDNA (B) Glucose
(C) Phospholipids (D) RNA

244. Alterations in DNA structure that produce permanent changes in the genetic information.
(A) Mutations
(B) Genetic transformation
(C) Evolution
(D) Gene expression

245. Which process carried on in several nucleotide bases which undergo spontaneous loss of their exocyclic amino groups.
(A) Phosphorylation (B) Methylation
(C) Acetylation (D) Deamination

246. In whose case, the nucleotide moiety of coenzyme A appears to be a binding "handle" that helps to pull the substrate (acetoacetyl-CoA) into the active site.
(A) Acetoacetyl CoA
(B) β-ketoacyl-CoA transferase
(C) superoxide dismutase
(D) Succinyl-CoA transferase

247. In DNA of eukaryotic cells, Methylation is most common at .... sequences on both strands of the DNA?
(A) C≡G (B) C=T
(C) A=T (D) A=G

248. Which arose during irradiation or as a byproduct of aerobic metabolism such as hydrogen peroxide, hydroxyl radicals, and superoxide radicals?
(A) Excited-hydrogen species
(B) Excited-oxygen species
(C) Excited-nitrogen species
(D) Excited-sulphur species

249. Which alkylating chemical agent can alter to methylate a guanine to yield O6-methylguanine which cannot base-pair with cytosine?
(A) Ethylsulfate (B) Methylsulfate
(C) Dimethylsulfate (D) Ethyl-methyl sulfate

250. Which of the following type of two prominent methylation systems found in *E. coli*?
(A) The part of a defense mechanism system
(B) The part of a repair mechanism system
(C) Both (A) and (B)
(D) None of the above

251. What would gradually lead to a decrease in GC base pairs and an increase in AU base pairs in the DNA of all cells?
(A) Cytosine phosphorylation
(B) Cytosine methylation
(C) Cytosine acetylation
(D) Cytosine deamination

252. Which of the following enzyme use template to form DNA strand?
(A) DNA ligase
(B) Reverse transcriptase
(C) DNA methylase
(D) RNA polymerase

253. Which process is involved the formation of *N-β-glycosyl* bond between the base and the pentose that occurs at a higher rate for purines than for pyrimidines?
(A) Acetylation (B) Methylation
(C) Hydrolysis (D) Deamination

254. Which steps is not involved in renaturation if the two strands are completely separated?
(A) First, finding the two strands each other by random collisions and form a short segment of complementary double helix
(B) Remaining unpaired bases successively base pairs
(C) Two strands "zipper" themselves together to form the double helix
(D) Breaking phosphodiester bonds

255. Which of the following physiological factor can selectively remove purines from DNA and result in a derivative called apurinic acid *in vitro*?
(A) pH 3 (B) Dilute acid
(C) UV light (D) Temperature

256. What is the angel between the stacked base pairs to the long axis of the double helix B-DNA?
(A) 60° (B) 36°
(C) 90° (D) 6°

257. Which of the following physiological factor can induce the condensation of two ethylene groups to form a cyclobutane ring for removal purine bases from DNA *in vitro*?
(A) UV light (B) Dilute acid
(C) pH 3 (D) Temperature

258. Which reaction between adjacent pyrimidine bases in nucleic acids form a cyclobutane ring as cyclobutane pyrimidine dimmers in the cell that reaction induced by UV light?
(A) Condensation of two amino groups
(B) Condensation of two ethylene groups
(C) Condensation of two hydroxyl groups
(D) Condensation of two methyl groups

259. Condensation of two ethylene groups forms a cyclobutane ring as cyclobutane pyrimidine dimmers in the cell that is induced by UV light. Which residues on the same DNA strand are most frequently involved in this reaction?
(A) Adjacent guanosine residues
(B) Adjacent adenosine residues
(C) Adjacent thymidine residues
(D) Adjacent cytidine residues

260. Condensation of two ethylene groups forms a cyclobutane ring as cyclobutane pyrimidine dimmers in the cell that is induced by UV light. Which carbon atoms of thymidine residues on the same DNA strand are most frequently involved in this reaction?
(A) C-6 and C-4 of adjacent pyrimidine residues
(B) C-5 and C-6 of adjacent pyrimidine residues
(C) C-3 and C-2 of adjacent pyrimidine residues
(D) C-6 and C-1 of adjacent pyrimidine residues

261. What is called 6-4 photoproduct during UV irradiation is formed with a linkage between C-6 of one pyrimidine and C-4 of its neighbor on the same DNA strand?
(A) Purine dimer (B) pyrimidine dimer
(C) Adinine dimer (D) Guanine dimer

262. Which is not non-ionizing radiation?
(A) UV irradiation (B) X-rays
(C) Gamma rays (D) Cosmic rays

263. Which radiation can cause ring opening and fragmentation of bases as well as breaks in the covalent backbone of nucleic acids.
(A) UV irradiation (B) X-rays
(C) Gamma rays (D) Both (B) and (C)

264. Which radiation can cause ring opening and fragmentation of bases as well as breaks in the covalent backbone of nucleic acids.
(A) Ionizing radiation
(B) UV radiation
(C) Non-ionizing radiation
(D) α and β radiations

265. Range of wavelength of UV radiation which makes up a significant portion of the solar spectrum?
(A) 1000 to 2000 nm
(B) 200 to 400 nm
(C) 100 to 200 nm
(D) 500 to 1000 nm

266. Which of the following radiation is known to cause pyrimidine dimer formation and other chemical changes in the DNA of bacteria and of human skin cells?
(A) UV irradiation (B) X-rays
(C) Gamma rays (D) Ionizing radition

267. Ionizing radiation in the form of cosmic rays which can penetrate deep into the earth, as well as radiation emitted from radioactive elements that are ....
(A) Radium (B) Plutonium
(C) Uranium (D) All of the above

268. In which form ionizing radiation can penetrate deep into the earth, as well as radiation emitted from radioactive elements such as radium, plutonium, uranium, radon, ^{14}C, and ^{3}H?
(A) UV irradiation (B) X-rays
(C) Gamma rays (D) Cosmic rays

269. Which is used in medical and dental examinations and in radiation therapy of cancer and other diseases are another form of ionizing radiation?
(A) UV irradiation
(B) X-rays
(C) Gamma rays
(D) Cosmic rays

270. What is the percentage of all DNA damage by UV and ionizing radiations?
(A) About 50% of all DNA
(B) About 100% of all DNA
(C) About 10% of all DNA
(D) About 80% of all DNA

271. Which prominent chemical agents may be DNA damaged by reactive chemicals which introduced into the environment as products of industrial activity and may be metabolized by cells into form?
(A) Deaminating agents
(B) Alkylating agents
(C) Aromatic agents
(D) Both a and b

272. Which is deaminating chemical agent may be DNA damaged by reactive chemicals that introduced into the environment as products of industrial activity and may be metabolized by cells into form?
(A) Nitrous acid (HNO_2)
(B) Nitric acid (HNO_3)
(C) Sulfuric acid (H_2SO_4)
(D) Hydrochloric acid (HCl)

273. Which is an organic precursor from which nitrous acid formed?
(A) β-Glucoronidase (B) Aspartic acid
(C) Glucosamines (D) Nitrosamines

274. Which chemical agent obtained from organic precursors such as nitrosamines, is a potent accelerator of the deamination of bases.
(A) Nitrate salt (B) Nitrite
(C) Nitrous acid (HNO_2)
(D) Urea (NH_2CONH_2)

275. Which chemical agents are used as preservatives in processed foods to prevent the growth of toxic bacteria?
(A) Sulfuric acid
(B) Bisulfite and Nitrous acid
(C) Benzene and Phenol
(D) Urea and Bisulfite

276. Which highly reactive alkylating chemical agent can alter certain bases of DNA?
(A) Dimethylsulfate
(B) Methylsulfate
(C) Ethylsulfate
(D) Ethyl-methyl sulfate

277. How dimethylsulfate can alter certain bases of DNA?
(A) Methylate a guanine to yield O^6-methylguanine
(B) Methylate a adinine to yield O^6-methylguanine
(C) Methylate a cytosine to yield O^6-methylguanine
(D) Methylate a thymine to yield O^6-methylguanine

278. What is the product of cytosine deamination?
(A) Thymine (B) Uracil
(C) Adenine (D) Guanine

279. Which is the most important source of mutagenic alterations in DNA?
(A) Hydroxylated damage
(B) Reductive damage
(C) Methylated damage
(D) Oxidative damage

280. Which is excited-oxygen species?
(A) Hydrogen peroxide
(B) Hydroxyl radicals
(C) Superoxide radicals
(D) All of the above

281. Under typical cellular conditions, what is the rate of deamination of adenine and guanine occurs of spontaneous events per day in a mammalian cell occur in 24 hours?
(A) About 1/10th (B) About 1/100th
(C) About 1/50th (D) About 1/200th

282. Which excited-oxygen species are responsible for most oxidative DNA damage?
(A) Hydrogen peroxide
(B) Hydroxyl radicals
(C) Superoxide radicals (D) Nitric acid

283. Which enzymes are involved in an elaborate defense system to destroy reactive oxygen species that convert reactive oxygen species to harmless products and are the source of mutagenic alterations in DNA in cells?
(A) Catalase
(B) Superoxide dismutase
(C) Both (A) and (B)
(D) Nitrous reductase

284. Which is responsible the damage to DNA occurs through the reaction ranging from oxidation of deoxyribose and base moieties to strand breaks?
(A) Excited-hydrogen species
(B) Excited-oxygen species
(C) Excited-nitrogen species
(D) Excited-sulphur species

285. Which are the following nucleotide bases in DNA molecules are more enzymatically methylated?

(A) Adenine and Cytosine
(B) Guanine and Thymine
(C) Adenine and Thymine
(D) Guanine and Cytosine

286. Which is used by all known DNA methylases as a methyl group donor?

(A) S-adenosylmethionine
(B) Methylsulfate
(C) Dimethylsulfate
(D) Ethyl-methyl sulfate

287. If DNA normally contained uracil and unrepaired uracils lead to permanent sequence changes as they were paired with ………… during replication.

(A) Thymine
(B) Uracil
(C) Adenine
(D) Guanine

288. How the prominent methylation system in *E. coli* helps in a defense mechanism

(A) Cell to distinguish its DNA from foreign DNA by marking its own DNA with methyl groups
(B) Destroying (foreign) DNA without the methyl groups
(C) A restriction-modification system
(D) All of the above

289. Which enzyme include to methylate adenosine residues within the sequence (5'-GATC-3') to N^6-methyladenosine repairs mismatched base pairs during DNA replication?

(A) cDam methylase
(B) Ram methylase
(C) Dam methylase
(D) Pam methylase

290. Dam is

(A) RNA adenine methylation
(B) DNA adenine methylation
(C) cDNA adenine methylation
(D) Deoxy adenine methylation

291. How many cytidine residues in DNA are methylated to 5-methylcytidine in eukaryotic cells?

(A) About 10% (B) About 20%
(C) About 50% (D) About 5%

292. At which position about 5% of cytidine residues in DNA of eukaryotic cells are methylated?

(A) 5-methylcytidine (B) 3-methylcytidine
(C) 4-methylcytidine (D) 2-methylcytidine

293. Under typical cellular conditions, how many spontaneous events per day in a mammalian cell occur for deamination of cytosine in DNA to uracil occurs in about one of every 107 cytidine residues in 24 hours?

(A) About 100
(B) About 1010
(C) About 4
(D) About 107

294. Migration of transposons is suppressed by ....

(A) Acetylation
(B) Methylation
(C) Alkylation
(D) Phosphorylation

295. Which is the following DNA form increased by making 5-methylcytosine through structurally methylations of cytosine in an alternating CpG sequence?

(A) A form (B) B form
(C) Z form (D) H form

296. Which of the following nucleoside triphosphates hydrolysis shifts the equilibrium of the overall process to favor product formation when coupled to a reaction with a positive free-energy change?

(A) Adenosine 5'-triphosphate (ATP)
(B) Guanosine 5'-triphosphate (GTP)
(C) Thymidine 5'-triphosphate (TTP)
(D) Cytosine 5'-triphosphate (CTP)

297. What is removed from acetoacetyl CoA which is the coenzyme A derivative of acetoacetate reduces and reduces its reactivity as a substrate for β-ketoacyl-CoA transferase by a factor of 106.

(A) 3'-phosphoguanosine diphosphate
(B) 3'-phosphocytosine diphosphate
(C) 3'-phosphoadenosine diphosphate
(D) 3'-phosphothymidine diphosphate

298. Which of the following enzyme involved in lipid metabolism?
(A) Acetoacetyl CoA
(B) β-ketoacyl-CoA transferase
(C) Superoxide dismutase
(D) Succinyl-CoA transferase

299. Under typical cellular conditions, which process of cytosine in DNA to uracil occurs in about one of every 107 cytidine residues in 24 hours?
(A) Phosphorylation (B) Deamination
(C) Acetylation (D) Methylation

300. What is called a single protein domain that binds adenosine can be used in a wide variety of enzymes and bind ATP and nucleotide cofactors?
(A) Protein-binding fold
(B) Enzyme-binding fold
(C) Nucleotide-binding fold
(D) Active site Nucleotide-binding fold

Answer-Key

Q	Ans	Q	Ans	Q	Ans	Q	Ans	Q	Ans	Q	Ans	Q	Ans	Q	Ans
1.	(B)	41.	(B)	81.	(C)	121.	(C)	161.	(B)	201.	(D)	241.	(D)	281.	(B)
2.	(D)	42.	(D)	82.	(B)	122.	(A)	162.	(D)	202.	(D)	242.	(D)	282.	(B)
3.	(D)	43.	(A)	83.	(C)	123.	(D)	163.	(C)	203.	(B)	243.	(D)	283.	(C)
4.	(A)	44.	(A)	84.	(C)	124.	(D)	164.	(D)	204.	(A)	244.	(A)	284.	(B)
5.	(A)	45.	(A)	85.	(D)	125.	(A)	165.	(A)	205.	(D)	245.	(D)	285.	(A)
6.	(D)	46.	(B)	86.	(C)	126.	(D)	166.	(D)	206.	(D)	246.	(B)	286.	(A)
7.	(A)	47.	(D)	87.	(A)	127.	(B)	167.	(A)	207.	(C)	247.	(A)	287.	(C)
8.	(C)	48.	(C)	88.	(A)	128.	(A)	168.	(A)	208.	(B)	248.	(B)	288.	(D)
9.	(A)	49.	(B)	89.	(D)	129.	(D)	169.	(C)	209.	(B)	249.	(B)	289.	(C)
10.	(D)	50.	(B)	90.	(A)	130.	(A)	170.	(A)	210.	(A)	250.	(C)	290.	(B)
11.	(A)	51.	(B)	91.	(B)	131.	(A)	171.	(B)	211.	(D)	251.	(D)	291.	(D)
12.	(B)	52.	(C)	92.	(A)	132.	(A)	172.	(A)	212.	(D)	252.	(B)	292.	(A)
13.	(C)	53.	(A)	93.	(B)	133.	(C)	173.	(B)	213.	(C)	253.	(C)	293.	(A)
14.	(D)	54.	(D)	94.	(B)	134.	(D)	174.	(A)	214.	(B)	254.	(D)	294.	(B)
15.	(A)	55.	(A)	95.	(D)	135.	(B)	175.	(D)	215.	(A)	255.	(A)	295.	(C)
16.	(A)	56.	(C)	96.	(A)	136.	(D)	176.	(A)	216.	(A)	256.	(D)	296.	(A)
17.	(A)	57.	(B)	97.	(D)	137.	(B)	177.	(C)	217.	(D)	257.	(C)	297.	(C)
18.	(C)	58.	(D)	98.	(D)	138.	(D)	178.	(A)	218.	(D)	258.	(B)	298.	(B)
19.	(D)	59.	(C)	99.	(D)	139.	(B)	179.	(B)	219.	(A)	259.	(C)	299.	(B)
20.	(A)	60.	(A)	100.	(D)	140.	(B)	180.	(C)	220.	(A)	260.	(A)	300.	(C)
21.	(B)	61.	(D)	101.	(D)	141.	(A)	181.	(A)	221.	(D)	261.	(B)		
22.	(B)	62.	(D)	102.	(B)	142.	(B)	182.	(C)	222.	(A)	262.	(A)		
23.	(A)	63.	(A)	103.	(A)	143.	(A)	183.	(A)	223.	(C)	263.	(D)		
24.	(A)	64.	(B)	104.	(C)	144.	(B)	184.	(A)	224.	(C)	264.	(C)		
25.	(D)	65.	(A)	105.	(A)	145.	(C)	185.	(C)	225.	(C)	265.	(B)		
26.	(C)	66.	(C)	106.	(B)	146.	(B)	186.	(A)	226.	(C)	266.	(A)		
27.	(A)	67.	(A)	107.	(C)	147.	(B)	187.	(A)	227.	(B)	267.	(D)		
28.	(A)	68.	(C)	108.	(B)	148.	(A)	188.	(C)	228.	(A)	268.	(D)		
29.	(A)	69.	(C)	109.	(C)	149.	(C)	189.	(B)	229.	(A)	269.	(B)		
30.	(B)	70.	(A)	110.	(A)	150.	(D)	190.	(A)	230.	(A)	270.	(C)		
31.	(B)	71.	(C)	111.	(D)	151.	(A)	191.	(A)	231.	(D)	271.	(D)		
32.	(C)	72.	(B)	112.	(B)	152.	(D)	192.	(A)	232.	(D)	272.	(A)		
33.	(A)	73.	(C)	113.	(D)	153.	(A)	193.	(D)	233.	(D)	273.	(D)		
34.	(C)	74.	(A)	114.	(D)	154.	(B)	194.	(D)	234.	(A)	274.	(C)		
35.	(D)	75.	(A)	115.	(A)	155.	(D)	195.	(B)	235.	(D)	275.	(B)		
36.	(A)	76.	(A)	116.	(C)	156.	(A)	196.	(A)	236.	(B)	276.	(A)		
37.	(A)	77.	(C)	117.	(D)	157.	(C)	197.	(A)	237.	(C)	277.	(A)		
38.	(D)	78.	(A)	118.	(D)	158.	(A)	198.	(C)	238.	(C)	278.	(B)		
39.	(D)	79.	(B)	119.	(A)	159.	(B)	199.	(A)	239.	(A)	279.	(D)		
40.	(B)	80.	(D)	120.	(D)	160.	(C)	200.	(B)	240.	(C)	280.	(D)		

3

Genes and Chromosomes

A gene is the segment of DNA also called as basic physical and functional unit of heredity. Genes are used to produce functional molecules called proteins. In living organism like human, plants and microorganisms, genes differ in size from a few hundred nucleotides to more than 2 million nucleotides. According to Human Genome Project (HGP), human contain more than 35,000 genes. Gene present in two copies which obtained from both parent. Alternative forms of gene are called alleles which found in chromosomes at specific locations with same order. Same genes are present on homologous chromosomes in the same loci which separated during meiosis. Non-Homologous chromosomes contain different alleles for different type of genes. For example in human, chromosome 1 has homologous chromosomes but one part from chromosome 1 and one from chromosome 2 are non- homologous chromosomes. In prokaryotes, most of all genes are monocistronic which means that gene contains genetic information for one specific polypeptide or protein or enzyme whereas in eukaryotes, genes may be polycistronic which code for more than one polypeptides or proteins or enzymes. Gene independently replicates in organism. In order to form proteins, the gene first transcribed into mRNA which is converted into proteins by translation process. Replication, transcription and translation collectively called as central dogma proposed by crick. Seymour Benzer (1956) proposed the concept of gene and fine structrure of gene working on bacteriophage with mutant *rII* genes. He also subdivided gene into three classes according to its function: cistron (unit of function), recon (unit of recombination) and muton (unit of mutation). Cistron exhibit cis-trans phenomenon as a unit of heredity. Mutations of genes lead important functions that may affect growth and development of the individual.

Chromosomes contain smaller units of genetic material called DNA. DNA is a sequence of letters that spell out the genetic code. The DNA is organized into words and sentences called genes. Humans have approximately 20,000 genes and each one influences a part of development. These genes usually come in pairs. One copy is inherited from our mother and one from our father. A change in the spelling of a DNA sequence or the gene is called a mutation. Every person's DNA contains mutations which are usually harmless. However, some mutations are responsible for causing specific disease.

Prokaryotes usually have one single circular chromosome, whereas most eukaryotes are diploid. Genetic material of prokaryotes presents in cytoplasm withour any covering membrane known as "nucleoid" and also not contains histone protein. Whereas chromosomes in eukaryotes are composed of chromatin fiber made of nucleosomes. Nucleosome is a histone octamer of H2A, H2B, H3 and H4. H1 histone works as linker protein between to nucleosome particles. Total length of DNA in nucleosome is 200bp which divided into two parts. A 146bp segment of DNA (1.75 times) wrapped around the histone octamer and 54bp of DNA is known as linker DNA used to connect two nucleosome molecules.

Human contains 46 (2n) chromosomes form 23 (n) pairs in which 22 are autosomes and remaining 1 pair (23rd, XX in female and XY in male) called as sex chromosomes. Different variation in chromosome structure and number shows many abnormalities which cause diseases in human being. Numerical aberration in chromosomes called as aneuploidy and occurs by increasing or decreasing of chromosome numbers. *Example:*

- Down syndrome: an extra copy of chromosome 21st known as Trisomy 21
- Turner syndrome: Missing of one X sex chromosome in female (contains only 45 chromosomes)
- Klinefelter syndrome: an extra 'X' chromosome in male (47, XXY)
- Edwards syndrome: caused by an extra chromosome 18 (Trisomy)
- Patau syndrome: caused by an extra chromosome 13 (Trisomy)

Chromosomal structural abnormalities occur due to deletion, addition, inversion, and translocation. A various types of diseases are caused by these chromosomal aberrations. *Example:*

- Cri-du-Chat syndrome: a deletion on the short arm of chromosome 5
- Williams syndrome: deletion by unequal recombination on chromosome 7,

Central dogma

Central dogma of molecular biology was given was first given by Francis Crick in 1958. Central dogma explains the flow of genetic information in biological systems.

According to Francis Crick, 1958

"The Central Dogma. This states that once 'information' has passed into protein it cannot get out again. In more detail, the transfer of information from nucleic acid to nucleic acid, or from nucleic acid to protein may be possible, but transfer from protein to protein, or from protein to nucleic acid is impossible. Information means here the precise determination of sequence, either of bases in the nucleic acid or of amino acid residues in the protein."

Revised central dogma in a Nature paper published in 1970 by Francis Crick

"The central dogma of molecular biology deals with the detailed residue-by-residue transfer of sequential information. It states that such information cannot be transferred back from protein to either protein or nucleic acid."

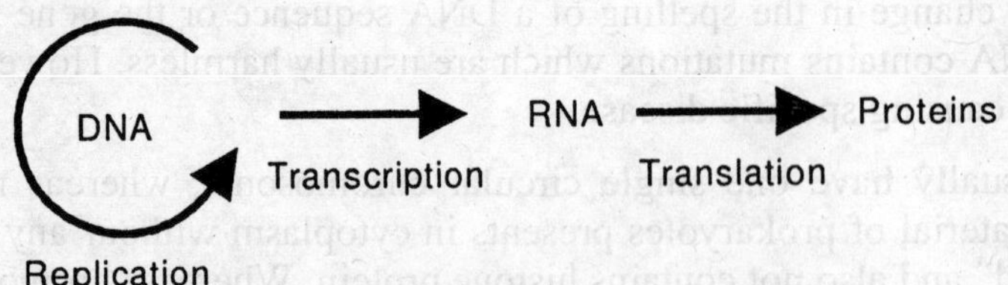

Diagrammatic representation of flow of genetic information

Key notes

- A gene is a locus or section of DNA made up of nucleotides and is the molecular unit of heredity.
- Term gene was given by Wilhelm Johannsen (1909)
- The transmission of genes from one generation to another is the base of the inheritance of traits.

- Genes can be mutated in their sequence, leading to diverse variations in the form of alleles responsible for different varieties of a protein, main cause of different phenotype traits.
- Genes evolve due to natural selection or survival of the fittest of the alleles.
- A chromosome contains of a single, lengthy DNA helix on which thousands of genes are pre-arranged.
- The section of the chromosome at which individual gene is positioned is called its locus. Each locus contains one allele of a gene.
- The chromosomes are crowded within the nucleus fused with proteins called histones to form a unit called a nucleosome.
- DNA wrapped and shortened in this way is called chromatin (coined by Walther Flemming.
- Prokaryotes (bacteria and archaea) naturally store their genomes on a single large, circular chromosome.
- Prokaryotes have extra small circles of DNA called plasmids which usually encode only a few genes and are transferable between individuals.
- A chromosome is a packed and structured unit comprising most of the DNA of a living organism.
- The chromosome gets replicated during S phase, resulting in an X-shaped structure called a metaphase chromosome. Both the original and the newly copied DNA are now called chromatids.
- The two "sister" chromatids are joined together at a protein junction called a centromere (forming the X-shaped structure).
- Chromosomes are generally noticeable under a light microscope only when the cell is enduring mitosis (metaphase).
- DNA and its linked proteins and macromolecules are jointly known as chromatin, which is further packaged along with its associated molecules into a distinct structure called a nucleosome.
- In prokaryotic cells, chromatin occurs free-floating in cytoplasm called nucleoids. Bacteria lack histones.
- The DNA-bound macromolecules are proteins, which serve to package the DNA and control its functions.
- Small circular genomes called plasmids are often found in bacteria and also in mitochondria and chloroplasts.
- Walter Sutton and Theodor Boveri independently developed the chromosome theory of inheritance in 1902.
- The number of human chromosomes was published in 1923 by Theophilus Painter through the microscopic study with counting 24 pairs mean 48 chromosomes. His error was copied by others and it was not until 1956 that the true number, 46, was determined by Indonesia-born cytogeneticist Joe Hin Tjio.
- Prokaryotes do not have nuclei. In its place, their DNA is organized into a structure called the nucleoid.
- In archaea, the DNA in chromosomes is even more organized, similar to eukaryotic nucleosomes.

- Prokaryotic chromosomes and plasmids are generally supercoiled like eukaryotic DNA.
- The structure of chromosomes and chromatin varies through the cell cycle.
- Eukaryotes have numerous huge linear chromosomes contained in the cell's nucleus. Each chromosome has one centromere.
- Chromosomes have a shorter arm called *p arms* and the longer arms called *q arms*.
- Chromosomes in humans can be divided into two types: autosomes (body chromosome(s) and allosome (sex chromosome(s).
- Human cells have 23 pairs of chromosomes (22 pairs of autosomes and one pair of sex chromosomes).
- Highest number of genes were estimated in human chromosome 18 (200 genes), lowest number in 'Y' sex chromosome (50 genes) and 'X' sex chromosome contains 800 genes.
- Gametes are produced by meiosis of a diploid germ line cell. Gametes, reproductive cells, are haploid (n) mean one set of chromosomes.
- Somatic cells are diploid (2n) having two sets of chromosomes.
- Polyploid cells possess more than two sets of homologous chromosomes. The more-common bread wheat is polyploid, having 42 (hexaploid) chromosomes.
- Prokaryotic species normally have one copy of each major chromosome, but most cells can easily survive with multiple copies.
- The number of plasmids in the cell completely determined by the rate of division of the plasmid (fast division causes high copy number).
- A karyotype is the number and appearance of chromosomes in the nucleus of a eukaryotic cell.
- For karyotyping, the cells are stained, photographed, and arranged into a *karyogram*, where chromosomes arranged in descending order of length.
- Chromosomal aberrations are disruptions in the normal chromosomal content of a cell such as Down syndrome (three copies of chromosome 21).
- Cri du chat, is caused by the deletion of portion of the short arm of chromosome 5. "Cri du chat" means "cry of the cat" in French; the condition was so-named because affected babies make high-pitched cries that sound like those of a cat. Affected individuals have wide-set eyes, a small head and jaw, moderate to severe mental health problems, and are very short.
- In Down's syndrome, decreased muscle tone, stockier build, asymmetrical skull, slanting eyes and mild to moderate developmental disability.
- Edwards syndrome, or trisomy-18, symptoms include motor retardation, developmental disability and numerous congenital anomalies causing serious health problems.
- In Klinefelter syndrome (XXY) men are usually sterile and tend to be taller and have longer arms and legs than their peers. Boys with the syndrome are often shy and quiet and have a higher incidence of speech delay and dyslexia.
- Platau Syndrome, also called D-Syndrome or trisomy-13.
- Triple-X syndrome (XXX). XXX girls to be likely to be tall and thin and have a complex incidence of dyslexia.
- Turner syndrome (X instead of XX or XY). In Turner syndrome, female sexual characteristics are present but underdeveloped.

- XYY syndrome boys are generally taller than their siblings. Like XXY boys and XXX girls, they are more likely to have learning difficulties.
- Wolf–Hirschhorn syndrome, which is produced by fractional removal of the short arm of chromosome 4. It is categorized by growth delay, hindered motor skills advancement, "Greek Helmet" facial features, and minor reflective mental health problems.

List of diploid chromosome number (2n) of important plants and animal

Organism	Species	chromosome number (2n)
Alfalfa	*Medicago sativa*	32
Barley	*Hordeum vulgare*	14
Broad bean	*Vicia faba*	12
Corn (maize)	*Zea mays*	20
Cotton	*Gossypium barbadense*	52
Garden pea	*Pisum sativum*	14
Guava	*Psidium guajava*	22
Kidney bean	*Phaseolus vulgaris*	22
Mango	*Mangifera indica*	40
Oats	*Avena sativa*	42
Onion	*Allium cepa*	16
Papaya	*Carica papaya*	18
Peanut	*Arachis hypogaea*	40
Pineapple	*Ananas comosus*	50
Potato	*Solanum tuberosum*	48
Rice	*Oryza sativa*	24
Rye	*Secale cereale*	14
Sorghum	*Sorghum vulgare*	20
Soybean	*Glycine max*	40
Sugar beet	*Beta vulgaris*	18
Sugar cane	*Saccharum officinarum*	80
Tobacco	*Nicotiana tabacum*	48
Tomato	*Lycopersicon esculentum*	24
Wheat, durum	*Triticum durum*	28
Wheat, common	*Triticum vulgare*	42
Man	*Homo sapiens*	46
Fruit fly	*Drosophila melanogaster*	8
Honeybee	*Apis mellifera*	32
House fly	*Musca domestica*	12
Rhesus monkey	*Macaca mulatta*	42

Multiple Choice Questions

1. Which is the fundamental unit of genetic information in living systems?
(A) Gene (B) Chromosome
(C) Histone (D) Protein

2. How can be defined a gene biochemically?
(A) As a segment of amino acids that encodes proteins
(B) As a segment of RNA that encodes DNA
(C) As a segment of DNA that encodes a functional biological product
(D) As a segment of protein that encodes other protein

3. What is the final product of gene which is defined as a segment of DNA that encodes a functional biological product?
(A) Gene (B) Chromosome
(C) RNA (D) Protein

4. Might a functional gene product also be one of several classes of RNA molecules?
(A) Sometimes
(B) Never
(C) A functional gene product is only DNA molecules
(D) A functional gene product is only RNA molecules

5. Who wrote the book 'the origin of species" on natural selection theory?
(A) Weisman
(B) J. B. Lamarck
(C) Hugo de Veries
(D) Charles Darwin

6. Who discovered the double-helical structure of DNA?
(A) Watson and Crick (1953)
(B) Sutton and Bovery (1954)
(C) Harshey and Chase (1952)
(D) Avery, McCleod and McCarty (1944)

7. How many energy is required for the formation of a peptide bond?
(A) Only about 14 kJ/mol
(B) Only about 100 kJ/mol
(C) Only about 21 kJ/mol
(D) Only about 125 kJ/mol

8. How many energy invested by the cell to synthesize a bond between two specific amino acids at a particular point in a polypeptide?
(A) Only about 14 kJ/mol
(B) Only about 125 kJ/mol
(C) Only about 21 kJ/mol
(D) Only about 100 kJ/mol

9. Which processes are shown by W, X, Y and Z in the following figure?

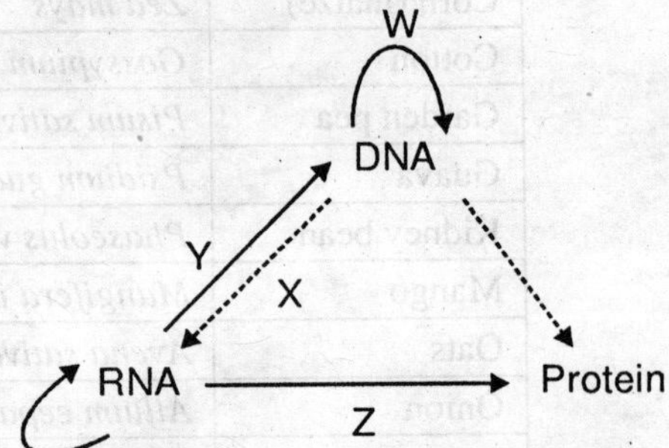

(A) W=Replication, X=Translation, Y=Trancription and Z= Reverse transcription
(B) W=Transcription, X=Replication, Y=Translation and Z= Reverse transcription
(C) W=Replication, X=Transcription, Y= Reverse transcription and Z=Translation
(D) W=Replication, X= Reverse transcription, Y=Translation and Z= Transcription

10. Who proposed central dogma?
(A) Sutton and Bovery
(B) Francis Crick and Jems Watson
(C) Erwin Chargaff
(D) Francis Crick

11. What is shown by central dogma?
(A) General pathways of lipid biosynthesis
(B) General pathways of nucleotide biosynthesis
(C) General pathways of respiration and photosynthesis
(D) General pathways of genetic information flow

12. Which continuous processes included in central dogma?
(A) Replication, Transcription and Translation
(B) Lipid Biosynthesis, Replication and Photosynthesis
(C) Replication, Transcription, Respiration and Translation
(D) Respiration, Replication, Photosynthesis, Transcription and Translation

13. Which process is not included in central dogma?
(A) Transcription (B) Translation
(C) Respiration (D) Translation

14. Which process is not included in central dogma?
(A) DNA to DNA (B) DNA to RNA
(C) Protein to RNA (D) RNA to protein

15. What is called the total complement of genetic material in a cell?
(A) Proteome (B) Gene
(C) Trascriptome (D) Genome

16. What is called condense portion in nucleus where gene are located?
(A) Chromosomes (B) Proteomes
(C) Transcriptomes
(D) Complimentary genomes

17. What are number of chromosomes (2n) in yeast *Saccharomyces cerevisiae*?
(A) 8 chromosomes
(B) 6 chromosomes
(C) 16 chromosomes
(D) 20 chromosomes

18. What is the molecular mass of the small genome of the yeast *Saccharomyces cerevisiae* that contains DNA molecules with 230,000 to 1,532,000 contiguous base pairs (bp).
(A) From 1×10^8 to 1.5×10^9 daltons
(B) From 1.5×10^8 to 1×10^9 daltons
(C) From 5×10^8 to 10^{10} daltons
(D) From $.5 \times 10^8$ to 2×10^9 daltons

19. What is genome size of human chromosomes?
(A) 5386 bp (B) 12×10^6 bp
(C) 3.2×10^9 bp (D) 4.8×10^7 bp

20. What is gene that code for RNAs and polypeptide chains?
(A) A segment of amino acids
(B) A segment of DNA
(C) A segment of Glucose
(D) A segment of Nucleosides

21. According to classical genetics, what is called a portion of a chromosome that determines or affects a single character or phenotype?
(A) Proteome (B) Gene
(C) Trascriptome (D) Genome

22. What is called visible character that determined by a portion of a chromosome known as gene?
(A) Phenotypes (B) Genotypes
(C) Alleles (D) Abnormalities

23. Who proposed a molecular definition of a gene in 1940 after exposing spores of the fungus *Neurospora crassa* to X-rays known to damage DNA and cause alterations in DNA sequence?
(A) Watson and Crick
(B) Jacob and Monad
(C) Beadle and Tatum
(D) Wilson and Franklin

24. Which of the following is not a major class of chromatin proteins?
(A) Globular proteins (B) SMC proteins
(C) Topoisomerases (D) SSBP

25. What is meaning of SMC proteins that is a major class of chromatin proteins?
(A) Structural maintenance proteins of chromosome
(B) Single stranded binding chromosomal proteins
(C) Simple mature cellular protein
(D) Simple managing chromosome proteins

26. Which of the following organism is used in the X-rays mutation experiment by George Beadle and Edward Tatum for causing alterations in DNA sequences?
(A) Bacterium *Escherichia coli*
(B) Yeast *Saccharomyces cerevisiae*
(C) Bacteriophage Ø X 174
(D) Fungus *Neurospora crassa*

27. What is the function of regulatory sequences of DNA that encode different processes in cell?
(A) Influence the transcription of genes
(B) Function as initiation points for replication
(C) Function as initiation points for recombination
(D) All of these

28. Which is not true about gene(s)?
(A) Some genes expressed to generate multiple gene products from one segment of DNA
(B) Regulatory sequences of DNA Influence the transcription of genes
(C) Genes are located on chromosomes
(D) Genome of human is totally made up of genes that code proteins

29. What is called a sequence of three consecutive nucleotides in a single strand of DNA which code for each amino acid of a polypeptide chain?
(A) Recon (B) Intron
(C) Codon (D) Muton

30. Which domains of SMC (Structural maintenance of chromosomes) proteins has a part of an ATP hydrolytic site, are connected by two regions of α-helical coiled-coil motifs?
(A) Hinge domains
(B) Amino- and carboxyl-terminal globular domains
(C) V-shaped complex domains
(D) Looped domains

31. What can be the size of DNA which encodes 333 amino acids?
(A) 900 (B) 100
(C) 333 (D) 1000

32. Which of the following the genome of prokaryotic organism first of all completely sequences?
(A) Mycoplasma (B) Archeabacteria
(C) *Escherichia coli* (D) BGA

33. Which organism has circular DNA molecule?
(A) *Escherichia coli* (B) *Homo sapience*
(C) *Neurospora crass*
(D) *Drosophilla melanogaster*

34. Which organism has linear DNA molecule?
(A) *Escherichia coli*
(B) *Pseudomonas putida*
(C) Mycoplasm gallisapticum
(D) Drosophilla melanogaster

35. Which of the following correct matching (Organism vs Genome size)?
i. *Escherichia coli* = 4.63×10^6 bp
ii. Ø × 174 bacteriophage = 5386 bp
iii. Human = 3.2×10^9 bp
iv. *Drosophila melanogaster* = 1.8×10^8 bp
(A) i, ii & iii (B) ii, iii & v
(C) ii & iii (D) All of the above

36. Which of the following correct matching (Organism vs Gene number)?
i. *Escherichia coli* = 4300
ii. Arabidopsis thaliana = 25500
iii. Human = 30000 – 35000
iv. *Drosophila melanogaster* = 13600
(A) i, ii & iii (B) ii, iii & v
(C) ii & iii (D) All of the above

37. Which of the following correct matching [Organism vs Chromosome number (2n)]?
i. *Escherichia coli* = 1
ii. *Arabidopsis thaliana* = 10
iii. Human = 46
iv. *Drosophila melanogaster* = 18
(A) i, ii & iii (B) ii, iii & v
(C) ii & iii (D) All of the above

38. Which of the following free-living organism?
(A) Mycoplasma (B) BGA
(C) Virus (D) Bacteria

39. Which of the following organism made up of only nucleic acid and proteins?
(A) Mycoplasma (B) Virus
(C) Bacteria (D) BGA

40. What is the genome of viruses?
(A) A single RNA and DNA molecule
(B) Only RNA (C) Only DNA
(D) A single RNA or DNA molecule

41. How genetic material of viruses is covered?
(A) In carbohydrate coat
(B) In phosphoprotein coat
(C) In protein coat
(D) In glycoprotein coat

42. Which of the following facts is not correct about the genome of HIV?
(A) Single stranded RNA
(B) About 9,000 nucleotides long
(C) It is retroviral genome which cause AIDS disease
(D) The primary targets of HIV infection are B cells

43. Which of the following virus group has the single stranded RNA genomes?
(A) Ø × 174 bacteriophage and λ virus
(B) HIV and bacteriophage Q*β*
(C) T4 bacteriophage and influenza virus
(D) TMV and FMD virus

44. What is called the specific type of circular viral DNA of λ virus during replication within a host cell?
(A) Replicative forms (B) Cruciform
(C) Hairpin form (D) Repetitive form

45. Which of the following virus has a circular double helix DNA in its replicative form inside cells during replication when infects *E. coli*?
(A) λ virus (B) HIV
(C) bacteriophage Q*β* (D) TMV

46. Which of the following true about different viruses?
i. ssRNA = HIV
ii. dsDNA = λ virus
iii. ssRNA = TMV
iv. ssDNA = Ø × 174 bacteriophage
(A) i, ii & iii (B) i, ii, iii & v
(C) ii & iii (D) i, ii & iv

47. What is the contour length of λ DNA that is a circular double-stranded contains 48,502 bp?
(A) 1.5 μm (B) 7.5 μm
(C) 5 μm (D) 17.5 μm

48. How many longer the DNA of bacteriophage T4 than the viral particle itself?
(A) About 290 times
(B) Same size both DNA and viral particle
(C) about 90 times
(D) about 190 times

49. Which of the following organism has a single double-stranded circular DNA?
(A) TMV (B) *E. coli*
(C) Human (D) Arabidopsis

50. A single *E. coli* cell contains a single double-stranded circular DNA molecule with 4,639,221 bp have a contour length of .... but some 850 times the length of the *E. coli* cell.
(A) About 7 mm
(B) About 1 mm
(C) About 107 mm
(D) About 1.7 mm

51. What is called the region of circular DNA chromosome of *E. coli* cell that are free in the cytosol?
(A) Nucleus (B) Nucleosome
(C) Nucleoid (D) Nucellus

52. What is called the circular extra-chromosomal elements of E. coli cell that are free in the cytosol?
(A) Plasmids (B) Protoplast
(C) Plastid (D) Phasmid

53. What is range of base pair of plasmids of bacteria?
(A) 10,000 – 1,00,000 bp
(B) 100 bp
(C) A few thousand – 10,000 bp
(D) Cannot determine the range of plasmid size

54. How daughter plasmids passed in daughter bacterial cells with genetic information after replication of plasmid?
(A) Through transcription
(B) Through mitosis
(C) Through cell division
(D) Through plasmid recombination

55. Which of the following organism does not contain extra-chromosomal genetic elements called plasmid?
(A) Yeast (A) Algae
(B) Fungi (C) Bacteri(A)

56. The enzyme's gene that confer resistance to β-lactam antibiotics such as penicillin and amoxicillin by plasmid?
(A) *β*-lactamase (B) *β*-lactase
(C) *β*-galactase (D) *β*-amylase

57. A yeast cell, one of the simplest eukaryotes has .... more DNA in its genome than an *E. coli* cell?
(A) 6 times (B) 2 times
(C) 26 times (D) 2.6 times

58. The cells of *Drosophila* (fruit fly) used contain more than .... as much DNA as E. coli cells, and human cells have .... as much.
(A) 700 times and almost 35 times
(B) 35 times and almost 700 times
(C) 300 times and almost 75 times
(D) 75 times and almost 300 times

59. Which type of cells are diploid?
(A) Pollen cells (B) Egg cells
(C) Somatic cells (D) Germ cells

60. How many different types of chromosomes are found in human that contain variable duplex DNA in length?
(A) 23 (B) 46
(C) 20 (D) 24

61. The DNA molecules in the 24 different types of human chromosomes contain ....
(A) 22 matching pairs of XX and two X and X sex chromosomes
(B) 22 matching pairs of XX and two X and Y sex chromosomes
(C) 23 matching pairs of XX and two X and Y sex chromosomes
(D) 46 matching pairs of XX and two X and Y sex chromosomes

62. What are number of autosomes and sex chromosomes in human?
(A) 44 autosomes (XX) and two X and Y sex chromosomes
(B) 46 autosomes (XX) and two X and Y sex chromosomes
(C) 23 autosomes (XX) and two X and Y sex chromosomes
(D) Two X and Y autosomes and 44 sex chromosomes (XX)

63. An adult human body contains .... cells?
(A) Approximately 10^{14}
(B) Approximately 10
(C) Approximately 10^{46}
(D) Approximately 3.2×10^{9}

64. What is the total DNA length of human
(A) 20 km (B) 10^{11} km
(C) 2×10^{11} km (D) 2 km

65. Which of the following cell organelles contain self circular DNA in eukaryotes?
(A) Mitochondria and Lysosomes
(B) Chloroplast and Mitochondria
(C) Chloroplast and Endoplasmic reticulum
(D) Ribosomes and Lysosomes

66. Which of the following cell organelle contain RNA in eukaryotes?
(A) Lysosomes
(B) Golgibody
(C) Endoplasmic reticulum
(D) Ribosomes

67. What is size of mitochondrial DNA (mtDNA) in human?
(A) 4630 bp (B) 5386 bp
(C) 3.2×10^{9} bp. (D) 16,569 bp

68. Which cell organelle does not contain circular DNA as genetic material?
(A) Mitochondria (B) Nucleus
(C) Chloroplast (D) Plasmid

69. How many copies of mtDNA normally found in each mitochondrion in human cell?
(A) Hundred copies (B) Only one copy
(C) Two to ten copies
(D) 10-100 copies

70. In which cells, hundreds copies of mtDNA contain?
(A) Differentiated embryonic cells
(B) Zygotic cell
(C) Endosperm
(D) Dedifferentiated embryonic cells

71. In which organism, contained the thousands of copies of mtDNA in each mitochondrion of its cell?
(A) Paramecium (B) Amoeba
(C) Trypanosomes (D) Hydra

72. What is called a complex and interlinked matrix which contain the thousands of copies of mtDNA in each mitochondrion of trypanosomes?
(A) Kinetochore (B) Centrosome
(C) Phragmoplast (D) Kinetoplast

73. Which of the following statement is not true about mitochondrial DNA?
(A) Plant cell mtDNA ranges in size from 200,000 to 2,500,000 bp
(B) mtDNA is replicated after and during division
(C) Mitochondrial DNA codes for the mitochondrial tRNAs and rRNAs
(D) Number of mtDNA copies in plant cell is more than animal cell

74. More than 95% of mitochondrial proteins are encoded by
(A) Nuclear DNA (B) mtDNA
(C) cpDNA (D) Plasmid

75. How many copies of chromosome contained in each cell of many bacterial species in nearly all cases?
(A) Two copies (B) 100 copies
(C) Only One copy (D) Infinite copies

76. How many copies of each gene contained in each chromosome of many bacterial species in nearly all cases?
(A) Only One copy (B) 100 copies
(C) Two copies (D) Infinite copies

77. Which types of RNA is coded by a very few genes that are repeated several times in bacterial cell?
(A) tRNAs (B) mRNAs
(C) hnRNAs (D) rRNAs

78. What is called eukaryotic nucleotide sequences that do not code for the amino acid sequence of the polypeptide product?
(A) Intervening sequences
(B) Transposon sequences
(C) Housekeeping gene sequences
(D) Pseudogene sequences

79. Which of the following is not true about the intervening segments of eukaryotic that are nontranslated inserts?
(A) Interrupt the collinear relationship between the coding gene and amino acids
(B) Nontranslated DNA segments in genes are called introns
(C) Few prokaryotic genes contain intervening segments
(D) In higher eukaryotes, gene has much more exon sequences than introns

80. What is called coding and non-coding DNA segments in an organism, respectively?
(A) Introns and Exons
(B) Exons and Introns
(C) Intervening segments and cistrons
(D) Trasnposons and SI elements

81. Which of the following gene coding for the single polypeptide chain of the avian egg protein that has the introns are much longer than the exons?
(A) Ovalbumin gene
(B) Globulin gene
(C) Histone gene
(D) Heamoglobin gene

82. Which of the following statement is not true?
(A) Seven introns make up 85% of the Ovalbumin gene's DNA for avian egg protein
(B) A single intron contains more than half of gene's DNA for hemoglobin β subunit
(C) The gene for the muscle protein titin is the intron champion with 178 introns
(D) Genes for histones appear to have one intron

83. Which of the following gene coding proteins have no introns in gene'DNA which encode the proteins?
(A) Ovalbumin (B) Globulin
(C) Histones (D) heamoglobin

84. How many percentage of total human DNA is coding or exon DNA that carry information for proteins or RNA products?
(A) Only about 5% (B) Only about 1.5%
(C) Only about 98.5%
(D) Only about 15%

85. In which form much of the nongene DNA exist in human genome?
(A) Repeated sequences
(B) Nonrepeated sequences
(C) Exons
(D) Transcripts

86. How many part of human genome made up of moderately repeated sequences that are derived from transposable elements?
(A) About 50% (B) only about 1.5%
(C) About 98.5% (D) only about 3%

87. What is not true regarding to transposones?
(A) DNA sequences that can move from one location to another in the genome
(B) About half the human genome is made up of transposons
(C) Transposable elements are a kind of molecular parasite
(D) Transposons have genes encoding proteins that are used in human cells

88. What is the role of transposable elements?
(A) Encode proteins that are used in human cells
(B) A major role in human evolution
(C) Transposons have no genes encoding proteins catalyzing transposition process
(D) Movement of transposons not lead to redistribution of other genomic sequences

89. What is percentage of the human genome consists of highly repetitive sequences referred to as simple sequence repeats (SSRs)?
(A) Almost 3% (B) About 50%
(C) only about 1.5% (D) About 98.5%

90. What may be the number of nucleotides in one repeat of Simple Sequence Repeats?
(A) Less than 10 bp (B) More than 10 bp
(C) 10 – 50 bp (D) 100 – 200 bp

91. The simple-sequence DNA has also been called .... because its unusual base composition causes it to migrate as "satellite" bands when fragmented cellular DNA samples are centrifuged in a cesium chloride density gradient.
(A) Satellite DNA (B) mtDNA
(C) Housekeeping DNA
(D) Circular ssDNA

92. Why highly repetitive DNA is important in human cellular metabolism?
(A) It is associated to centromeres and telomeres of eukaryotic chromosomes
(B) It is associated to mtDNA of eukaryotic mitochondria
(C) It is associated to cpDNA of plant chloroplast
(D) It is associated to protein synthesis of eukaryotes

93. What is the function of centromere present in eukaryotic chromosomes?
(A) As an attachment point for proteins to the recombination
(B) As an attachment point for proteins to the crossing over
(C) Help in the stabilization of chromosomes
(D) As an attachment point for proteins to the mitotic spindle

94. What is the length of sequences essential to centromere function are about 130 bp long that are very rich in AT pairs.
(A) About 200 bp long
(B) About 130 bp long
(C) About 190 bp long
(D) About 150 bp long

95. The sequences which are essential to centromere function of yeast (*Saccharomyces cerevisiae*) that are about 130 bp long and very rich in ....
(A) GT pairs
(B) CT pairs
(C) CG pairs
(D) AT pairs

96. What is the size of thousands of tandem copies of one or a few short sequences in the same orientation in the centromeric sequences of higher eukaryotes that generally contain simple-sequence DNA?
(A) 50 to 100 bp
(B) 5 to 10 bp
(C) 100 to 500 bp
(D) 10 to 100 bp

97. Which of the following organism' centromeres have been isolated and studied with finding the about 130 bp long and are very rich in AT pairs in centromeric sequences that essential to centromere function?
(A) Yeast (*Saccharomyces cerevisiae*)
(B) Ø × 174 bacteriophage
(C) Bacteria *Escherichia coli*
(D) *Arabidopsis thaliana*

98. What is the meaning of *telo* in term 'telomeres' in Greek language?
(A) *telos* = "end" (B) *telos* = "start"
(C) *telos* = "remain" (D) *telos* = "middle"

99. What is called the sequences at the ends of eukaryotic chromosomes that help stabilize the chromosome?
(A) Centromeres (B) Telomeres
(C) Chromomeres (D) Chromonema

100. What is the role of the sequences at the ends of eukaryotic chromosomes?
(A) Help in the stabilization of the chromosome
(B) As an attachment point for proteins to the mitotic spindle
(C) Help in the movement of chromatids during cell division
(D) Help in the crossing over and recombination

101. What is the length of yeast telomeres end of imprecisely repeated sequences of the form (5')(T*x*G*y*)*n* and (3')(A*x*C*y*)*n* where *x* and *y* are generally between 1 and 4?
(A) About 100 bp (B) 5 to 10 bp
(C) 100 to 500 bp (D) About 10 bp

102. Which enzyme id involved in addition of repeated telomeric sequences to eukaryotic chromosome ends?
(A) Telomerase (B) Centromerase
(C) DNA polymerase I (D) Topoisomerase II

103. Which of the following components does not require for a stable artificial linear chromosome?
(A) A centromere
(B) Telomeres at each end
(C) Sequences for the initiation of DNA replication
(D) Sequence for antibiotic resistant gene

104. Which of the following artificial chromosome have not been developed as a research tool in biotechnology?
(A) Yeast artificial chromosomes (YACs)
(B) Human artificial chromosomes (HACs)
(C) Arabidopsis artificial chromosomes (AACs)
(D) Bacterial artificial chromosomes (BACs)

105. Which are celled as a kind of molecular parasite?
(A) Transposable elements
(B) Exons (C) Pseudogenes
(D) Teleomeres

106. Why folding mechanism require for cellular DNA?
(A) Packaging of DNA in to chromosomes
(B) Access to the information in the DNA
(C) due to more length of DNA than chromosome length
(D) All of the above

107. Which is an important property of cellular DNA structure in eukaryotes?
(A) Only supercoiling
(B) Supresscoiling
(C) Linearity without supercoiling
(D) Linearity with supercoiling

108. What is the mean of supercoiling of DNA?
(A) Coiling of a coil
(B) Recoiling of a coil
(C) Despoiling of a coil
(D) Discoiling of a coil

109. What happen in DNA supercoling?
(A) DNA coiling of axis upon itself after coiled DNA with both strands
(B) Open coiling of coiled DNA with both strands
(C) DNA tightly linked to axis with hydrogen bonds
(D) No net bending of the DNA axis upon itself

110. What happen in relaxed state of DNA?
(A) DNA coiling of axis upon itself after coiled DNA with both strands
(B) One time both strands of the DNA coiling around an axis
(C) No net bending of the DNA axis upon itself
(D) DNA tightly linked to axis with hydrogen bonds

111. Which process are affected by DNA supercoiling because these processes require a separation of DNA strands?
(A) Replication
(B) Translation
(C) Transcription
(D) both (B) and (C)

112. In which state many circular DNA molecules remain after extractaction and purification from protein and other cellular components.

(A) Highly supercoiled
(B) Relaxed state (C) Unwind state
(D) Only in circular state without coiling

113. Which structure of DNA is shown by supercoiling after they are extracted and purified, freed from protein and other cellular components?
(A) Primary structure
(B) Tertiary structure
(C) Secondary structure
(D) Relaxed structure

114. What is called the conformational changes of DNA due to thermal motion or an interaction with proteins or other molecules?
(A) DNA continuous deformations
(B) DNA discontinuous deformations
(C) DNA supercoiling
(D) DNA melting

115. What happen in discontinuous deformations of DNA?
(A) DNA strand breakage
(B) Open supercoiling
(C) DNA supercoiling
(D) DNA strand binding

116. Which of the following one reason is not considered for the conformational changes of DNA?
(A) Due to thermal motion
(B) An interaction with proteins or other molecules
(C) Due to changing in pH and salt concentration
(D) DNA supercoling

117. Why topological properties are changed?
(A) Only by breakage of the backbone of one or both DNA strands
(B) Only by rejoining of the backbone of one or both DNA strands
(C) Only by breakage and rejoining of the backbone of one or both DNA strands
(D) Only by breakage and rejoining of the backbone of one DNA strands

118. Which of the following are not the examples of small circular DNAs?
(A) Plasmids
(B) Small viral DNAs
(C) Human DNA
(D) Eukaryotic mtDNAs and cpDNAs

119. What will not happen, if the DNA of a closed-circular molecule conforms closely to the B-form structure with one turn of the double helix per 10.5 bp.
(A) The DNA is not relaxed rather than supercoiled
(B) The DNA is relaxed rather than supercoiled
(C) The DNA replicates as eukaryotic DNA
(D) The DNA transcribed as eukaryotic DNA

120. Which process is induced after subjecting DNA to some form of structural strain whereas purified closed-circular DNA is rarely relaxed, if the DNA of a closed-circular molecule conforms closely to the B-form structure with one turn of the double helix per 10.5 bp?
(A) Supercoiling (B) Replication
(C) Unwinding (D) Melting

121. How many base pairs per turn contained by the DNA of a closed-circular molecule if it conforms closely to the B-form structure?
(A) 10.5 bp (B) 20 bp
(C) 12 bp (D) 11 bp

122. How many double helical turns contained by an 84 bp segment of a circular DNA in the relaxed state, if the DNA of a closed-circular molecule conforms closely to the B-form structure with one turn of the double helix per 10.5 bp?
(A) 10 (B) 10.5
(C) 8 (D) 12

123. How many double helical turns contained, if one of eight turns was removed from an 84 bp segment of the closed-circular DNA molecule that conforms closely to the B-form structural features?
(A) 12.0 bp per turn
(B) 10.0 bp per turn
(C) 10.5 bp per turn
(D) 11.0 bp per turn

124. Which is a deviation from the most stable DNA form?
(A) Circular DNA (B) Z – DNA
(C) A – DNA (D) B – DNA

125. Why underwinding is reuired for demage in isolated closed-circular DNA and it is generally accommodated by supercoiling rather than strand separation?

(A) Coiling the axis of the DNA requires less energy than breaking hydrogen bonds

(B) Coiling the axis of the DNA requires more energy than breaking hydrogen bonds

(C) The underwinding of DNA in vivo makes it not easier to separate DNA strands

(D) Underwinding induced by without enzatic activity

126. What is the role of the underwinding of DNA in cell?

(A) Important to enzymes of DNA metabolism about strand separation

(B) In gene regulation

(C) In post-translation mechanism

(D) In post transcription mechanism

127. What will be happened? If a break occurred in one strand of an isolated protein-free circular DNA?

(A) Strands are not free to rotate about each other

(B) The number of helical turns cannot be changed

(C) Strands not remain in the relaxed state

(D) Strands are free to rotate about each other

128. DNA underwinding is defined by ....

(A) Topological linking number

(B) Number of base pairs per helical turns

(C) Without breaking in one strand of DNA

(D) State of DNA supercoiling

129. Which is a topological property of double-stranded DNA that does not vary when the DNA is bent or deformed as long as both DNA strands remain intact?

(A) Base pair number per helical turn

(B) Diameter of DNA helix

(C) Linking number

(D) Length of DNA helix

130. Which statement is not true about the linking number of DNA double helix?

(A) Linking number is the number of times the second strand pierces this surface

(B) Topological linking number defines DNA underwinding

(C) The linking number for a closed-circular DNA is always an integer

(D) Dependent of the length of the DNA molecule

131. What is the linking number for a closed-circular DNA?

(A) Always an integer

(B) A decimal number

(C) Always one

(D) Always zero

132. What is the linking number, if the two DNA strands are interwound in a right-handed helix?

(A) Negative integer

(B) Positive integer

(C) Always zero

(D) Always one

133. What is the linking number, if the two DNA strands are interwound in a left-handed helix?

(A) Negative integer (B) Positive integer

(C) Always zero (D) Always one

134. What is the linking number to a closed-circular DNA with 2,100 bp in relaxed state of B-form?

(A) 200 (B) 198

(C) Not defined (D) Zero

135. What is the formula of linking number of a closed-circular DNA in relaxed state of B-form?

(A) Sum of total number of base pairs and the number of base pairs per turn

(B) Difference between total numbers of base pairs and the number of bp per turn

(C) Total number of base pairs divided by the number of base pairs per turn

(D) Multiplication of Total number of base pairs and the number of base pairs per turn

136. How many linking number of a circular closed DNA with 10500 bp with breaking at each 10.5 bp apart?

(A) 200 (B) 1000

(C) 100 (D) Zero

137. How many base pairs is found in a circular closed B-DNA with breaking at each 10.5 bp apart that has ten linking number?

(A) 1 (B) 105
(C) 1050 (D) Zero

138. What is linking number, if strands of a circular DNA molecule with 2100 bp do not contain a break?

(A) 1 (B) An integer
(C) Undefined (D) Zero

139. What is linking number, If a circular closed DNA has a break in either strand

(A) 1 (B) An integer
(C) Undefined (D) Zero

140. Which is true about DNA underwinding?

(A) It changes in the linking number
(B) It does not help in supercoiling
(C) it is not depend on topological property
(D) Positive linking number is due to supercoiling of the DNA

141. What is the symbole of linking number for relaxed DNA?

(A) *Lk0* (B) *Lk*
(C) *Lk1* (D) *Lkα*

142. If a B-DNA molecule has Linking number (*Lk*0) = 200, What is the value of *Lk*, if two turns are removed from this molecule?

(A) $Lk = 105$ (B) $Lk = 1$
(C) $Lk = 2$ (D) $Lk = 198$

143. What is the formula of changing linking number if any turn is removed from DNA molecule?

(A) $\Delta Lk = Lk - Lk0$ (B) $\Delta Lk = Lk + Lk0$
(C) $\Delta Lk = Lk/Lk0$ (D) $\Delta Lk = Lk \times Lk0$

144. What is the formula of specific linking difference (σ) or superhelical density

(A) $\sigma = Lk/Lk0$ (B) $\sigma = \Delta Lk/Lk0$
(C) $\sigma = \Delta Lk \times Lk0$ (D) $\sigma = Lk - Lk0$

145. What called a measure of the number of turns removed relative to the number present in relaxed DNA?

(A) Superhelical density
(B) Supercoiling Activity
(C) Unwinding density
(D) Superhelical specificity

146. If a B-DNA molecule has linking number (*Lk0*) = 200, what is the value of superhelical density (σ), if two turns are removed from this molecule?

(A) – 0.01 (B) + 0.01
(C) + 0.001 (D) – 1.98

147. If a B-DNA molecule has Linking number (Lk0) = 200, What is the percent value of superhelical density (σ), if two turns are removed from this molecule?

(A) 10% (B) 2%
(C) 1% (D) 98%

148. Generally, what is the degree of underwinding in cellular DNAs?

(A) Range of 5% to 7%
(B) Range of 10% to 15%
(C) Only 1%
(D) Range of 50-100%

149. Generally, what is the degree of underwinding in cellular DNAs

(A) $\sigma = 0.05$ to 0.07
(B) $\sigma = -0.05$ to -0.07
(C) $\sigma = 5$ to 7
(D) $\sigma = -5$ to -7

150. Supercoiling is not a .... process>

(A) Continuous
(B) Constant
(C) Random
(D) Specific pointable

151. The path of the supercoiling is largely prescribed through the torsional strain imparted to the DNA by .... relative to B-DNA.

(A) Decreasing the linking number
(B) Increasing the linking number
(C) Decreasing or increasing the linking number
(D) First increasing then decreasing the linking number

152. Linking number can be changed by .... by breaking one DNA strand, rotating one of the ends 360° about the unbroken strand and rejoining the broken ends.

(A) 2 (B) 1
(C) Zero (D) Infinite

153. What are called two forms of a circular DNA that differ only in a topological property (linking number) ?
(A) Topoisomers
(B) Isoschizomers
(C) Centrosomers
(D) Topoisomerasers

154. What are the structural components of linking numbers?
(A) Dextrorotatory (*D*) and Laevorotatory (*L*)
(B) Right (*R*) and Left (*L*)
(C) Writhe (*Wr*) and Twist (*Tw*)
(D) N-terminal (*N*) and C-terminal (*C*)

155. Which one of the component of linking number may be thought of as a measure of the coiling of the helix axis and second as determining DNA facilitates a number of structural changes in the molecule, respectively?
(A) Twist (*Tw*) and Writhe (*Wr*)
(B) Writhe (*Wr*) and Twist (*Tw*)
(C) Right (*R*) and Left (*L*)
(D) Left (*L*) and Right (*R*)

156. What facilitates by underwinding of a right-handed DNA helix?
(A) The formation of short stretches of left-handed A-DNA
(B) The formation of short stretches of left-handed B-DNA
(C) The formation of short stretches of left-handed H-DNA
(D) The formation of short stretches of left-handed Z-DNA

157. Which enzymes increase or decrease the extent of DNA underwinding?
(A) Teloemerases
(B) Topoisomerases
(C) Transcriptases
(D) DNA polymerases

158. Which enzymes changes in the Linking Number of DNA?
(A) Teloemerases
(B) DNA polymerases
(C) Transcriptases
(D) Topoisomerases

159. What is the function of enzemes topoisomerases?
(A) Increase or decrease the extent of DNA underwinding
(B) Initiation of transcription
(C) Cut the DNA at specific base sequences
(D) RNA splicing

160. In which processes enzyme topoisomerases play important role?
(A) Cut the DNA at specific base sequences
(B) Elongation of transcription
(C) DNA replication and packaging
(D) RNA splicing

161. Which are the major types of enzyme topoisomerases in eukaryotes?
(A) Type I topoisomerases and Type II topoisomerases
(B) Topoisomerases α and Topoisomerases β
(C) Topoisomerases III and Topoisomerases IV
(D) Negative topoisomerases and Positive topoisomerases

162. Which type of topoisomerases act by transiently breaking one of the two DNA strands with passing the unbroken strand through the break and rejoining the broken ends?
(A) Topoisomerases type IIα
(B) Topoisomerases type IIβ
(C) Topoisomerases type I
(D) Topoisomerases type II

163. Which type of topoisomerases change Lk in increments of 1 with acting by transiently breaking one of the two DNA strands?
(A) Topoisomerases type IIα
(B) Topoisomerases type IIβ
(C) Topoisomerases type I
(D) Topoisomerases type II

164. Which formula shows the value of linking number?
(A) $Lk = Tw + Wr$ (B) $Lk = Tw - Wr$
(C) $Lk = Tw \times Wr$ (D) $Lk = Tw / Wr$

165. What are the values of twist (*Tw*) and writhe (*Wr*) for finding the linking number?
(A) *Tw* and *Wr* need not be integers
(B) *Tw* and *Wr* are always zero or 1
(C) *Tw* and *Wr* are always integers
(D) *Tw* and *Wr* need be integers

166. Which statement is not correct about twist (*Tw*) and writhe (*Wr*) for finding the linking number?

(A) *Tw* and *Wr* need not be integers
(B) *Tw* and *Wr* are geometric rather than topological properties
(C) *Tw* and *Wr* may be changed by deformation of a closed-circular DNA molecule
(D) *Tw* and *Wr* are always integers

167. Which type of topoisomerases change *Lk* in increments of 2 with by transiently breaking both DNA strands?

(A) Topoisomerases type III
(B) Topoisomerases type IV
(C) Topoisomerases type I
(D) Topoisomerases type II

168. Which types of enzyme topoisomerases is not found in E.coli?

(A) Topoisomerases I
(B) Topoisomerases II
(C) Topoisomerases III
(D) Topoisomerases V

169. What happen when topoisomerases I and III relax DNA by removing negative supercoils?

(A) Increase linking number
(B) Decrease linking number
(C) First increase followed by decrease linking number
(D) First decrease followed by increase linking number

170. Which type of enzyme topoisomerase is called DNA gyrase in bacterium E.coli?

(A) Topoisomerases I
(B) Topoisomerases II
(C) Topoisomerases III
(D) Topoisomerases V

171. Which type of enzyme topoisomerase is called DNA gyrase in bacterium E.coli that introduces negative supercoils with decreasing linking number?

(A) Topoisomerases I
(B) Topoisomerases III
(C) Topoisomerases II
(D) Topoisomerases V

172. Which type of enzyme topoisomerase is called DNA gyrase in bacterium E.coli that introduces negative supercoils by cleaving both strands of DNA?

(A) Topoisomerases I
(B) Topoisomerases IV
(C) Topoisomerases III
(D) Topoisomerases II

173. The net activity of which types of topoisomerases maintain the degree of supercoiling of bacterial DNA by regulation?

(A) Topoisomerases I and III
(B) Topoisomerases II and III
(C) Topoisomerases I and II
(D) Topoisomerases I and IV

174. Which types of enzyme topoisomerases is not found eukaryotic cells?

(A) Topoisomerases III and IV
(B) Topoisomerases I and IV
(C) Topoisomerases I and II
(D) Topoisomerases I, II, III and IV

175. What are the forms of type I topoisomerase are found in eukaryotic cells?

(A) Topoisomerases I and III
(B) Topoisomerases I and IV
(C) Topoisomerases I and II
(D) Topoisomerases Iα and Iβ

176. What is the function the eukaryotic type II topoisomerases?

(A) Underwind DNA introduce negative supercoils
(B) Relax both positive and negative supercoils
(C) Underwind DNA introduce positive supercoils
(D) Both rewind and underwind DNA introduce negative supercoils

177. What is reuire for the DNA compaction?

(A) DNA unwinding
(B) DNA relaxing
(C) DNA supercoiling
(D) Longer base pair sequences

178. Which type of coiling exists in a negatively supercoiled DNA molecule with multiple branches?

(A) Left-handed supercoils
(B) Right-handed supercoils
(C) Dextrorotatory supercoil
(D) Random supercoil

179. At the superhelical densities, the length of the supercoil axis including branches is about 40% of the length of the DNA. This type of supercoiling is referred to as
(A) Plectonemic (B) Palindromic
(C) Randomized (D) Chromonemic

180. What is meaning of greek term "plectonemic"
(A) *plektos* = writhe and *nema* = *thread*
(B) *plektos* = left and *nema* = *thread*
(C) *plektos* = twisted and *nema* = *thread*
(D) *plektos* = coloured and *nema* = *thread*

181. Which is the second form of supercoiling required due to unsufficient compaction by plectonemic supercoiling to package DNA in the cell?
(A) Solenoidal (B) Palindromic
(C) Cylindrical (D) Circular

182. Which are the two forms of negative supercoiling that can be taken up by the same segment of underwound DNA.
(A) Cylindrical and circular
(B) Palindromic and solenoidal supercoiling
(C) Palindromic and Plectonemic
(D) Plectonemic and solenoidal supercoiling

183. Which form of supercoiling of DNA is more stable in solution?
(A) Cylindrical and circular
(B) Plectonemic
(C) Palindromic
(D) Solenoidal supercoiling

184. The solenoidal form of supercoing can be stabilized by .... and is the form found in chromatin that provides a much greater degree of compaction?
(A) Protein binding (B) Underwinding
(C) Lipid binding (D) Sugar binding

185. What is called the densely colored bodies seen in the nuclei of dye-stained under a light microscope in a virus, a bacterium, a eukaryotic cell, or an organelle?
(A) Centrosomes (B) Chromosomes
(C) Nucleosomes (D) Ribosomes

186. What is mostly chromatin?
(A) DNA-sugar complex
(B) DNA-lipid complex
(C) DNA-protein complex
(D) RNA-protein complex

187. In which process changes in eukaryotic chromosome structure?
(A) Cell cycle
(B) Respiration
(C) Protein synthesis
(D) Lipid synthesis

188. Which is the non-dividing phase of eukaryotic cell cycle?
(A) G1 (B) S
(C) G0 (D) G2

189. Which phase is not involed in interphage of eukaryotic cell cycle?
(A) G1 (B) S
(C) M (D) G2

190. What is called the chromosomal material that is amorphous and appears to be randomly dispersed in certain parts of the nucleus?
(A) Chromatin
(B) Centromere
(C) Chromomere
(D) Telomere

191. In which phase of interphase DNA is replicated and each chromosome produces two sister chromosomes called sister chromatids?
(A) G1 (B) S
(C) M (D) G2

192. Which is not the component of chromatin?
(A) DNA (B) Lipid
(C) RNA (D) Protein

193. What is the relation between protein and DNA of chromatin fibers?
(A) Both are in approximately equal mass
(B) Protein is more than DNA
(C) DNA is more than protein
(D) Protein Inversely proportion to DNA

194. Which protein is very tightly associated with DNA in the chromatin of eukaryotic chromosomes?
(A) Hemoglobin (B) Histones
(C) Tubulines (D) SSBP

195. What is called the structural unit of chromatin that are made up of very tightly association of histone proteins and DNA?
(A) Centrosomes (B) Telomeres
(C) Centioles (D) Nucleosomes

196. What is the role of nonhistone proteins present in chromatin of eukaryotic cell?
(A) Maintain chromosome structure and regulate the expression of specific genes
(B) Stabilizing of chromosomes
(C) Movement of chromosomes
(D) Division of chromosomes

197. Histones are ....
(A) Basic proteins
(B) Acidic proteins
(C) Neutral protein
(D) H1 is histone basic and others are acidic proteins

198. Which types of histones is found in eukaryotic chromosomes?
(A) H1 (B) H2A and H2B
(C) H3 and H4 (D) All of the above

199. Which is called linker histone protein?
(A) H1 (B) H2A and H2B
(C) H3 (D) H4

200. What is the molecular weight (in daltons) of histone protein?
(A) Between 1000 and 2000
(B) Between 1100 and 2100
(C) Between 11,000 and 21,000
(D) Between 110 and 210

201. Which amino acid residues found in histones than other amino acid residues?
(A) Tryptophan and Tyrosin
(B) Phenilalanin and Serine
(C) Methionine and proline
(D) Arginine and lysine

202. Which types of proteins are very rich in the basic amino acids arginine and lysine that make up together about one-fourth of the amino acid residues?
(A) Glycophorin A
(B) Histones
(C) Single stranded binding proteins
(D) GFP proteins

203. How many major histones are found in eukaryotic cells?
(A) Four (B) Three
(C) Two (D) Five

204. Which if the following histones show less sequence similarity among eukaryotic species?
(A) H3 and H4
(B) H1, H2A and H2B
(C) H1, H2A, H2B, H3 and H4
(D) H1

205. Only 2 of 102 amino acid residues of which histones differ between peas and cows and only 8 differ between humans and yeast.
(A) H1 (B) H2A and H2B
(C) H3 (D) H4

206. Which of the following enzymatic modifications in the amino acid side chains of each type of histone that has variant forms?
(A) Methylation
(B) ADP-ribosylation and glycosylation
(C) Phosphorylation and acetylation
(D) All of the above

207. Which properties of histones effect by different types of modifications?
(A) Net electric charge
(B) Shape
(C) The structural and functional properties of the chromatin
(D) All of the above

208. In which process histones do not play any role?
(A) In the regulation of transcription
(B) Restriction digestion
(C) DNA supercoiling
(D) DNA replication

209. Which is the fundamental organizational unit of chromatin?
(A) Centrosomes (B) Chromosomes
(C) Nucleosomes (D) Ribosomes

210. What is the diameter of cell nucleus in which the a DNA molecule of eukaryotic chromosome compacted about 10^5 μm long
(A) 5 to 10 μm (B) 50 to 100 μm
(C) 0.5 to 1 μm (D) 0.05 to 0.01 μm

211. Which type of structure is seen by the association of DNA and proteins in chromosomes?
(A) Domb-shape structure
(B) Beads-like structure
(C) Cone-shape structure
(D) Sleeper-shape structure

212. How many histones are presnt in the bead of each nucleosome?
(A) Six (B) Ten
(C) One (D) Eight

213. What is the ratio of histones in the bead of each nucleosome?
(A) H2A : H2B : H3 : H4 = 2 : 2 : 1 : 1
(B) H2A : H2B : H3 : H4 = 1 : 1 : 1 : 1
(C) H2A : H2B : H3 : H4 = 2 : 2 : 2 : 2
(D) H2A : H2B : H3 : H4 = 1 : 1 : 2 : 2

214. Which type of histone is not found in the bead of each nucleosome?
(A) H1 (B) H2A and H2B
(C) H3 (D) H4

215. What is called the combination of eight histones in the bead of each nucleosome?
(A) Histone octamer
(B) Nucleosome centre
(C) Histone centre
(D) Core space centre

216. What is the size of DNA which wrapped on histone core (Histone octamer) in a nucleosome?
(A) 200 bp (B) 54 bp
(C) 146 bp (D) 400 bp

217. What is the size of DNA which wrapped on a complete nucleosome with linker histone H1?
(A) 200 bp (B) 54 bp
(C) 146 bp (D) 400 bp

218. What is the size of DNA which connected to linker histone H1?
(A) 200 bp (B) 54 bp
(C) 146 bp (D) 400 bp

219. Which type of histone is not found in the bead of each nucleosome and this histone serves as linker DNA between nucleosome beads?
(A) H1 (B) H2A and H2B
(C) H3 (D) H4

220. Which of the following process provided the information about nucleosomes structure and different sizes of DNA?
(A) Nuclease treatment
(B) Site direct mutagenesis experiment
(C) Autoradiography
(D) Protein determination method

221. Which of the following process reveals a particle made up of the eight histone molecules with the DNA wrapped around it in the form of a left-handed solenoidal supercoil?
(A) Nuclease treatment
(B) X-ray diffraction analysis
(C) DNA sequencing
(D) Autoradiography

222. What requires for the tight wrapping of DNA around the histone core in nucleosome of eukaryotes?
(A) Removal of H1 histone from nucleosome
(B) Completely relax DNA
(C) Removal of about one helical turn in the DNA
(D) Insertion a nick by topoisoemarse

223. Where do histone cores bind to DNA at certain locations.
(A) GC base pairs region
(B) AC base pairs region
(C) GT base pairs region
(D) AT base pairs region

224. What happen if the minor groove of the helix compressed at AT base pair points makes this compression more likely?
(A) Tightly wrapped of the DNA around the nucleosome's histone core
(B) Loose wrapping of the DNA around the nucleosome's histone core
(C) DNA is completely relaxed
(D) Remove H1 histone from nucleosome

225. What is the function of nonhistone proteins which bind to a specific DNA sequence in nucleosome?
(A) Facilitate the formation of a nucleosome core
(B) Loose wrapping of the DNA around the nucleosome's histone core
(C) completely relax DNA
(D) Easily provide a position for cut to DNA

226. How many folds is done by wrapping of DNA around a nucleosome core which compacts the DNA length?
(A) About eleven fold
(B) About ten fold
(C) About eight fold
(D) About seven fold

227. Which is the first organized structure of nucleosome cores in chromosomes which isolated by very gentle methods?
(A) 30 nm fiber
(B) Scaffold structure
(C) Solanoid structure
(D) 100-fold compaction structure

228. Which histone require for the 30 nm fibers that is first organized structure of nucleosome cores in chromosomes which isolated by very gentle methods?
(A) H3 (B) H2A and H2B
(C) H1 (D) H4

229. Which property is found in 30 nm fibers is punctuateed by regions bound by sequence-specific (nonhistone) DNA binding proteins.
(A) The translation activity of the particular region of DNA
(B) The transcriptional activity of the particular region of DNA
(C) The gene regulation of the particular region of DNA
(D) The Post-translation activity of the particular region of DNA

230. What is the second level of chromatin organization which provided by The 30 nm fiber?
(A) An approximately 100-fold compaction of the DNA
(B) Scaffold structure
(C) Solanoid structure
(D) Nucleoid structure

231. Which is the main characteristics of the scaffold-associated regions that are separated by it?
(A) Nucloidal structure
(B) Solanoidal shape
(C) 30 nm fibers
(D) Loops of DNA with 20 to 100 kbp

232. Where are found complete sets of histone-coding genes in *Drosophila* that are bonded to scaffold attachment sites during chromatin organization?
(A) In mitochondrial genome
(B) In loops of DNA with 20 to 100 kbp
(C) In 30 nm fibers
(D) In Plasmamembrane

233. Which enzyme contained in the scaffold with several proteins and large amounts of histone H1 located in the interior of the fiber?
(A) Topoisomerase II
(B) Topoisomerase I
(C) DNA polymerase
(D) DNA ligase

234. What are called several drugs that used in cancer chemotherapy and promote strand breakage but not the resealing of the breaks?
(A) Topoisomerase I inhibitors
(B) DNA ligase inhibitors
(C) DNA polemerase II inhibitor
(D) Topoisomerase I inhibitors

235. Three-Dimensional packaging of nuclear chromosomes condensed chromosome structures are maintained by ....
(A) Globular proteins
(B) SMC proteins
(C) Tubulin proteins
(D) SSBP

236. Who proposed one gene–one enzyme hypothesis after exposing spores of the fungus *Neurospora crassa* to X-rays known to damage DNA and cause alterations in DNA sequence?
(A) Watson and Crick
(B) Jacob and Monad
(C) Wilson and Franklin
(D) Beadle and Tatum

237. What is called the sudden and heritable alterations in DNA sequences?
(A) Mutations (B) Evolution
(C) Abnormalities (D) Genotypes

238. How many amino acids required for forming a polypeptide chain that coded by 1,050 bp DNA in size?
(A) 1050 (B) 350
(C) 300 (D) 575

239. Three-Dimensional packaging of nuclear chromosomes condensed chromosome structures are maintained by SMC (Structural Maintenance of Chromosomes) proteins that are .... proteins.
(A) Monomeric
(B) Polymeric
(C) Dimeric
(D) Tetrameric

240. In which protein, V-shaped complex is formed that maintain three-dimensional packaging of nuclear chromosomes condensed chromosome structures?
(A) Histone proteins
(B) SMC proteins
(C) Topoisomerases
(D) SSBP

241. In which organism is not found SMC (Structural Maintenance of Chromosomes) proteins that maintain three-dimensional packaging of nuclear chromosomes condensed chromosome structures?
(A) Bacterial
(B) BGA
(C) Virus
(D) Human

242. Which are the major types of eukaryotic SMC (Structural Maintenance of Chromosomes) proteins that maintain three-dimensional packaging of nuclear chromosomes condensed chromosome structures?
(A) Histones and SSBP
(B) Cohesins and Condensins
(C) Topoisomerases I and Topoisomerases II
(D) Tubulin and Actin

243. Which is the major type of eukaryotic SMC (Structural Maintenance of Chromosomes) proteins play a substantial role in linking together sister chromatids immediately after replication and keeping them together as the chromosomes condense to metaphase.
(A) Histones and SSBP
(B) Condensins
(C) Topoisomerases II
(D) Cohesins

244. Which is the major type of eukaryotic SMC (Structural Maintenance of Chromosomes) proteins are essential to the condensation of chromosomes as cells enter mitosis.
(A) Histones and SSBP
(B) Condensins
(C) Topoisomerases II
(D) Cohesins

245. Condensin proteins bind to DNA in a manner that creates .... that causes the DNA to become overwound.
(A) Positive supercoils
(B) Negative supercoils
(C) Random supercoils
(D) DNA relaxing

246. What is called a structure in which the bacterial DNA is compacted?
(A) Nucleosome (B) Nucleus
(C) Nucleoid (D) Nucleolus

247. Which type of structure of *E. coli* DNA appears to organize the circular chromosome into a series of looped domains?
(A) V-shape complex structure
(B) A scaffold like structure
(C) Solanoid structure
(D) Nucleosome structure

248. Which is the histone like proteins are abundant in *E. coli* that is a two-subunit protein with Mr 19,000 dalton?
(A) HU
(B) Histones and SSBP
(C) Condensins and Cohesins
(D) Topoisomerases II

249. In the following figure is shown different parts of nucleosome structure. What are represented by X, Y and Z?

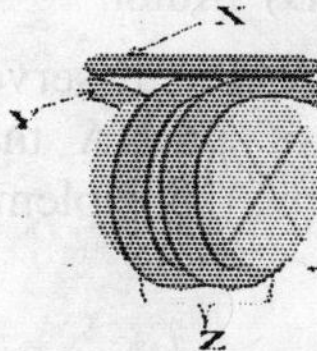

(A) Core DNA, Linker histone, Linker DNA
(B) Linker DNA, Linker histone, Core DNA
(C) Linker histone, Linker DNA, Core DNA
(D) Linker histone, Core DNA, Linker DNA

250. In the following figure shown the phosphate group covalently linked at the 5' hydroxyl of a nucleotide. What are X, Y and Z, respectively?

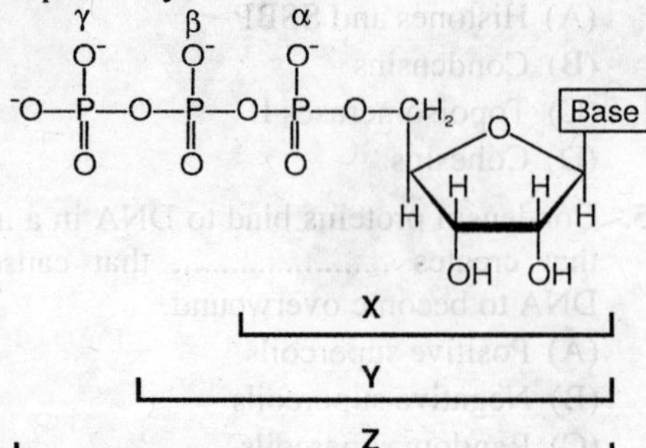

(A) NDPs, NTPs, and NMPs
(B) NTPs, NMPs, and NDPs
(C) NMPs, NDPs, and NTPs
(D) NTPs, NDPs, and NDPs

251. Which is not a part of an undamaged T2 bacteriophage particle?
(A) Protein coated-head
(B) A tail
(C) Single linear DNA
(D) Single linear RNA

252. What is the function of the tail of the bacteriophage T2 during bacterial infection?
(A) Help in attachment itself to the outer surface of a bacterial cell
(B) Help in DNA replication
(C) Help in bacteriophage division
(D) Help in protein transfer to a bacterial cell

253. What is called the triplets of nucleotide units in DNA determine the amino acids in a protein through the intermediary mRNA.
(A) Exon (B) Intron
(C) Codon (D) Muton

254. Which strand of the DNA strands serves as a template for synthesis of mRNA that has nucleotide triplets (codons) complementary to those of the DNA?
(A) Sense strand
(B) Antisense strand
(C) Semi-strand
(D) Both strands

255. What is called one strand of the duplex DNA that serves as a template for synthesis of mRNA which has nucleotide triplets (codons) complementary to those of the DNA?
(A) Sense strand
(B) Both strands synthesis mRNA
(C) Semi-strand
(D) Template

256. How many nucleotides require for forming 4 polypeptide bonds?
(A) 12 (B) 4
(C) 15 (D) 5

257. How many amino acids require for forming 4 polypeptide bonds?
(A) 12 (B) 4
(C) 15 (D) 5

258. How many peptide bonds are formed by 72 nucleotides?
(A) 72 (B) 23
(C) 24 (D) 22

259. How many amino acids coded by 72 nucleotides?
(A) 72 (B) 23
(C) 24 (D) 22

260. Four core histones show at least five types of covalent modification?
(A) Acetylation and covalent linkage (H2A only) to ubiquitin
(B) Methylation and ADP ribosylation
(C) Phosphorylation
(D) All of the above

261. In the following figure W, X, Y and Z are shown, respectively.

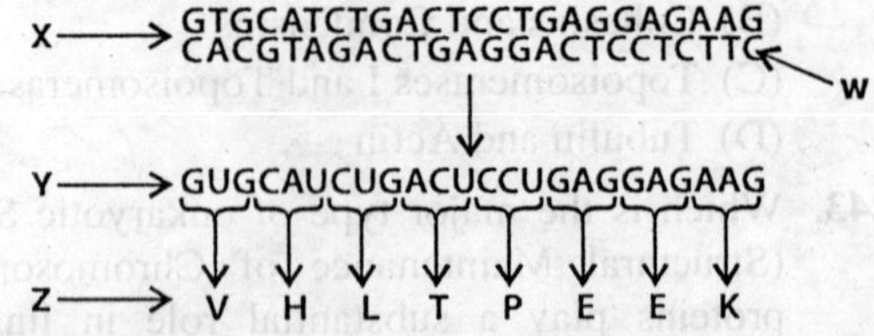

(A) Amino acids chain, DNA chain, Template DNA, RNA chain
(B) RNA chain, Template DNA, DNA chain, amino acids chain
(C) RNA chain, DNA chain, Template DNA, amino acids chain
(D) Template DNA, DNA chain, RNA chain, amino acids chain

262. What are the length of the E. coli chromosome in linear form and the length of a typical E. coli cell?
(A) 17 mm and 2 µm
(B) 1.7 mm and 2 µm
(C) 1.7 mm and 0.2 µm
(D) 2 µm and 1.7 mm

263. What is the number of a complete set of chromosomes in every normal human somatic cell?
(A) 46 chromosomes
(B) 23 chromosomes
(C) 44 chromosomes
(D) 46 pairs chromosomes

264. Wild yeast strains generally are
(A) Hexaploid (B) Tetraploid
(C) Octoploid (D) Diploid

265. How many number of chromosomes is found in male nematodes (Caenorhabditis elegans) in all whereas females have two X chromosomes?
(A) 11 chromosomes (X but no Y)
(B) 11 chromosomes (no X but Y)
(C) 12 chromosomes (X and Y)
(D) 11 chromosomes (X but no Y)

266. Which of the following match is correct?

S. No.	Types of sequences in the human genome	Percentage
i	Genes	30%
ii	LINEs	21%
iii	SINEs	13%
iv	Transposons	45%
v	Introns and noncoding segments	28.5%

(A) i, ii, iii and iv (B) i, ii, iii and v
(C) i, ii and iii (D) i, ii, iii, iv and v

267. Which of the following match is correct?

S. No.	Types of sequences in the human genome	Percentage
i	Long interspersed elements (LINEs)	6 – 8 kbp long
ii	Short interspersed elements (SINEs)	100 – 300 bp long
iii	Retrolike transposons	1.5 – 11 kbp long
iv	HIV	Retrolike transposons

(A) i, ii, iii and iv (B) i, ii and iii
(C) Only ii and iii (D) Only I and ii

268. How many helical turn will be present in a segment of DNA within a closed-circular molecule of 84 bp long?
(A) 7 (B) 8
(C) 8.4 (D) 10

269. In the following figure, each blue ribbon represents one strand of a double-stranded DNA molecule. If the linking number is defined as the number of times the second strand pierces the surface, what are the linking number (*Lk*) in X and Y, respectively?

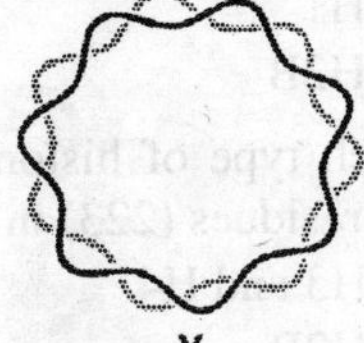
X Y

(A) X = 1 and Y = 1
(B) X = 6 and Y = 1
(C) X = 6 and Y = 6
(D) X = 1 and Y = 6

270. Suppose that a closed-circular DNA with 2,100 bp which shown in three conditions in the figure has different the linking number that is simply the number of base pairs divided by the number of base pairs per turn, which is close to 10.5 bp. Find the values of X, Y and Z, respectively?

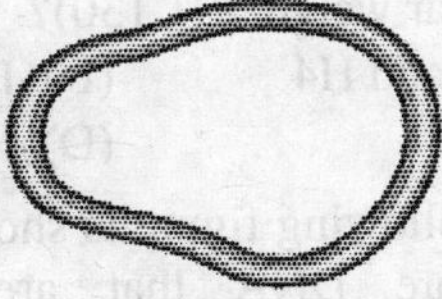

(A) 200, 202 and 198 (B) 198, 202 and 200
(C) 202, 198 and 200 (D) 200, 198 and 202

271. Which bond is broken in DNA by activity of topoisomerases?
(A) N – glycosidic bond
(B) Phosphodiester bond
(C) Hydrogen bond
(D) Ionic bonds

272. Which type of histone has maximum lysine amino acid residues (29.5%)?
(A) H1 (B) H2A and H2B
(C) H3 (D) H4

273. How many peptide bonds present in protein containing 1000 amino acids?
(A) 1000 (B) 999
(C) 333 (D) 1001

274. Which type of histone has maximum arginine amino acid residues (19.3%)?
(A) H1 (B) H2A
(C) H2B (C) H3 and H4

275. Which type of histone has maximum amino acid residues (223) in its polypeptide chain?
(A) H3 and H4 (B) H2A
(C) H2B (D) H1

276. Which type of histone has minimum amino acid residues (102) in its polypeptide chain?
(A) H3 (B) H4
(C) H2A and H2B (D) H1

277. Which type of histone collectively has maximum content of arginine and lysine amino acid residues?
(A) H1 (B) H2A and H2B
(C) H3 (D) H4

278. Which type of histone has maximum molecular weight (21,130)?
(A) H3 and H4 (B) H2A
(C) H2B (D) H1

279. In the following figure is shown Linker DNA and Core DNA that are connected to nucleosome structure in human. What are represented by X and Y?

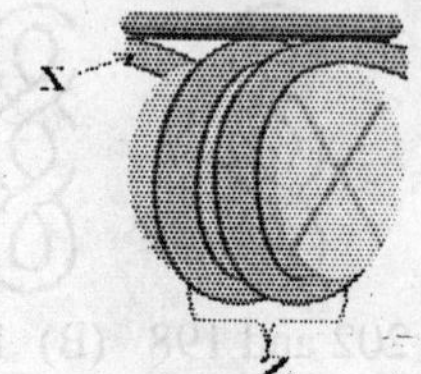

(A) X = Core DNA (200 bp) and Z = Linker DNA (54 bp)
(B) X = Linker DNA (54 bp) and Z = core DNA (146 bp)
(C) X = Linker DNA (200 bp) and Z = Core DNA (146 bp)
(D) X = Core DNA (54 bp) and Z = Linker DNA (146 bp)

280. Which of the following enzymes are shown by W, X, Y and Z involved in different biological processes?

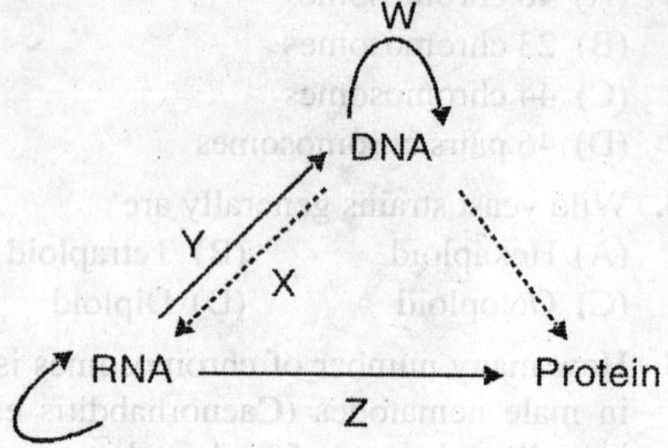

(A) W=DNA polymerase III, X= Reverse transcriptase, Y= RNA polymerase and Z= Peptidyl transferase
(B) W=RNA polymerase III, X=DNA polymerase, Y=Reverse transcriptase and Z= Peptidyl transferase
(C) W=DNA polymerase III, X= Peptidyl transferase, Y= RNA polymerase and Z= Reverse transcriptase
(D) W=DNA polymerase III, X=RNA polymerase, Y=Reverse transcriptase and Z= Peptidyl transferase

281. .... superhelical turns of DNA are wrapped around the surface of the histone octamer with covering .... base pairs of DNA.
(A) 1.75 and 146
(B) 1.46 and 175
(C) 1.50 and 146
(D) 1.50 and 200

282. A mutation in a germ cell transmitted to offspring called .... of hereditary disease.
(A) Transverse transmission
(B) Horizontal transformation
(C) Vertical transformation
(D) Lateral transformation

283. A nucleosome core particle consists of two copies each of histones namely .... to form a histone octamer.
(A) Only H1
(B) H2B, H3 and H4
(C) H2A, H2B and H4
(D) H2A, H2B, H3 and H4

284. AC repeats in microsatellite sequences are very useful in constructing ....
(A) RFLP
(B) Cytogenetic map
(C) Genetic linkage maps
(D) Genetic transformation

285. .... of core histones is associated with chromosomal assembly during DNA replication.
(A) Acetylation
(B) Methylation
(C) Phosphorylation
(D) Hydroxylation

286. ADP-ribosylation of histones is associated with ....
(A) DNA recombination
(B) DNA repair
(C) Mutation
(D) DNA demage

287. AGA and AGG are standard codons for Arg but Human mitochondrial DNA read as ....
(A) Initiation codons
(B) Sense codons
(C) Overaapping codons
(D) Stop codons

288. All mitochondria are contributed by the ovum during zygote formation that is transmitted by .... inheritance.
(A) Paternal nonmendelian
(B) Maternal mendelian
(C) Maternal nonmendelian
(D) Paternal mendelian

289. Alu family belongs to .... in the human genome.
(A) SINEs
(B) LINEs
(C) Tn5
(D) Viral transposon

290. Alu family belongs to SINEs is present in .... copies per haploid genome and accounts for at least 5–6% of the human genome.
(A) About 500,000 (B) About 500
(C) About 5000 (D) About 50,00,000

291. At metaphase, mammalian chromosomes with the identical duplicated sister chromatids connected at a ....
(A) Centrosome (B) Chromomere
(C) Centromere (D) Telomere

292. Before joining the ends of a linear duplex DNA into a closed circle, one end is twisted relative to the other to introduce some strain into the molecule. Such coiling of the DNA helix upon itself is called ....
(A) Reverse coiling (B) Supercoiling
(C) Parallel coiling (D) Simple coiling

293. What are the components of chromatin?
(A) DNA, histone proteins and RNA
(B) Histone proteins and RNA
(C) DNA and RNA
(D) Only DNA

294. Chromatin has dense spherical particles called .... which are approximately 10 nm in diameter and connected by DNA filaments.
(A) Centrosomes (B) Chromosome
(C) Centromere (D) Nucleosomes

295. Closed-circular DNA of prokaryotic cells is .... supercoiled.
(A) Negatively (B) Positively
(C) Normal (D) No coiling

296. Constitutive heterochromatin is always condensed and inactive in the regions near the chromosomal centromere and at chromosomal ends called....................
(A) Centrosome (B) Chromomere
(C) Centromere (D) Telomere

297. DNA in active chromatin contains large regions (about 100,000 bases long) that are sensitive to digestion by a nuclease such as.
(A) RNase I
(B) DNase I
(C) Restriction endonuclease
(D) Exonuclease

298. DNase I makes .... in any segment of DNA.
(A) Double-strand cuts
(B) No cuts
(C) Single-strand cuts
(D) Cut at specific locus by double cuts

299. Acetylation of histones H3 and H4 is associated with the activation or inactivation of ....
(A) Gene translation
(B) Gene transcription
(C) Gene replication
(D) Gene transformation

300. .... is the enzyme responsible for telomere synthesis.
(A) Restriction endonuclease
(B) Polymerase
(C) Replicase
(D) Telomerase

Answer-Key

Q	Ans	Q	Ans	Q	Ans	Q	Ans	Q	Ans	Q	Ans	Q	Ans	Q	Ans
1.	(A)	41.	(C)	81.	(A)	121.	(A)	161.	(A)	201.	(D)	241.	(C)	281.	(A)
2.	(C)	42.	(D)	82.	(D)	122.	(C)	162.	(C)	202.	(B)	242.	(B)	282.	(B)
3.	(D)	43.	(B)	83.	(C)	123.	(A)	163.	(C)	203.	(D)	243.	(D)	283.	(D)
4.	(A)	44.	(A)	84.	(B)	124.	(D)	164.	(A)	204.	(B)	244.	(B)	284.	(C)
5.	(D)	45.	(A)	85.	(A)	125.	(A)	165.	(A)	205.	(D)	245.	(A)	285.	(A)
6.	(A)	46.	(B)	86.	(A)	126.	(A)	166.	(D)	206.	(D)	246.	(C)	286.	(B)
7.	(C)	47.	(D)	87.	(D)	127.	(D)	167.	(D)	207.	(D)	247.	(B)	287.	(D)
8.	(B)	48.	(A)	88.	(B)	128.	(A)	168.	(D)	208.	(B)	248.	(A)	288.	(C)
9.	(C)	49.	(B)	89.	(A)	129.	(C)	169.	(A)	209.	(C)	249.	(C)	289.	(A)
10.	(D)	50.	(D)	90.	(A)	130.	(D)	170.	(B)	210.	(A)	250.	(C)	290.	(A)
11.	(D)	51.	(C)	91.	(A)	131.	(A)	171.	(C)	211.	(B)	251.	(D)	291.	(C)
12.	(A)	52.	(A)	92.	(A)	132.	(B)	172.	(D)	212.	(D)	252.	(A)	292.	(B)
13.	(C)	53.	(C)	93.	(D)	133.	(A)	173.	(C)	213.	(C)	253.	(C)	293.	(A)
14.	(C)	54.	(C)	94.	(B)	134.	(A)	174.	(C)	214.	(A)	254.	(B)	294.	(D)
15.	(D)	55.	(B)	95.	(D)	135.	(C)	175.	(A)	215.	(A)	255.	(D)	295.	(A)
16.	(A)	56.	(A)	96.	(B)	136.	(B)	176.	(B)	216.	(C)	256.	(C)	296.	(D)
17.	(C)	57.	(D)	97.	(A)	137.	(C)	177.	(C)	217.	(A)	257.	(D)	297.	(B)
18.	(A)	58.	(B)	98.	(A)	138.	(C)	178.	(B)	218.	(B)	258.	(B)	298.	(C)
19.	(C)	59.	(C)	99.	(B)	139.	(C)	179.	(A)	219.	(A)	259.	(C)	299.	(B)
20.	(B)	60.	(D)	100.	(A)	140.	(A)	180.	(C)	220.	(A)	260.	(D)	300.	(D)
21.	(B)	61.	(B)	101.	(A)	141.	(A)	181.	(A)	221.	(B)	261.	(D)		
22.	(A)	62.	(B)	102.	(A)	142.	(D)	182.	(D)	222.	(C)	262.	(B)		
23.	(C)	63.	(A)	103.	(D)	143.	(A)	183.	(B)	223.	(D)	263.	(A)		
24.	(D)	64.	(C)	104.	(C)	144.	(B)	184.	(A)	224.	(A)	264.	(C)		
25.	(A)	65.	(B)	105.	(A)	145.	(A)	185.	(B)	225.	(A)	265.	(D)		
26.	(D)	66.	(D)	106.	(D)	146.	(A)	186.	(C)	226.	(D)	266.	(D)		
27.	(D)	67.	(D)	107.	(D)	147.	(C)	187.	(A)	227.	(A)	267.	(A)		
28.	(D)	68.	(B)	108.	(A)	148.	(A)	188.	(C)	228.	(C)	268.	(B)		
29.	(C)	69.	(C)	109.	(A)	149.	(B)	189.	(C)	229.	(B)	269.	(D)		
30.	(B)	70.	(A)	110.	(D)	150.	(C)	190.	(A)	230.	(A)	270.	(D)		
31.	(D)	71.	(C)	111.	(D)	151.	(C)	191.	(B)	231.	(D)	271.	(B)		
32.	(C)	72.	(D)	112.	(A)	152.	(B)	192.	(B)	232.	(B)	272.	(A)		
33.	(A)	73.	(B)	113.	(B)	153.	(A)	193.	(A)	233.	(A)	273.	(B)		
34.	(D)	74.	(A)	114.	(A)	154.	(C)	194.	(B)	234.	(C)	274.	(B)		
35.	(D)	75.	(C)	115.	(A)	155.	(B)	195.	(D)	235.	(B)	275.	(D)		
36.	(D)	76.	(A)	116.	(D)	156.	(D)	196.	(A)	236.	(D)	276.	(B)		
37.	(D)	77.	(D)	117.	(C)	157.	(B)	197.	(A)	237.	(A)	277.	(A)		
38.	(C)	78.	(A)	118.	(C)	158.	(D)	198.	(D)	238.	(B)	278.	(D)		
39.	(B)	79.	(D)	119.	(A)	159.	(A)	199.	(A)	239.	(C)	279.	(B)		
40.	(D)	80.	(B)	120.	(A)	160.	(C)	200.	(C)	240.	(B)	280.	(D)		

4

DNA Replication

Deoxyribonucleic acid (DNA) contains the genetic instructions for the biological development of a cellular form of life including some viruses. DNA is an antiparallel double helix molecule with sugar-phosphate backbone on the outer side and nitrogen bases in the inner side. The bases are paired specifically, also known as complementary pairing, Adenine (A) with Thymine (T), and Guanine (G) with Cytosine (C) by two and three hydrogen bonds, respectively. DNA is a long polymer of nucleotides (a polynucleotide) that transfers the information to RNA (mRNA) that in turn controls the sequence of amino acid residues in proteins, using the genetic code. James Watson and Francis Crick proposed the DNA double helix structure in 1953 and shared the the Noble Prize in 1962 with Maurice Wilkins. During cell division, each DNA molecule precisely forms its identical copy which moves to each of the two daughter cells called DNA replication in which both strands of DNA separate from each other at a specific point called origin of centre (*Ori*C) and both strands serve as a template for synthesis of new complementary strands.

Models for DNA Replication

A. **Conservative replication** according to which the original strand is unchanged and a completely new DNA molecule is synthesized.

(B) **Semi-conservative replication** describes a way in which each DNA strand of a double helical molecule serves as a template for the synthesis of two new DNA molecules each with one new and one old strand.

C. **Dispersive replication** exhibits more-or-less random interspersion of parental and new segments in daughter DNA molecules.

Among the replication models, semi-conservative replication is widely accepted and experimentally proved by Meselson and Stahl (1958). The basic mechanisms of DNA replication are similar in all the living organisms.

Mechanism of DNA Replication

The basic steps involved in DNA synthesis are: Initiation, Elongation and Termination.

Initiation

In *E.coli*, DNA replication gets started at a discreet region of the chromosome called origin of replication (*oriC*). As per Cairns model of DNA replication, both parental strands act as templates simultaneouly for synthesis of two daughter complementary DNA strands. This type of synthesis generates what is called a replication fork as it resembles a two-pronged fork. With the local unwinding of dsDNA, two template strands now become available for the initiation of replication. All known DNA polymerases have a common property i.e. they cannot initiate DNA replication or in other words all of them require the presence of a short stretch of nucleotides, called as primer, on to which incoming deoxynucleotides are added. During DNA synthesis, an RNA polymerase primase or primase adds a short stretch (10-60 nucleotides long) of RNA nucleotides called as primer, complementary to one of the parental DNA strands.

Proteins/enzymes involved in initiation step of DNA replication in *Escherichia coli*

Protein/enzyme	Function
DnaA protein	Recognition of *oriC* region in DNA; denaturation of *oriC* region
Helicase (DnaB protein)	Unwinding of dsDNA
DnaC protein	Assists helicase to bind to DNA
HU	Helps DnaB protein in denaturation of DNA
Primase (DnaG protein)	Synthesis of RNA primers
Single strand DNA binding proteins (SSBP)	Binding to ssDNA
RNA polymerase	Promotes DnaA protein activity
DNA gyrase (DNA topoisomerase II)	Relieves topological strain generated by DNA unwinding
Dam methylase	Methylates (5′) GATC palindromic sequences present in *oriC* locus

Elongation

Once the primer has been formed, DNA polymerase starts adding new nucleotides and thus the DNA daughter strands start to grow by adding nucleotides at 3' free OH group. This process continues as the longer template stretch is made availably by progressive unwinding of the dsDNA. In order to maintain the 5′- 3′ direction of DNA synthesis on two antiparallel template strands, Reiji Okazaki and his colleagues proposed that one of the DNA strands is synthesized in small stretches called Okazaki fragments, ranging in length from few hundred to few thousand nucleotides. This strand is termed as lagging strand. The direction of synthesis of lagging strand is opposite to the direction of fork movement. The other DNA daughter strand is synthesized continuously, called the leading strand, in which the 5′-3′ synthesis of new strand moves in the same direction as that of the replication fork. DNA Polymerase III molecule is main enzyme which synthesizes both leading strand and synthesizes lagging strand. Primer sequence from leading as well as individual Okazaki fragments is excised and replaced by DNA polymerase I. The nicks are sealed by DNA ligase. The ligation is carried out in the presence of NAD^+ in bacteria and ATP in case of viruses and eukaryotes.

Role of proteins in the elongation of DNA strands during DNA replication

Protein	Function
SSB	SSB proteins coat ssDNA and prevents its degradation by nucleases
DnaB (helicase)	Unwinding of dsDNA
DnaG (primase)	Synthesis of RNA primer
DNA polymerase III	Synthesis of DNA strands
DNA polymerase I	Filling of gaps in DNA strands; excision of RNA primers
DNA ligase	Ligation
DNA gyrase (Topoisomerase II)	Supercoiling of DNA

Termination

In this process, the *Tus* (terminus utilization substance) protein binds to the *Ter* sequence and forms a *Tus-Ter* complex. This complex halts the replication fork which ever enters first at the termination site. *Ter* sequence prevents the over replication in case when one of the replication fork gets delayed due to DNA damage or some other reason by stopping the first replication fork. At the end of theta mode of replication, it generates a topologically intertwined two circular molecules of chromosomal DNA, which are released by topoisomerase IV later on and the ligase closes them up.

Key notes

- DNA replication is a process of generating two duplicate copies of DNA from one original DNA molecule in all living organisms.
- During replication, both strands of DNA are separated from each other called denaturation.
- Each strand of the DNA molecule works as a template for the production of its complement, a procedure denoted to as semiconservative replication.
- DNA replication begins at specific locations called origins of replication (*oriC*) in the genome and directed by initiator proteins 'DnaA' in *E. coli.*
- Unwinding of DNA at the origin and synthesis of new strands results in replication forks growing bi-directionally from the origin.
- DNA polymerase III synthesizes the complementary strand by addition of nucleotides .
- DNA replication happens during the S-stage of interphase.
- DNA replication, proceeds in three enzymatically catalyzed steps: initiation, elongation and termination.
- DNA polymerase has 5'-3' polymerase activity. All DNA replication methods need a free 3' hydroxyl group before synthesis.
- DNA template is read in 3' to 5' direction whereas a new strand is synthesized in the 5' to 3' direction.
- Four different mechanisms for initiation of synthesis are acknowledged:
  1. All living organisms and many DNA viruses, phages and plasmids use a primase to synthesize a short RNA primer with a free 3' OH group which is successively elongated by a DNA polymerase.
  2. Retroviruses employ a transfer RNA that primes DNA replication.
  3. Adenoviruses and the φ29 family of bacteriophages, get the 3' OH group from side chain of an amino acid of the genome attached protein (the terminal protein).
  4. Single stranded DNA viruses, (circoviruses, geminiviruses, parvoviruses and many phages and plasmids) use the rolling circle replication (RCR) mechanism.
- The RCR endonuclease creates a nick in the genome strand (single stranded viruses) or one of the DNA strands (plasmids). The 5′ end of the nicked strand is transferred to a tyrosine residue on the nuclease and the free 3′ OH group is then used by the DNA polymerase to synthesize the new strand.
- In eukaryotes, the leading strand receives one RNA primer while the lagging strand receives several.
- The leading strand is constantly extended from the primer with high processivity, while the lagging strand is prolonged disjointedly from each primer making Okazaki fragments.
- A single nick on the leading strand and several nicks on the lagging strand can be found Ligase works to fill these nicks to complete the newly duplicated DNA molecule.

- Bacteria use a primase belonging to the DnaG protein in replication
- Multiple DNA polymerases perform diverse roles in the DNA replication process.
- In *E. coli*, DNA Pol III is the polymerase responsible for DNA replication.
- DNA Pol I is the enzyme responsible for replacing RNA primers with DNA.
- DNA Pol I have a 5' to 3' exonuclease activity in addition to its polymerase activity, and use its exonuclease activity to degrade the RNA primers.
- DNA Pol I is less processive than Pol III because its primary function in DNA replication is to create many short DNA regions rather than a few very long regions.
- In eukaryotes, the low-processive enzyme, Pol α, supports to initiate replication. The high-processive extension enzymes are Pol δ and Pol ε.
- In bacteria, which have a single origin of replication on their circular chromosome, this process creates a "theta structure" (θ).
- Eukaryotes have longer linear chromosomes and initiate replication at multiple origins within these.
- DNA is always synthesized in the 5' to 3' direction.
- Leading and lagging strand templates are oriented in opposite directions at the replication fork.
- The polymerase responsible for leading strand synthesis is DNA polymerase III (DNA Pol III) in prokaryotes.

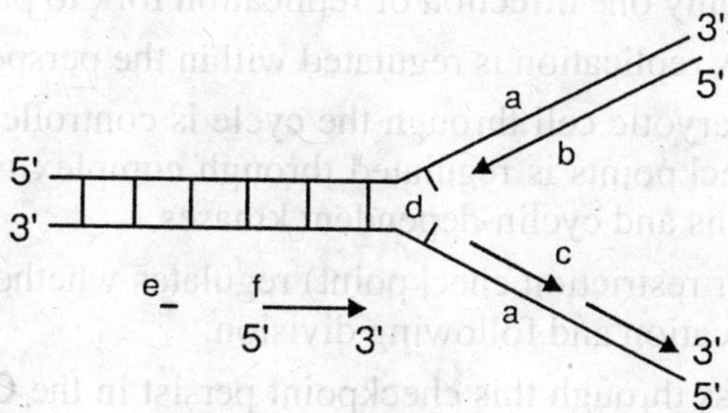

Scheme of the replication fork: a: template, b: leading strand, c: lagging strand, d: replication fork, e: primer, f: Okazaki fragments

- The lagging strand is the strand of DNA whose direction of synthesis is opposite to the direction of the growing replication fork.
- The lagging strand is synthesized in short, separated segments called Okazaki fragments.
- DNA polymerase III (in prokaryotes) and combination of FEN1 and RNaseH factors (in eukaryotes) are responsible for extension of the primers added during replication of the lagging strand.
- Primer removal is performed by DNA polymerase I (in prokaryotes) and combination of FEN1 and RnaseH factors (in eukaryotes).
- First enzyme act in DNA replication is helicase which unwinds DNA at the replication fork.
- Topoisomerases (including DNA gyrase) are enzymes that temporarily break the strands of DNA and release the tightness caused by unwinding the two strands of the DNA helix by adding negative supercoils.

- At the replication fork, many replication enzymes assemble on the DNA into a complex molecular machine called the replisome.
- DNA replication ends at the telomere region of repetitive DNA close to the ends. This shortens the telomere of the daughter DNA strand.
- Within the germ cell line, DNA permits to the next generation, telomerase extends the repetitive sequences of the telomere region to prevent degradation.
- Shortening of the telomeres is a normal process in somatic cells. As a result, cells can only divide a certain number of times before the DNA loss prevents further division, this is known as the Hayflick limit.
- Increased telomerase activity is one of the hallmarks of cancer.
- Termination requires that the progress of the DNA replication fork must stop or be blocked.
- Termination at a specific locus, involves the interaction between two components: (1) a termination site sequence in the DNA, and (2) a protein which binds to this sequence to physically stop DNA replication.
- In various bacterial species, this is named the DNA replication terminus site-binding protein, or Ter protein.
- Because bacteria have circular chromosomes, termination of replication occurs when the two replication forks meet each other on the opposite end of the parental chromosome.
- *E. coli* controls this procedure through the use of termination sequences that, when bound by the Tus protein, enable only one direction of replication fork to pass through.
- Within eukaryotes, DNA replication is regulated within the perspective of the cell cycle.
- The progress of the eukaryotic cell through the cycle is controlled by cell cycle checkpoints. Progression through checkpoints is regulated through complex interactions between various proteins, including cyclins and cyclin-dependent kinases.
- The G1/S checkpoint (or restriction checkpoint) regulates whether eukaryotic cells arrive the procedure of DNA replication and following division.
- Cells that do not progress through this checkpoint persist in the G_0 stage and do not replicate their DNA.
- Replication of chloroplast and mitochondrial genomes happens individually of the cell cycle, through the process of D-loop replication.
- ATP builds up when the cell is in a rich medium, triggering DNA replication once the cell has reached a specific size. ATP competes with ADP to bind to DnaA, and the DnaA-ATP complex is able to initiate replication.
- *In vitro*, DNA commonly replicates using the polymerase chain reaction (PCR) discovered by Karry mullis (1983).
- In PCR, a thermo stable *Taq* DNA polymerase uses to replicate the targeted DNA region through multiple cycles amplification. At the start of each cycle, the mixture of template and primers is heated, denatured the newly synthesized molecule and template.
- In PCR the number of copies of the target region doubles each round, increasing exponentially.
- DNA replication can also be done *in vitro* (artificially, outside a cell) through polymerase chain reaction (PCR), a common laboratory technique using *Taq* polymerase.

Multiple Choice Questions

1. Which of the following gene is responsible for *resistance* to the damaging effects of *UV* radiation present in DNA molecule of *E. coli*?

(A) dna (B) uvr
(C) *rec* (D) *ter*

2. Which of the following gene is responsible for resistance to the DNA adenine methylation and origin of replication present in DNA molecule of *E. coli*?

(A) *dam* and *ter* (B) *dam* and *rpo*
(C) *rec* and *lig* (D) dam and ori

3. Which of the following gene and its relative protein are responsible for DNA repair present in DNA molecule of *E. coli*?

(A) *dnaQ* and *DnaQ*
(B) *uvrB* and *UvrB*
(C) *dnaN* and *DnaN*
(D) *DnaQ* and *dnaQ*

4. What is called the strand of duplex DNA that replicated to produce new DNA strand during replication?

(A) Codon (B) Template
(C) Sense strand (D) Filament

5. What is the function of template strand of DNA during replication?

(A) Provide nucleotide sequence for new strand
(B) Require for binding of DNA polymerase
(C) Formation of Okajaki fragments
(D) Help in Unwinding ds DNA

6. The sequence of newly synthesized DNA strand during replication is……….

(A) Complementary to template strand
(B) Complementary to sense strand
(C) Identical to Template
(D) Complementary to mRNA strand

7. A structure that would allow molecules to be lined up in a specific order and joined, to create a macromolecule with a unique sequence and function.

(A) Codon (B) Sense strand
(C) Template (D) Filament

8. Which strand of DNA is act as template during DNA replication?

(A) Only one strand
(B) 3'→5' DNA strand
(C) 5'→3' DNA strand
(D) Both b and c

9. Which method of DNA replication is done in an organism?

(A) Conservative replication
(B) Semi-conservative replication
(C) Dispersive replication
(D) Discontinuous replication

10. In which method of DNA replication, each DNA strand synthesis a new strand producing two new DNA molecules with one new strand and one old strand?

(A) Semi-conservative replication
(B) Conservative replication
(C) Dispersive replication
(D) Discontinuous replication

11. In which process each DNA strand serves as a template for the synthesis of a new strand producing two new DNA molecules?

(A) Transcription (B) Translation
(C) Transformation (D) Replication

12. Which method of replication is proposed by Watson and Crick?

(A) Discontinuous replication
(B) Conservative replication
(C) Dispersive replication
(D) Semi-conservative replication

13. Who proposed the hypothesis of semi-conservative replication of DNA after describing the structure of DNA in 1957?

(A) Meselson and Stahl
(B) Watson and Crick
(C) Hershey and Chase
(D) Chargaff and Franklin

14. Which type of histone has minimum molecular weight (11,236)?

(A) H3 and (B) H2A and H2B
(C) H4 (D) H1

15. Which method of replication was proved by Meselson and Stahl in 1957?

(A) Discontinuous replication
(B) Conservative replication

(C) Dispersive replication
(D) Semi-conservative replication

16. Which organism was used for proving the hypothesis of semi-conservative replication of DNA by Matthew Meselson and Franklin Stahl in 1957?
(A) *Streptococcus pneumoniae*
(B) *Tobacco mosaic virus*
(C) *Bacteriophase* T2
(D) *Escherichia coli*

17. Which of the following was used as nitrogen source in the medium for proving semi-conservative mode of replication of DNA by Meselson and Stahl in 1957?
(A) $(NH_4)_2SO_4$ (B) $NaNO_3$
(C) NH_4Cl (D) $(NH_4)_2CO_3$

18. Which types of nitrogen isotopes were used for labeling of DNA in the experiment of semi-conservative mode of replication of DNA by Meselson and Stahl in 1957?
(A) Heavy isotope (^{14}N) and Light isotope (^{15}N)
(B) Heavy isotope (^{17}N) and Light isotope (^{15}N)
(C) Heavy isotope (^{15}N) and Light isotope (^{14}N)
(D) Heavy isotope (^{16}N) and Light isotope (^{15}N)

19. Why nitrogen isotopes were used for labeling of DNA in the experiment of semi-conservative mode of replication of DNA by Meselson and Stahl in 1957?
(A) Because DNA does not contain nitrogen
(B) Because DNA contain nitrogen
(C) Because DNA obtain easily nitrogen from medium
(D) Because DNA has heavy weight with ^{14}N isotopes than ^{15}N

20. In which medium, Meselson and Stahl grew *E. coli* cells for many generations in their experiment of semi-conservative mode of replication of DNA instead of the normal more abundant light isotope, ^{14}N of NH_4Cl medium?
(A) H_3PO_4 medium contained heavy isotope of phosphorus (^{15}P)
(B) $CaCl_2$ medium contained heavy isotope of calcium (^{20}Ca)
(C) CCl_4 medium contained heavy isotope of crbon (^{12}C)
(D) NH_4Cl medium contained heavy isotope of nitrogen (^{15}N)

21. How much density found in the DNA isolated from growing cells in medium containing ^{15}N than that of normal ^{14}N DNA in Meselson and Stahl experiment.
(A) Similar density both ^{14}N DNA and ^{15}N DNA
(B) About 1% greater
(C) About 10% greater
(D) About 1% less

22. Which technique was used for the separation of a mixture of heavy ^{15}N DNA and light ^{14}N DNA in Meselson and Stahl?
(A) Cesium chloride density gradient centrifugation
(B) Autoradiography
(C) DNA fingerprinting
(D) Paper chromatography

23. How many bands found when the *E. coli* cells grown in the ^{15}N medium then transferred to a fresh medium containing only the ^{14}N isotope and where just doubled cell population obtained in the first generation in the CsCl gradient at a position that indicate the double helical DNA molecules of the daughter cells?
(A) Double bands
(B) A Single band
(C) A light and a heavy band
(D) Smear type band

24. Which types of double helical DNA found when the *E. coli* cells grown in the ^{15}N medium then transferred to a fresh medium containing only the ^{14}N isotope and where just doubled cell population obtained in the first generation in the CsCl gradient at a position that indicate the double helical DNA molecules of the daughter cells?
(A) Containing Heavy isotope (^{15}N)
(B) Containing Light isotope (^{14}N)
(C) Containing both Heavy isotope (^{15}N) and Light isotope (^{14}N)
(D) Obtaining both types of ^{14}N dsDNA and ^{15}N dsDNA

25. Which types of double helical DNA found when the *E. coli* cells grown in the ^{15}N medium then transferred to a fresh medium containing only the ^{14}N isotope and where just doubled cell population obtained in the first generation in the CsCl gradient at a position that indicate the double helical DNA molecules of the daughter cells?

(A) Containing one new ^{14}N strand and one parent ^{15}N strand
(B) Containing both parent ^{15}N strand
(C) Containing both new ^{14}N strand
(D) Containing the segments of new ^{14}N strand and parent ^{15}N strand in daughter strands

26. In which mode of replication, daughter DNA has both strands identical to parental DNA?

(A) Discontinuous replication
(B) Conservative replication
(C) Dispersive replication
(D) Semi-conservative replication

27. In which mode of replication, daughter DNA has the segments of both parental DNA and newly synthesized DNA?

(A) Discontinuous replication
(B) Conservative replication
(C) Dispersive replication
(D) Semi-conservative replication

28. What is the ratio of ^{15}N and ^{14}N DNA strands in daughter DNA (hybrid dsDNA) in first generation after growing the *E. coli* cells in the ^{15}N medium then transferred to a fresh medium containing only the ^{14}N isotope?

(A) 3 : 2 (B) 1 : 4
(C) 1 : 1 (D) 1 : 3

29. What percent of the types of ^{15}N DNA and ^{14}N DNA in daughter DNAs during first generation after growing the *E. coli* cells in the ^{15}N medium then transferred to a fresh medium containing only the ^{14}N isotope?

(A) ^{15}N DNA = 50 and ^{14}N DNA = 50
(B) ^{15}N-^{14}N DNA = 100
(C) ^{14}N DNA = 100
(D) ^{15}N DNA = 100

30. Which types of daughter DNA were found in second generation after growing the *E. coli* cells in the ^{15}N medium then continue two time transferred to a fresh medium containing only the ^{14}N isotope?

(A) Only ^{14}N DNA
(B) Only ^{15}N-^{14}N DNA
(C) ^{15}N-^{14}N DNA and ^{14}N DNA
(D) Only ^{15}N DNA

31. What is the ratio of daughter ^{15}N- and ^{14}N-DNA were found in second generation after growing the *E. coli* cells in the ^{15}N medium then continue two times transferred to a fresh medium containing only the ^{14}N isotope?

(A) ^{15}N DNA = 50 and ^{14}N DNA = 50
(B) ^{15}N-^{14}N DNA = 100
(C) ^{15}N-^{14}N DNA = 50 and ^{14}N DNA = 50
(D) ^{15}N-^{14}N DNA = 50 and ^{15}N DNA = 50

32. How many bands found after centrifugation in the CsCl gradient of daughter DNAs of second generation which obtained through growing the *E. coli* cells in the ^{15}N medium then transferred to a fresh medium containing only the ^{14}N isotope?

(A) Double bands
(B) A Single band
(C) A light and a heavy band
(D) Smear type band

33. How many bands found after centrifugation in the CsCl gradient of daughter DNAs of second generation which obtained through growing the *E. coli* cells in the ^{15}N medium then transferred to a fresh medium containing only the ^{14}N isotope, if this process is considered as conservative mode of replication ?

(A) Double bands
(B) A Single band
(C) A light and a heavy band
(D) Smear type band

34. How many bands found when the *E. coli* cells grown in the ^{15}N medium then transferred to a fresh medium containing only the ^{14}N isotope and where just doubled cell population obtained in the first generation in the CsCl gradient at a position that indicate the double helical DNA molecules of the daughter cells, if this process is considered as conservative mode of replication?

(A) Double bands (B) A Single band
(C) A light and a heavy band
(D) Smear type band

35. In the following figure is shown bands of ^{15}N DNA, ^{14}N DNA and ^{15}N-^{14}N DNA after a experiment in which the *E. coli* cells grown in the ^{15}N medium then transferred to a fresh medium containing only the ^{14}N isotope and where just doubled cell population obtained in the first generation in the CsCl gradient at a position that indicate the double helical DNA molecules of the daughter cells.

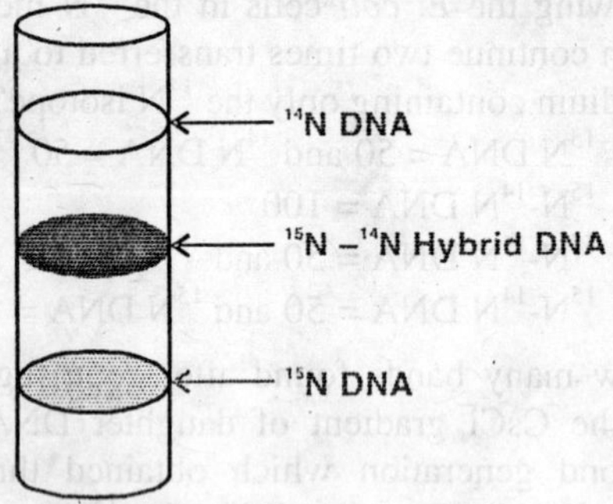

Which of the following figure shows if replication is considered as conservative mode?

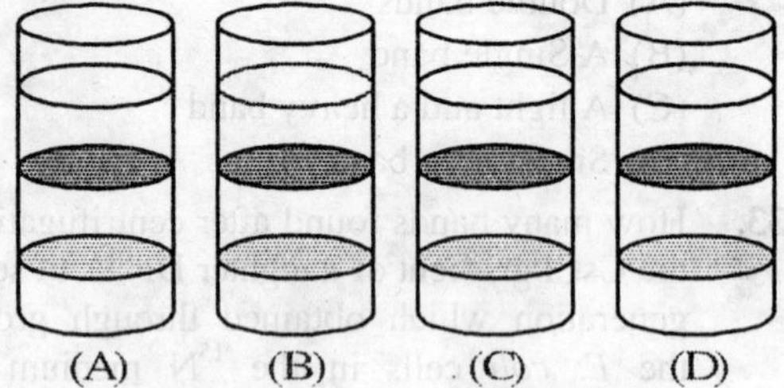

36. DNA replication of in an organism is .... process.
(A) Unidirectional (B) Bidirectional
(C) Random (D) Multidirectional

37. Who suggested that replication is a highly coordinated process in which the parent strands are simultaneously unwound and replicated was provided using autoradiography?
(A) John Cairns
(B) F. Sanger
(C) Meselson and Stahl
(D) Watson and Crick

38. Which technique is used for suggesting that replication is a highly coordinated process in which the parent strands are simultaneously unwound and replicated was provided by John Cairns?
(A) Cesium chloride density gradient centrifugation
(B) Autoradiography
(C) X-ray crystallography
(D) Paper chromatography

39. Which isotopes was used for labeling to *E. coli* DNA by growing cells in a medium containing thymidine to suggest that replication is a highly coordinated process in which the parent strands are simultaneously unwound and replicated was provided by John Cairns?
(A) Deuterium (^2H) (B) Proteium (^1H)
(C) Tritium (^3H) (D) Phosphorus (^{32}P)

40. What was revealed by the tracks of silver grains generated by the radioactive thymidine residues after isolation, spreading and overlaying the DNA with a photographic emulsion for several weeks when labeled DNA of *E. coli* cells grown in a medium containing thymidine with tritium (^3H) I John Cairns experiment?
(A) Segmented chromosome of *E. coli*
(B) Intact chromosome of *E. coli* is a single huge circle with 1.7 mm long
(C) Intact chromosome of *E. coli* is a linear with 1.7 mm long
(D) Intact chromosome of *E. coli* without tritium (^3H)

41. What is called the dynamic points where parent DNA is unwounded and separated strands quickly replicated?
(A) Replication forks
(B) Replication termination points
(C) Replication bubble
(D) Replication fizz

42. Who proposed that replication of bacterial chromosomes is bidirectional?
(A) Ross Inman (B) F. Sanger
(C) Meselson and Stahl
(D) John Cairns

43. Who gave the concept of replication fork from where parent DNA is unwounded and separated strands quickly replicated?
(A) Ross Inman (B) John Cairns
(C) Meselson and Stahl
(D) F. Sanger

44. Who developed a denaturation mapping technique by which determined that the replication loops originate at a unique point in the DNA?
(A) Ross Inman
(B) John Cairns
(C) Meselson and Stahl
(D) F. Sanger

45. Which technique was developed by Ross Inman for determination the replication loops that originate at a unique point in the DNA?
(A) Genetic mapping
(B) Chromosomal mapping
(C) Denaturation mapping
(D) Linkage mapping

46. Which organism was used by Ross Inman for denaturation mapping and showed that DNA selectively denatured at sequences unusually rich in AT base pairs generating a reproducible pattern of single-strand bubbles?
(A) 48,502 bp chromosome of bacteriophage λ
(B) Intact chromosome of *E. coli* with 1.7 mm long
(C) 40,000 bp DNA molecule of *E. coli*
(D) 300 bp DNA chromosomal segment of *Bacillus Subtilis*

47. Which information is not provided through denaturation mapping technique developed by Ross Inman?
(A) DNA selectively denatured at sequences unusually rich in AT base pairs
(B) Replication loops always initiate at a unique point which termed an origin
(C) Two replication forks form at a point in opposite to the origin in circular DNA
(D) Parental DNA strands are simultaneously unwound and replicated

48. In which direction a new strand of DNA is always synthesized with the free 3' OH
(A) 5' to 3' direction
(B) 3' to 5' direction
(C) In random direction
(D) 5' to 2' direction

49. In which direction template DNA strand read because the two DNA strands are antiparallel?
(A) 5' to 3' direction (B) 3' to 5' direction
(C) In random direction
(D) 5' to 2' direction

50. Who solved the problem that if both strands of DNA synthesized continuously while the replication fork moved and one strand would have to undergo 3' to 5' synthesis?
(A) Ross Inman (B) John Cairns
(C) Reiji Okazaki (D) F. Sanger

51. What solution was given by Okazaki for solving the problem that if both strands of DNA synthesized continuously while the replication fork moved and one strand would have to undergo 3' to 5' synthesis?
(A) Both strands of DNA synthesized continuously
(B) One of the new DNA strands is synthesized in short pieces
(C) Both strands of DNA synthesized discontinuously
(D) One strand is synthesized in 5' to 3' and the other in 3' to 5'

52. Which of the following conclusion was not given by Okazaki from their experiment?
(A) Both strands of DNA always synthesized in 5' to 3' direction
(B) One of the new DNA strands is synthesized in short pieces
(C) One strand is synthesized continuously and the other discontinuously
(D) One strand is synthesized in 5' to 3' and the other in 3' to 5'

53. What are called the one of the DNA strands synthesized continuously and the other discontinuously, respectively?
(A) Continuously = Lagging strand and the discontinuously = Leading strand
(B) Continuously = Leading strand and the discontinuously = Lagging strand
(C) Continuously = Lagging strand and the discontinuously = Logging strand
(D) Both strands were not defined

54. What is called the DNA strand that is synthesized in short pieces during replication?
(A) Lagging strand (B) Leading strand
(C) Logging strand
(D) Strand cannot be defined

55. What is called the short pieces of DNA that synthesized during replication?
(A) Inman fragments
(B) Cairns fragments
(C) Okazaki fragments
(D) Leading fragments

56. Which strands is synthesized in 5' to 3' direction opposite to the direction of replication fork movement?
(A) Lagging strand
(B) Leading strand
(C) Logging strand
(D) Strand cannot be defined

57. Which enzyme is responsible to degrade DNA?
(A) DNase (B) RNase
(C) DNA polymerase (D) DNA ligase

58. DNase and RNase enzyme have
(A) Polymerase activity
(B) Nuclease activity
(C) Ligation activity
(D) Isomerase activity

59. Which of the following two broad classes of nuclease enzymes are contained in every cell?
(A) Exonucleases and endonucleases
(B) Isomerases and Polymerases
(C) Topoisomerases and ligases
(D) Kinases and Catalses

60. Which of the following enzyme degrade nucleic acids from one end of the molecule at only the 5' to 3' or the 3' to 5' direction with removing nucleotides only from the 5' or the 3' end, respectively, of one strand of a double stranded nucleic acid or of a single-stranded DNA?
(A) Endonucleases (B) Helicase
(C) Topoisomerases (D) Exonucleases

61. Which of the following enzyme can degrade at specific internal sites in a nucleic acid strand or molecule with reducing it to smaller and smaller fragments?
(A) Endonucleases
(B) Helicase
(C) Topoisomerases
(D) Exonucleases

62. Exonucleases and endonucleases degrade ....
(A) Only double-stranded DNA
(B) Only single-stranded DNA
(C) Only circular-stranded DNA
(D) All of the above

63. What are called the enzymes that cleave DNA only at specific nucleotide sequences?
(A) Endonucleases
(B) Helicases
(C) Restriction endonucleases
(D) Exonucleases

64. Who discovered DNA polymerase I enzyme which responsible for the synthesis of DNA?
(A) Ross Inman
(B) John Cairns
(C) Reiji Okazaki
(D) Arthur Kornberg

65. Which of the following enzyme was purified and characterized from *E. coli* by Arthur Kornberg in 1955 for the synthesis of DNA?
(A) DNA polymerase II
(B) RNA polymerase II
(C) DNA polymerase I
(D) DNA polymerase III

66. Which of the following statement is not related to DNA polymerase I which
(A) Purified and characterized from *E. coli* by Arthur Kornberg in 1955
(B) Help in the synthesis of DNA
(C) A double-polypeptide enzyme
(D) It has *Mr* 103,000 and encoded by the *polA* gene

67. Which group is transferred by DNA polymerase I for the synthesis of DNA?
(A) A phosphoryl group
(B) A amino acid group
(C) A phosphoenol group
(D) A hydroxyl group

68. At which position of the nucleotide new coming nucleotides are added to growing DNA strand?
(A) 5'-hydroxyl group of sugar
(B) 3'-hydroxyl group of sugar
(C) 5'-phosphate group of sugar
(D) 2'-hydroxyl group of sugar

69. At which position of the nucleotide new coming nucleotides are added to growing DNA strand?
(A) 5'-hydroxyl group of sugar
(B) 3'-hydroxyl group of sugar
(C) 5'-phosphate group of sugar
(D) 2'-hydroxyl group of sugar

70. Which of the following group of the nucleotide act as nucleophile for new coming nucleotides that added to growing DNA strand during DNA replication?
(A) 3'-hydroxyl group of sugar
(B) 5'-hydroxyl group of sugar
(C) 5'-phosphate group of sugar
(D) 2'-hydroxyl group of sugar

71. Which of the following group of the nucleotide act as nucleophile for new coming nucleotides that added to growing DNA strand during RNA transcription?
(A) 2'-hydroxyl group of sugar
(B) 5'-hydroxyl group of sugar
(C) 5'-phosphate group of sugar
(D) 3'-hydroxyl group of sugar

72. Which of the following molecule is release from binding new nucleotide to 3'-hydroxyl group of sugar of growing DNA strand during DNA replication?
(A) A phosphoryl group
(B) A Inorganic pyrophosphate
(C) A phosphoenol group
(D) A base

73. Which reaction shows phosphodiester bond formation during DNA replication? Where dNMP and dNTP are deoxynucleoside 5'-monophosphate and 5'-triphosphate, respectively.

(A) $(NMP)n + NTP \rightarrow (NMP)_{n+1} + PPi$
DNA — Lengthened DNA

(B) $(dNMP)n + dNTP \rightarrow (dNMP)_{n+1} + PPi$
DNA — Lengthened DNA

(C) $(NMP)n + NTP \rightarrow (dNMP)_{n+1} + PPi$
DNA — Lengthened DNA

(D) $(dNMP)n + dNTP \rightarrow (NMP)_{n+1} + PPi$
DNA — Lengthened DNA

74. Which types of interaction provide stabilization to the lengthened DNA product relative to the free nucleotide during replication?
(A) Noncovalent base-stacking and base-pairing interactions
(B) Hydrogen bonds and glycosidic bonds
(C) Only noncovalent base-stacking
(D) Only base-pairing interactions

75. Which enzyme is responsible for hydrolysis of the pyrophosphate product during DNA replication?
(A) Topoisomerase (B) Helicases
(C) Pyrophosphatase (D) DNA polymerase

76. How much energy generated in the subsequent hydrolysis of the pyrophosphate product by the enzyme pyrophosphatase?
(A) 50 kJ/mol (B) 19 kJ/mol
(C) 100 kJ/mol (D) 39 kJ/mol

77. Which of the following require the polymerization reaction in DNA replication according to the base-pairing rules predicted by Watson and Crick?
(A) Codon (B) Primer
(C) Template DNA (D) Both b and c

78. Which of the following require to guide for the polymerization reaction in DNA replication according to the base-pairing rules predicted by Watson and Crick?
(A) Codon
(B) Primer
(C) Template DNA
(D) Both b and c

79. Which of the following statement is not correct about primer?
(A) It is complementary to the template
(B) It is a single stranded oligonucleotide
(C) It has a free 3'-hydroxyl group
(D) It has also 5'-hydroxyl group

80. What is called the free 3' end of the primer where new coming nucleotide is added that complementary to growing DNA strand?
(A) Primer shuffling origin
(B) Primer origin
(C) Primer terminus
(D) Primer slippage origin

81. What is the frequency of a mistake for adding of nucleotides during DNA replication in *E. coli*?
(A) Only once for every 10^9 to 10^{10} nucleotides
(B) Only once for every 1000 to 100000 nucleotides
(C) Only once for every 10 to 100 nucleotides
(D) Only once for every 10 nucleotides

82. An error occurs only once per how many replications for the *E. coli* chromosome of ~4.6 X 10^6 bp?
(A) 1 to10 replications
(B) 1,000 to 10,000 replications
(C) 10^9 to 10^{10} replications
(D) Only 1 replication

83. DNA polymerases insert one incorrect nucleotide for .... correct nucleotides during DNA replication in *E. Coli*?
(A) Only 1 (B) Every 1,000 to 10,000
(C) Every 10^9 to 10^{10} (D) Every 10^4 to 10^5

84. Why occur the mistakes of adding incorrect nucleotides to growing to DNA strand during replication?
(A) Due to an unusual tautomeric form
(B) Due to lose the activity of DNA polymerase I
(C) Due to form incorrect primer
(D) Due to form replication fork at unusual point

85. How reduced the error rate of adding incorrect nucleotide during replication?
(A) 5' to 3' exonuclease activity of DNA polymerase I
(B) 3' to 5' exonuclease activity of DNA polymerase II
(C) 3' to 5' exonuclease activity of DNA polymerase I
(D) 3' to 5' exonuclease activity of DNA polymerase III

86. Which of the following activity permits to remove a newly added nucleotide replace of incorrect nucleotide in growing DNA strand and is highly specific for mismatched base pairs?
(A) 3' to 5' exonuclease activity of DNA polymerase I
(B) 3' to 5' exonuclease activity of DNA polymerase II
(C) 5' to 3' exonuclease activity of DNA polymerase I
(D) 3' to 5' exonuclease activity of DNA polymerase III

87. What will happen, if the DNA polymerase I add the wrong nucleotide to growing DNA strand during replication?
(A) Translocation of enzyme where the next nucleotide addition is inhibited
(B) Lose of 3' to 5' activity of polymerase enzyme
(C) Lose of 5' to 3' activity of polymerase enzyme
(D) Degrade the polymerase enzyme

88. What is called the 3' to 5' exonuclease activity of DNA polymerase I to removes the mis-paired nucleotide and the polymerase begins again replication through adding correct nucleotide to growing DNA strand.
(A) Nick translation (B) Processivity
(C) Polymerization (D) Proofreading

89. Why proofreading is not simply the reverse of the DNA polymerization reaction?
(A) Because pyrophosphate is not involved
(B) Because DNA polymerase enzyme lost its 3' to 5' activity
(C) Because DNA polymerase enzyme degraded
(D) Because DNA polymerase enzyme lost its 5' to 3' activity

90. How many folds proofreading improve the inherent accuracy of the polymerization reaction?
(A) 10^9- to 10^{10}-folds
(B) 1000- to 10000-folds
(C) 10^2- to 10^3-folds
(D) Only one folds

91. Which of the following statement is not true about the polymerizing and proofreading activities of DNA during replication?
(A) Both have separate active sites within same polypeptide in monomeric DNA polymerase I

(B) Proofreading improve inherent accuracy of the polymerization reaction 10^2- to 10^3-folds
(C) Pyrophosphate involve in proofreading but not in DNA polymerization
(D) 3' to 5' activity is proofreading and 5' to 3' activity is polymerization

92. When base selection and proofreading are combined, DNA polymerase leaves behind one net error for .... bases added during replication in *E. coli*?
(A) Every 10^6 to 10^8
(B) Every 1000 to 10000
(C) Every 10^2- to 10^3-folds
(D) Every one

93. How many DNA Polymerases mostly found in *E. coli*?
(A) At Least Five DNA Polymerases
(B) Only three DNA Polymerases
(C) Only one DNA Polymerases
(D) Various types of DNA Polymerases during each replication

94. What is the rate of addition of nucleotides of DNA polymerase I to growing DNA strand during replication in *E. coli*?
(A) Only one nucleotides/min
(B) 10^6 to 10^8 nucleotides/min
(B) 1000 to 10000 nucleotides/min
(D) 600 nucleotide/min

95. Which of the following evidence of DNA polymerase I does not prove that it is not suited for replication of the large *E. coli* chromosome?
(A) The rate of polymerization (600 nucleotides/min) is too slow in the bacterial cell
(B) DNA polymerase I has a relatively low processivity
(C) DNA polymerase I clearly act alone
(D) DNA polymerase I produce an inactive enzyme according to J. Cairns

96. Who isolated a bacterial strain with altered gene for DNA polymerase I in 1956 that produced an inactive enzyme?
(A) Ross Inman (B) John Cairns
(C) Reiji Okazaki (D) Arthur Kornberg

97. In which organism DNA polymerase II and DNA polymerase III were discovered in the early 1970s?
(A) Yeast (B) Arabidopsis
(C) Bacterophage T_2 (D) *E. coli*

98. In which mechanism DNA polymerase II is involved?
(A) DNA polymerization
(B) DNA repair
(C) Formation of Okazaki fragments
(D) DNA Methylation

99. *E. Coli* has at least five DNA Polymerases but which of the following is not found in *E. coli*?
(A) DNA Polymerase I and II
(B) DNA Polymerase V
(C) DNA Polymerase III and IV
(D) DNA Polymerase α and β

100. Which of the following enzymes of *Escherichia coli* are involved in DNA repair mechanism?
(A) DNA Polymerase I and II
(B) DNA Polymerase V
(C) DNA Polymerase III and IV
(D) All of the above

101. Which is the principal replication enzyme in *E coli*?
(A) DNA Polymerase I and II
(B) DNA Polymerase IV and V
(C) DNA Polymerase III
(D) All of the above

102. How separated a structural domain of DNA polymerase I that has 5' to 3' exonuclease activity?
(A) Mild DNase treatment
(B) Mild protease treatment
(C) By inactivation of 5' to 3' exonuclease activity through mutation
(D) Mild RNase treatment

103. What is called the fragment that remains after separation of a structural domain of DNA polymerase I that has 5' to 3' exonuclease activity by mild protease treatment?
(A) Okazaki fragment
(B) Small fragment
(C) Restriction fragment
(D) Klenow fragment

104. What is molecular weight (*Mr*) of Klenow fragment that remains after separation of a structural domain of DNA polymerase I that has 5' to 3' exonuclease activity by mild protease treatment?
(A) 68,000 (B) 1,00,000
(C) 90,000 (D) 1,000

105. What are the forms of type II topoisomerase are found in eukaryotic cells?
(A) Topoisomerases I and III
(B) Topoisomerases I and IV
(C) Topoisomerases I and II
(D) Topoisomerases IIα and IIβ

106. Which of the following activities retain in Klenow fragment that obtained after separation of a structural domain of DNA polymerase I that has 5' to 3' exonuclease activity by mild protease treatment?
(A) Polymerization activities
(B) Proofreading activities
(C) Both a and b
(D) Any one activity as required condition during replication

107. What is called the process in which the 5' to 3' exonuclease activity of intact DNA polymerase I can be replaced a segment of DNA (or RNA) with pairing to the template strand?
(A) DNA repair
(B) Polymerization
(C) Nick translation
(D) Proofreading

108. Which of the following enzyme has a 5' to 3' exonuclease activity in *E. Coli*?
(A) DNA Polymerase II and III
(B) DNA Polymerase IV and V
(C) DNA Polymerase I
(D) All of the above

109. What is defined by the average number of nucleotides added to growing DNA strand during replication before a polymerase dissociates from template
(A) Processivity of primer
(B) Processivity of SSBP
(C) Processivity of DNA polymerase
(D) Processivity of topoisomerase

110. How many subunits contained in DNA polymerase III that is much more complex than DNA polymerase I?
(A) Only one (B) Five
(C) Ten (D) DNA polymerase III has no subunit

111. Which of the following subunits of DNA polymerase III have polymerization and proofreading activities?
(A) Polymerization = β subunit and proofreading activities = γ subunit
(B) Polymerization = δ subunit and proofreading activities = δ' subunit
(C) Polymerization = λ subunit and proofreading activities = ψ subunit
(D) Polymerization = α subunit and proofreading activities = ε subunit

112. Which of the following subunits of DNA polymerase III form a core polymerase that can polymerize DNA but limited processivity?
(A) θ subunit associates with α and ε subunits
(B) ψ subunit associates with α and ε subunits
(C) δ subunit associates with α and ε subunits
(D) β subunit associates with α and ε subunits

113. Which of the following subunits of DNA polymerase III form a clamp-loading complex?
(A) τ_2, γ, δ, δ' and β subunits
(B) λ, θ, τ, γ, and β subunits
(C) χ_2, ω, δ, δ' and α subunits
(D) α, β, ψ, δ, and β subunits

114. Which of the following subunit help in linking between core polymerases and a clamp-loading complex?
(A) β subunit (B) τ subunit
(C) χ subunits (D) δ subunit

115. DNA polymerase III polymerizes DNA but with a much lower processivity therefore, which subunit provide necessary increase in processivity?
(A) θ subunit (B) τ subunit
(C) α subunits (D) β subunit

116. Which of the following subunit of DNA polymerase III found four in number with DNA clamp required for optimal processivity?
(A) θ subunit (B) β subunit
(C) α subunits (D) τ subunit

117. Which of the following subunit of DNA polymerase III associate in pairs to form donut-shaped structures that encircle the DNA and act like clamps?
(A) θ subunit (B) β subunit
(C) α subunits (D) τ subunit

118. Which of the following prevents the dissociation of DNA polymerase III from DNA dramatically increasing processivity to greater than 500,000?
(A) Core polymerase
(B) Clamp-loading (γ) complex
(C) β sliding clamp
(D) τ subunit

119. What is rate of polymerization DNA polymerase I and III in *E. coli*, respectively?
(A) 40 and 250 – 1000 nucleotide per second
(B) 16 – 20 and 40 nucleotide per second
(C) 250 – 1000 and 16 – 20 nucleotide per second
(D) 16 – 20 and 250 – 1000 nucleotide per second

120. Which of the following mach is not correct between polymerase enzyme and its coding gene?
(A) DNA Polymerase I = *polA*
(B) DNA Polymerase II = *polB*
(C) DNA Polymerase III = *polE*
(D) DNA Polymerase III = *polC* or *dnaE*

121. Which of the following subunit of DNA polymerase III is generated by a translational frame shifting mechanism that leads to premature translational termination?
(A) θ subunit (B) β subunit
(C) γ subunits (D) τ subunit

122. Which of the following enzyme does not participate in DNA replication?
(A) DNA polymerase
(B) DNA ligase
(C) Topoisomerase
(D) DNA methylase

123. What is called the entire complex of different enzymes and protein factors that perform a specific task in replication in *E. coli*?
(A) Replisome (B) Spliciosome
(C) Centrosome (D) Nucleosome

124. Which enzyme requires for separation of the two parent DNA strands and moves along the DNA and separates the strands using chemical energy from ATP?
(A) Topoisomerase (B) Primase
(C) Ligase (D) Helicase

125. Which enzyme requires to create topological stress in the helical DNA structure?
(A) Topoisomerase (B) Primase
(C) Ligase (D) Helicase

126. Which of the following helps in stabilizing of the separated strands during replication?
(A) Ligase
(B) Topoisomerase
(C) DNA-binding proteins
(D) Helicase

127. What is the function of single stranded binding proteins during DNA replication?
(A) To stabilize of the separated strands
(B) To create topological stress in the helical DNA structure
(C) To separate of the two parent DNA strands
(D) To synthesize short segments of RNA

128. Which of the following enzyme synthesize a short segment of RNA as a primer during replication?
(A) Ligase (B) Topoisomerase
(C) Primase (D) Helicase

129. Which of the following enzyme remove and replace from DNA during replication?
(A) DNA polymerase II
(B) DNA polymerase I
(C) Primase
(D) DNA polymerase III

130. Which of the following enzyme seal the nicks that remained in the DNA backbone through broken phosphodiester bonds after removing RNA primers during replication?
(A) DNA ligase (B) Topoisomerase
(C) Primase (D) Helicase

131. Which type of structure is produced in the replication of a circular chromosome?
(A) Greek letter epsilon (ε)
(B) Greek letter omega (ω)
(C) Greek letter theta (θ)
(D) Greek letter delta (δ)

132. A new DNA strand is always synthesized in .... direction during replication whereas the template is read in the opposite direction ....?
(A) 3' to 5' direction and 5' to 3' direction
(B) 5' to 3' direction and 3' to 5' direction
(C) 5' to 2' direction and 2' to 5' direction
(D) 2' to 5' direction and 5' to 2' direction

133. In the following figure, DNA replication process is shown. where X, Y and Z, respectively indicate ....?

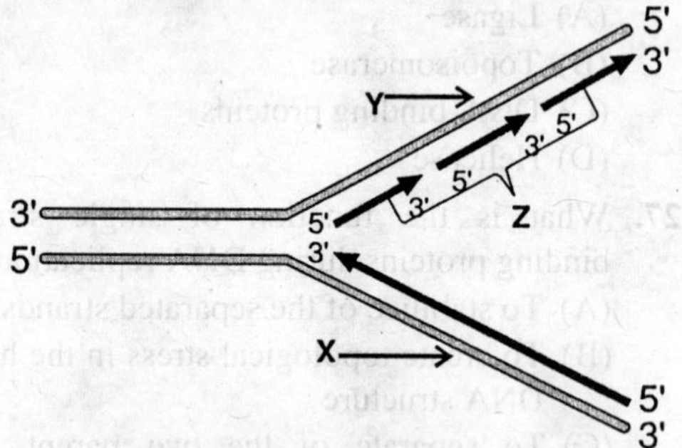

(A) Lagging strand, Leading strand and Okazaki Fragments
(B) Leading strand, Okazaki Fragments, and Lagging strand
(C) Discontinuous strand, Continuous strand and Okazaki Fragments
(D) Leading strand, Lagging strand and Okazaki Fragments

134. Which enzyme helps in joining of Okazaki fragments that are synthesized discontinuously in short pieces of the lagging strand?
(A) DNA ligase
(B) Topoisomerase
(C) Primase
(D) DNA polymerase

135. What is the size of Okazaki fragments that are synthesized discontinuously in short pieces of the lagging strand in bacterial replication?
(A) Almost 150 to 200 nucleotides long
(B) Almost 1,000 to 2,000 nucleotides long
(C) Almost 10 to 20 nucleotides long
(D) Only 100 nucleotides long

136. What is the size of Okazaki fragments that are synthesized discontinuously in short pieces of the lagging strand in eukaryotic replication?
(A) Almost 150 to 200 nucleotides long
(B) Almost 1,000 to 2,000 nucleotides long
(C) Almost 10 to 20 nucleotides long
(D) Only 100 nucleotides long

137. What require for DNA polymerase I activity for replication?
(A) Single unpaired strand and as template and a primer with 3' free hydroxyl group
(B) Double unpaired strand and as template and a primer with 3' free hydroxyl group
(C) A single paired strand and as template and a primer with 5' free hydroxyl group
(D) Double paired strand and as template and a primer with 3' free hydroxyl group

138. Which of the following ions involve in the catalytic mechanism coordinated to the phosphate groups of the incoming nucleotide triphosphate and to three Asp residues, two of which are highly conserved in all DNA polymerases?
(A) Two Mn^{2+} ions
(B) Only one Mg^{2+} ion
(C) Two Mg^{2+} ions
(D) Many Mg^{2+} ions

139. What is function of two Mg^{2+} ions during DNA replication and transcription?
(A) One Mg^{2+} ion facilitates attack of the 3'-hydroxyl group of the primer on the α phosphate of the nucleotide triphosphate and Second Mg^{2+} ion facilitates displacement of the pyrophosphate
(B) Both ions stabilize the structure of the pentacovalent transition state
(C) Both ions are also used in RNA polymerases during transcription in same fashion
(D) All of these

140. In the following structure which type of bond is shown by X?

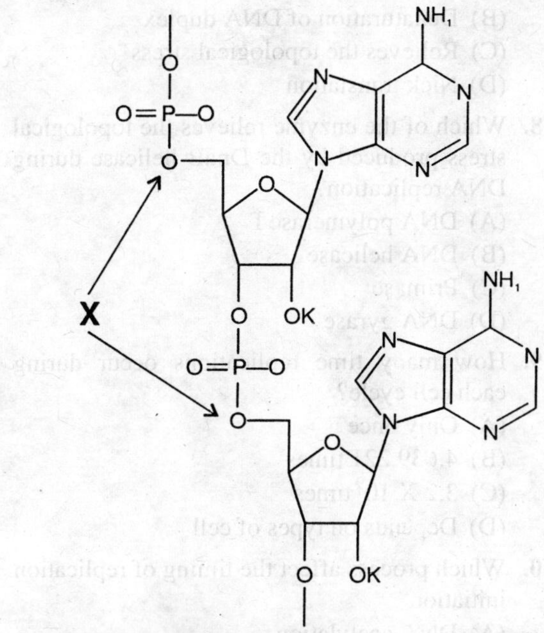

(A) Glycocidic bond
(B) Phosphodiester bond
(C) Ionic bond
(D) Hydrogen bond

141. In the following which type of bond is shown by dot lines?

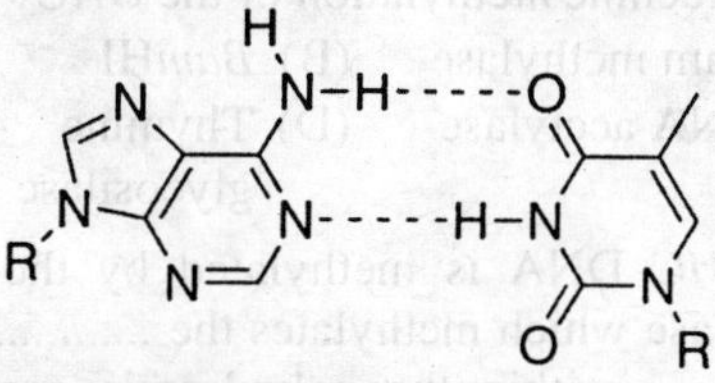

(A) Glycocidic bond
(B) Phosphodiester bond
(C) Ionic bond
(D) Hydrogen bond

142. In which process an RNA or DNA strand paired to a DNA template is simultaneously degraded by the 5' to 3' exonuclease activity of DNA polymerase I and replaced by the polymerase activity of the same enzyme?

(A) Nick translation
(B) DNA replication
(C) Mutation
(D) Translation

143. Which of the following enzyme extends the non-template DNA strand and moves the nick along the DNA that is called nick translation?

(A) DNA ligase
(B) DNA polymerase I
(C) DNA polymerase III
(D) DNA polymerase II

144. How many enzymes or proteins participate in the initiation phase of DNA replication?

(A) Twelve
(B) Five
(C) Nine
(D) Three

145. Which of the protein first binds to the four 9 bp repeats in the origin of DNA replication?

(A) DnaB molecules
(B) DnaA molecules
(C) DnaE molecules
(D) DnaC molecules

146. At which region DNA first denature after binding DnaA protein to the four 9 bp repeats in the origin of DNA replication?

(A) First the region of the four 9 bp then three 13 bp repeats
(B) The region of the four 9 bp repeats
(C) Both regions of the four 9 bp and three 13 bp repeats
(D) The region of the three 13 bp repeats

147. What is main characteristic of the region of the three 13 bp repeats in origin of DNA replication?

(A) Rich in AT pairs
(B) Rich in GC pairs
(C) Rich in both AT and GC pairs
(D) Rich in AU pairs

148. What requires for the danaturation of DNA template during *E. coli* replication?

(A) Primase enzyme
(B) DNA polymerase I, II and III
(C) ATP and the bacterial histone like protein HU
(D) Single stranded binding proteins

149. Which of the protein has main characteristic of two ring shaped hexamers act as helicases for unwinding the DNA bidirectionally and creating two potential replication forks during DNA replication?

(A) DnaB (B) DnaA
(C) DnaE (D) DnaC

150. Which of the following enzyme act as DNA gyrase during replication?
(A) DNA topoisomerase III
(B) DNA topoisomerase II
(C) DNA topoisomerase I
(D) Helicase

151. Which of the protein act as helicase that requires for separation of the two parent strands during replication.
(A) DnaC (B) DnaA
(C) DnaB (D) DnaE

152. Which of the following stages are involved in the synthesis of a DNA molecule?
(A) Initiation (B) Elongation
(C) Termination (D) All of the above

153. What is called the origin of the *E. coli* replication origin that consists of 245 bp?
(A) *oriC* (B) *oriB*
(C) *oriD* (D) *oriA*

154. What is size of the origin of the *E. coli* replication is called *oriC*?
(A) 245 bp (B) 1000 bp
(C) 9 bp (D) 13 bp

155. Which of the following bears DNA sequence elements that are highly conserved among bacterial DNA and help in replication initiation?
(A) Anticodons
(B) Promoter
(C) Replication origin (*oriC*)
(D) Replication termination points

156. Which of the following highly conserved sequences found in replication origin (*oriC*) of *E. coli*?
(A) Four repeats of a 13 bp sequence and three repeats of a 9 bp sequence
(B) Three repeats of a 13 bp sequence and four repeats of a 9 bp sequence
(C) Five repeats of a 13 bp sequence and ten repeats of a 9 bp sequence
(D) Three repeats of a 9 bp sequence and four repeats of a 13 bp sequence

157. What is the function of single strand binding proteins during replication?
(A) Stabilizing the separated strands and preventing renaturation
(B) Denaturation of DNA duplex
(C) Relieves the topological stress
(D) Nick translation

158. Which of the enzyme relieves the topological stress produced by the DnaB helicase during DNA replication?
(A) DNA polymerase I
(B) DNA helicase
(C) Primase
(D) DNA gyrase

159. How many time replications occur during each cell cycle?
(A) Only once
(B) 4,639,221 times
(C) 3.2 X 10^9 times
(D) Depends on types of cell

160. Which process affect the timing of replication initiation?
(A) DNA acetylation
(B) DNA methylation
(C) interactions with the bacterial plasma membrane
(D) Both b and c

161. Which of the following enzyme involves in DNA *a*denine *m*ethylation of the *OriC*?
(A) Dam methylase (B) *Bam*HI
(C) DNA acetylase (D) Thymine glycosilase

162. The *oriC* DNA is methylated by the Dam methylase which methylates the .... within the palindromic sequence (5') GATC.
(A) *N*6 position of cytocine
(B) *N*6 position of guanine
(C) *N*6 position of adenine
(D) *N*6 position of thymine

163. Which sequence is highly enriched in the *oriC* region of *E. coli*?
(A) GATC sequences
(B) GA sequences
(C) TC sequences
(D) TCGA sequences

164. What is the average frequency of GATC in the *OriC* of *E. coli* chromosome as a whole?
(A) 1 in 245 bp (B) 1 in 256 bp
(C) 1 in 247 bp (D) 1 in 243 bp

165. What is the meaning of the hemimethylated DNA?
(A) Parent strands have not methylated *oriC* sequences but newly synthesized strands do
(B) Both parent and newly synthesized strands have methylated *oriC* sequences
(C) Parent strands have methylated *oriC* sequences but newly synthesized strands not
(D) Both parent and newly synthesized strands have not methylated *oriC* sequences

166. Regulation of replication initiation involves the slow hydrolysis of ATP by DnaA protein which cycles the protein between active and inactive forms on a timescale of 20 to 40 minutes. What is the meaning of active and inactive forms, respectively?
(A) Active = with bound ADP and inactive = with bound ATP
(B) Active = with bound ATP and inactive = with bound ADP
(C) Both active and inactive = with bound ATP
(D) Both active and inactive = with bound ADP

167. In which pahse of DNA replication includes both leading strand synthesis, and lagging strand synthesis?
(A) Initiation (B) Elongation
(C) Termination (D) Denaturation

168. Which of the following operations involved in elongation phase of DNA replication?
(A) Leading strand synthesis
(B) Lagging strand synthesis
(C) Unwinding of DNA strands
(D) Both a and b

169. Which of the following enzyme first act in DNA replication?
(A) DNA polymerase I
(B) DNA helicase (C) Primase
(D) DNA gyrase

170. Which of the following process first is done by enzyme during DNA replication?
(A) Unwinding of DNA
(B) Binding of SSBP
(C) Polymerization
(D) Formation of replication fork

171. Which of the following protein act as enzyme primase for synthesizing a short (10 to 60 nucleotide) RNA primer at the replication origin?
(A) DnaC protein (B) DnaA protein
(C) DnaG protein (D) DnaB protein

172. What is size of RNA primer that synthesized at the replication origin by primase?
(A) Only 60 nucleotides
(B) 10 to 60 nucleotides
(C) 50 to 200 nucleotides
(D) More than 60 nucleotides

173. Which of the following enzyme add deoxyribonucleotides to DNA template for polymerization during replication?
(A) DNA polymerase III
(B) DNA helicase (C) Primase
(D) DNA gyrase

174. Which of the following proteins does not act at the *E. coli* Replication Fork?
(A) SSBP (B) Pyrophosphatase
(C) DNA helicase (D) DNA gyrase

175. How many nucleotides per seconds added to each strand (leading and lagging) during rapid DNA synthesis promoted by the replisome?
(A) ~100 nucleotides/s
(B) ~10 nucleotides/s (C) ~1 nucleotides/s
(D) ~1,000 nucleotides/s

176. Which of the following enzyme remove RNA primer after completing an Okazaki fragment in lagging strand?
(A) DNA polymerase III
(B) DNA polymerase I
(C) DNA ligase
(D) DNA polymerase II

177. Which of the following enzyme seal nick that remains after removing RNA primer by DNA polymerase I with completing an Okazaki fragment in lagging strand?
(A) DNA polymerase III
(B) DNA polymerase I
(C) DNA ligase
(D) DNA gyrase

178. How DNA ligase seal nick that remains after removing RNA primer by DNA polymerase I with completing an Okazaki fragment in lagging strand?
(A) Catalyzes the formation of a phosphodiester bond
(B) Catalyzes the formation of a glycosidic bond
(C) Catalyzes the formation of a hydrogen bond
(D) Catalyzes the formation of a ionic bond

179. DNA ligase catalyzes the formation of a phosphodiester bond between .... at the end of one DNA strand and .... at the end of another strand, respectively?
(A) 3' hydroxyl and 5' phosphate
(B) 5' phosphate and 3' hydroxyl
(C) 3' hydroxyl and 5' hydroxyl
(D) 5' phosphate and 3' phosphate

180. DNA ligase uses as energy for the formation of a phosphodiester bond between 3' hydroxyl at the end of one DNA strand and 5' phosphate at the end of another strand?
(A) GTP (B) ATP
(C) NADP (D) CTP

181. What is called the region containing multiple copies of a 20 bp sequence where the two-replication forks of the circular *E. coli* chromosome meet?
(A) Terminus region (*ter*)
(B) Replication origin (*oriC*)
(C) Promotor region (D) Regulator region

182. What is the meaning of *Tus* protein where found binding sites for the *Ter* sequences containing multiple copies of a 20 bp sequence in the circular *E. coli* chromosome meet?
(A) Terminus utilization sequence
(B) Terminus utilization solution
(C) Terminus utilization substance
(D) Terminus utilization stuff

183. What is the fuction of termination sequences (*Ter* sequences) containing multiple copies of a 20 bp sequence in the circular *E. coli* chromosome meet?
(A) As binding sites for DNA ligase
(B) As binding sites for a Tus protein
(C) As binding sites for replication origin (*oriC*)
(D) As termination sites for transcription termination

184. What is the fuction of the Tus-Ter complex in replication process of *E. coli*?
(A) Arrest a replication fork from two directions
(B) Arrest a replication fork from only one direction
(C) Continue a replication fork from only one direction
(D) Continue a replication fork from two directions

185. How many Tus-Ter complexes function per replication cycle for arresting a replication fork from only one direction?
(A) Two
(B) According to DNA length
(C) Only one
(D) Generally ten

186. Which of the following step is completed in DNA replication?
(A) Ligation of Okazaki fragments
(B) Removing RNA primer from lagging strand
(C) Replication of final few hundred base pairs of DNA between Tus-Ter complexes
(D) Removing all enzymes from DNA duplex

187. What is called two topologically interlinked circular chromosomes after replication of few hundred base pairs of DNA between Tus-Ter large protein complexes?
(A) Catenanes (B) Carotines
(C) Primosomes (D) Relisomes

188. Which of the following enzyme involved in the separation of the two topologically interlinked DNA circles (catenanes) in *E. coli* after replication of few hundred base pairs of DNA between Tus-Ter large protein complexes
(A) Topoisomerase I
(B) Topoisomerase III
(C) Topoisomerase II
(D) Topoisomerase IV

189. In the following figure of oriC, what is shown by 'X'?

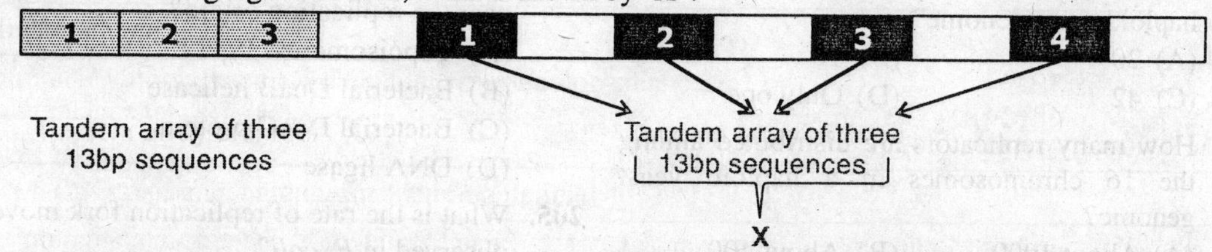

(A) Binding sites for DnaC protein
(B) Binding sites for DnaA protein
(C) Binding sites for DnaB protein
(D) Binding sites for DNA gyrase

190. In the following figure, which enzymes are involved in removing RNA primer (X) and joining the fragments of DNA (Y), respectively?

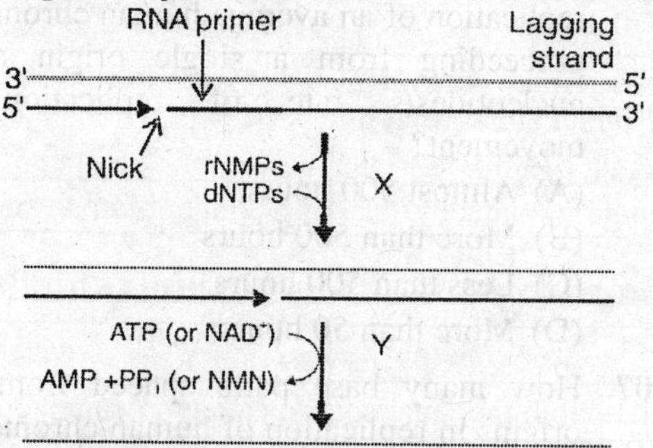

(A) DNA ligase and Topoisomerase I
(B) DNA polymerase I and DNA ligase
(C) DNA polymerase III and DNA ligase
(D) DNA polymerase III and DNA helicase

191. Where is situated replication factory for the replication of a circular and highly organized bacterial chromosome?

(A) Mitochondria
(B) Nucleoid
(C) Inner membrane of plasmalemma
(D) Outer membrane of plasmalemma

192. A circular and highly organized bacterial chromosome has two replisomes that are joined together to one point on the ....

(A) Bacterial inner membrane
(B) Bacterial outer membrane
(C) Bacterial cell wall
(D) Nucleoid

193. In which direction replication of circular bacterial chromosome is initiated?

(A) Unidirection
(B) Bidirection
(C) Only in 3' to 5' direction
(D) Random direction

194. Which of the following statement is not true about the replication of circular bacterial chromosome?

(A) It is performed in bidirectional at the origin
(B) Two replisomes involve in this process
(C) Two replisomes travel away from each other along the DNA in opposite direction
(D) Replisomes are linked together to one point on the bacterial inner membrane

195. Which of the following statement is not true about the replication of circular bacterial chromosome?

(A) Bacterial inner membrane is called replication factory
(B) Each newly synthesized replication origins is partitioned into one half of the cell
(C) Bacterial homologs of the SMC proteins and topoisomerases do not maintain the newly replicated chromosomes
(D) After termination of replication, cell divides

196. What are replicators?

(A) Autonomously replicating sequences
(B) Origin recognition complex
(C) Clamp loading complex
(D) Core polymerases

197. What is the length of Yeast replicators that contain several essential conserved sequences?

(A) About 10 bp
(B) About 400 bp
(C) About 1000 bp
(D) About 150 bp

198. How many chromosomes contained in a haploid yeast genome?
(A) 20 (B) 16
(C) 42 (D) Only one

199. How many replicators are distributed among the 16 chromosomes in a haploid yeast genome?
(A) About 1000 (B) About 400
(C) About 10 (D) About 150

200. What is called a multisubunit protein that requires for the initiation of replication in all eukaryotes and binds to several sequences within the replicator?
(A) ARS
(B) Topoisomerases
(C) Origin *recognition* complex
(D) Clamp-loading complex

201. Which of the following proteins bind to *Origin Recognition Complex* (*ORC*) and mediate the loading of a heterohexamer of *mini*chromosome *maintenance* proteins during DNA replication and involved in control of the eukaryotic cell cycle?
(A) Cell division cycle 6 and Cdc10-dependent transcript 1
(B) MCM2 and MCM7
(C) Cell division cycle 10 and Cdc6-dependent transcript 1
(D) MutL-MutS protein complex

202. Which of the following process is controlled by CDC6 and CDT1 (*Cdc10-dependent transcript 1*) proteins in eukaryotic cell?
(A) Cell fusion
(B) Mutation
(C) Cell division cycle
(D) Protoplast fusion

203. Which of the following enzyme is analogous to *mini*chromosome *maintenance* protein complex (MCM2 to MCM7) that is a ring-shaped replicative helicase in eukaryotic cell?
(A) Topoisomerase type I
(B) Bacterial DnaB helicase
(C) bacterial DnaC protein
(D) DNA ligase

204. Which of the following enzyme has a role comparable to CDC6 and CDT1 proteins that control cell division cycle in eukaryotic cell and load the MCM helicase onto the DNA near the replication origin?
(A) Topoisomerase type I
(B) Bacterial DnaB helicase
(C) Bacterial DnaC protein
(D) DNA ligase

205. What is the rate of replication fork movement observed in *E. coli*?
(A) ~100 nucleotides/s
(B) ~500 nucleotides/s
(C) ~10 nucleotides/s
(D) ~50 nucleotides/s

206. How many times would take for the replication of an average human chromosome proceeding from a single origin at ~50 nucleotides/s rate of replication fork movement?
(A) Almost 500 hours
(B) More than 500 hours
(C) Less than 500 hours
(D) More than 50 hours

207. How many base pairs spaced from many origins in replication of human chromosomes that proceeds bidirectionally?
(A) Only 300,000 bp apart
(B) 30,000 to 300,000 bp apart
(C) 30 to 300 bp apart
(D) Only 30,000 bp apart

208. How many origin of replication is found in bacterial and eukaryotic chromosomes, respectively?
(A) Single and multiple origins
(B) Multiple and single origins
(C) Single origin in both chromosomes
(D) Multiple origins in both chromosomes

209. Which of the following activity is not found in a multisubunit enzyme DNA polymerase α?
(A) Primase activity
(B) Polymerization activity
(C) Proofreading activity
(D) 5' to 3' activity

210. Which of the following activity found in largest subunit of DNA polymerase α?
(A) Primase activity
(B) 3' to 5' exonuclease activity

(C) Proofreading activity
(D) Polymerization activity

211. Which of the following activity is not found in a multisubunit enzyme DNA polymerase α in eukaryotic cell?
(A) 3' to 5' exonuclease activity
(B) Polymerization activity
(C) Primase activity
(D) 5' to 3' activity

212. Which of the following related to polymerization and proofreading activity of DNA polymerase enzyme, respectively?
(A) 3' to 5' exonuclease activity and 5' to 3' exonuclease activity
(B) Helicase activity and Primase activity
(C) Primase activity and helicase activity
(D) 5' to 3' activity and 3' to 5' exonuclease activity

213. Which of the following enzyme function only in the synthesis of short primers (containing either RNA or DNA) for Okazaki fragments on the lagging strand in eukaryotes?
(A) DNA polymerase β
(B) DNA polymerase δ
(C) DNA polymerase α
(D) DNA polymerase I

214. Which of the following enzyme is associated with and stimulated by a protein called *p*roliferating *c*ell *n*uclear *a*ntigen (PCNA; *Mr* 29,000) that found in large amounts in the nuclei of proliferating cells in eukaryotes?
(A) DNA polymerase β
(B) DNA polymerase δ
(C) DNA polymerase α
(D) DNA polymerase I

215. Which of the following multisubunit enzyme extend primers that synthesized by DNA polymerase α for Okazaki fragments on the lagging strand in eukaryotes.
(A) DNA polymerase β
(B) DNA polymerase I
(C) DNA polymerase α
(D) DNA polymerase δ

216. Which of the following enzyme subunit is a function analogous to the three-dimensional structure of a protein called *p*roliferating *c*ell *n*uclear *a*ntigen (PCNA) that found in large amounts in the nuclei of proliferating cells in eukaryotes?
(A) *β* subunit of *E. coli* DNA polymerase α
(B) *β* subunit of *E. coli* DNA polymerase III
(C) *β* subunit of *E. coli* DNA polymerase I
(D) *β* subunit of *E. coli* DNA polymerase II

217. Which of the following activity is found in a multisubunit enzyme DNA polymerase δ in eukaryotic cell?
(A) 3' to 5' exonuclease activity
(B) Polymerization activity
(C) Primase activity
(D) 5' to 3' activity

218. Which of the following enzyme replaces DNA polymerase δ in DNA repair and may also function at the replication fork with playing a role analogous to the bacterial DNA polymerase I in removing the primers of Okazaki fragments on the lagging strand?
(A) DNA polymerase α
(B) DNA polymerase β
(C) DNA polymerase δ
(D) DNA polymerase ε

219. Which of the following enzyme play a important role in removing the primers of Okazaki fragments on the lagging strand in eukaryotes?
(A) DNA polymerase ε
(B) DNA polymerase β
(C) DNA polymerase δ
(D) DNA polymerase α

220. Which of the following inhibit the activity of the DNA polymerase of the herpes simplex virus?
(A) Acyclovir
(B) Allopurinol
(C) Fluorouracil
(D) Aminopterin

221. Who discovered Acyclovir that consists of guanine attached to an incomplete ribose ring and inhibits the activity of the DNA polymerase of the herpes simplex virus?
(A) J. B. Collip
(B) G. Robert Greenberg
(C) Gertrude Elion
(D) Charles Best

222. What is the property of Acyclovir that inhibits the activity of the DNA polymerase of the herpes simplex virus?
(A) Consists of guanine attached to an incomplete ribose ring
(B) Consists of guanine attached to an incomplete deoxyribose ring
(C) Consists of thymine attached to an incomplete ribose ring
(D) Consists of thymine attached to an incomplete deoxyribose ring

223. Which of the following virally encoded enzyme does phosphorylation of acyclovir that inhibits the activity of the DNA polymerase of the herpes simplex virus?
(A) Phosphorylase kinase
(B) Peptidyl transferase
(C) Thymidine kinase
(D) Protein phosphatases

224. What is the function of enzyme cellular kinases, which is both an inhibitor and a substrate of DNA polymerases and competitively inhibits the herpes DNA polymerase more strongly than cellular DNA polymerases?
(A) Inhibit the joining between Okazaki fragments
(B) Specific repairing for mismatches after replication
(C) Control of the eukaryotic cell cycle
(D) Conversion of acyclo-GMP to acyclo-GTP

225. Which of the following enzymes act as an inhibitor of DNA polymerase?
(A) Phosphorylase kinase
(B) Peptidyl transferase
(C) Cellular kinases
(D) Protein phosphatases

226. What is the function of acyclo-GTP that results from acyclo-GMP by catalyzing cellular kinases during viral replication and lacks a 3' hydroxyl?
(A) Act as an inhibitor for joining Okazaki fragments
(B) Act as a chain terminator when incorporated into DNA
(C) Specific repairing for mismatches after replication
(D) Act as precursor when incorporated into DNA

227. Which of the following is a eukaryotic single-stranded DNA–binding protein equivalent in function to the *E. coli* SSB protein?
(A) Replication factor C (RFC)
(B) MutL protein
(C) MutH protein
(D) Replication protein A (RPA)

228. Which of the following protein is a clamp loader for proliferating cell nuclear antigen (PCNA) and facilitates the assembly of active replication complexes in eukaryotic DNA replication?
(A) Replication factor C (RFC)
(B) MutL protein
(C) MutH protein
(D) Replication protein A (RPA)

229. Which of the following special structure is synthesized at the each ends of chromosomes after termination of replication on linear eukaryotic chromosomes?
(A) Centromeres
(B) Chromomeres
(C) Telomeres
(D) Chromatids

230. Which of the following involved in maintain the fidelity of DNA replication?
(A) Base selection by the DNA polymerase
(B) A 3' to 5' proofreading exonuclease activity
(C) Specific repair systems for mismatches after replication
(D) All of the above

231. What is the number of telomere repeats (n) for most single-celled eukaryotes and generally more than 1,500 in mammals?
(A) In the range of 200 to 1000
(B) in the range of 2 to 10
(C) in the range of 50 to 200
(D) in the range of 20 to 100

232. What is called the study of the properties of an object that do not change under continuous deformations?
(A) Supercoiling (B) Relaxing of DNA
(C) Topology (D) Melting study

233. What is a topological property for circular DNA molecules?
(A) Unaffected by deformations of the DNA strands
(B) Unaffected by deformations of the DNA strands
(C) DNA supercoiling
(D) Continuous breakage in DNA strands

234. What is the linking number, if the two DNA strands are interwound in a right-handed helix?
(A) Negative (B) Positive
(C) Always zero (D) Always one

235. Which strand is accomplished in short Okazaki fragments during DNA replication?
(A) Leading strand
(B) Lagging strand
(C) Discontinuous strand
(D) Strand that read in 3' to 5' direction

236. Both strands of DNA are produced by .... which is accomplished by looping the DNA of the lagging strand bringing together the two points of polymerization.
(A) A single symmetric DNA polymerase III dimer
(B) A single asymmetric DNA polymerase II dimer
(C) A single asymmetric DNA polymerase III dimer
(D) A single asymmetric DNA polymerase I dimer

237. Which of the following proteins act as helicase and primase, respectively?
(A) DnaB and DnaG (B) DnaC and DnaA
(C) DnaG and DnaC (D) DnaE and DnaG

238. Which of the following enzyme unwinds the DNA at the replication fork as it travels along the lagging strand template in the 5' to 3' direction?
(A) DnaG (B) DnaA
(C) DnaC (D) DnaB

239. Which of the following activity is found in a multisubunit enzyme DNA polymerase δ that carry out both leading and lagging strand synthesis in eukaryotic cell in a complex comparable to the dimeric bacterial DNA polymerase III?
(A) Proofreading activity
(B) Polymerization activity
(C) Primase activity
(D) 5' to 3' activity

240. Which of the following enzyme first act in DNA replication?
(A) DNA polymerase III
(B) DNA helicase
(C) Primase
(D) DNA gyrase

241. Which is pentose sugar in the following figure?

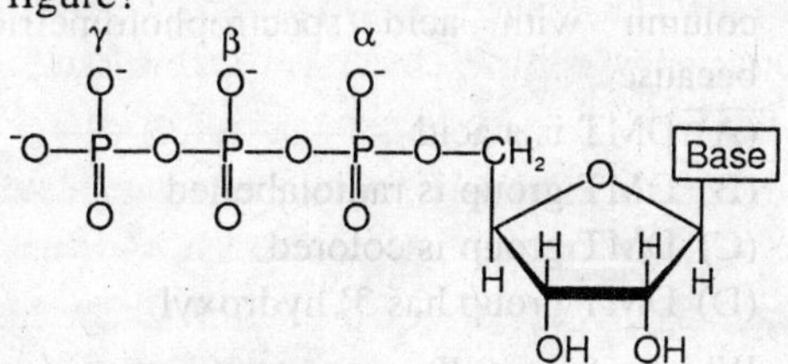

(A) 2'-D-ribose
(B) 2'-deoxy-D-ribose
(C) 2'-oxy-D-ribose
(D) 2'-hydroxy-D-ribose

242. Which bonds (X, Y and Z) are shown in following figure of ATP?

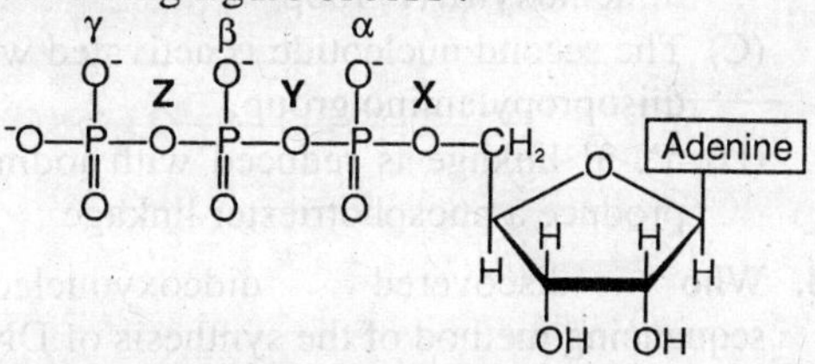

(A) X = phosphoanhydride ester and Y, Z = Phosphate bonds of ATP
(B) X , Y = ester bond and Z = phosphoanhydride bonds of ATP.
(C) X = ester bond and Y, Z = phosphoanhydride bonds of ATP.
(D) X , Y = Phosphate bonds and Z = phosphoanhydride ester of ATP.

243. What is used as solid support to build up oligonucleotide in automated DNA synthesis?
(A) Calcite (B) Silica
(C) Gelrite (D) Agarose

244. At which position of the first nucleoside is attached to the silica support through a linking group in automated DNA synthesis?

(A) 3' hydroxyl (B) 5' hydroxyl
(C) 2' hydroxyl (D) 4' hydroxyl

245. At which position of the first nucleoside is protected at the 5' hydroxyl with an acid-labile dimethoxytrityl group (DMT) to the silica support in automated DNA synthesis?
(A) 3' hydroxyl (B) 5' hydroxyl
(C) 2' hydroxyl (D) 4' hydroxyl

246. In automated DNA synthesis, the first nucleoside is protected at the 5' hydroxyl with an acid-labile dimethoxytrityl group (DMT) which is removed by washing the column with acid spectrophotometrically because
(A) DMT is a acid
(B) DMT group is radiolabelled
(C) DMT group is colored
(D) DMT group has 3' hydroxyl

247. Which of the following steps is not involved in automated DNA synthesis?
(A) The first nucleoside is attached to the silica support at the 3.hydroxyl.
(B) The first nucleoside is protected at the 5' hydroxyl with an acid-labile dimethoxytrityl group
(C) The second nucleotide is activated with a diisopropylamino group.
(D) 5', 3' linkage is reduced with iodine to produce a phosphotriester linkage

248. Who discovered dideoxynucleotide sequencing method of the synthesis of DNA?
(A) H. Gobind Khorana
(B) Maxam-Gilbert
(C) F. Sanger
(D) Merrifield

249. In which method used a fluorescent molecule that gives all the fragments terminating in that nucleotide a particular color?
(A) H. G. Khorana nucleic acids synthesizing method
(B) Maxam-Gilbert sequencing method
(C) F. Sanger sequencing method
(D) Merrifield sequencing method

250. Which of the following technique has been used for separation of coloured DNA fragments in the automating DNA sequencing reactions?
(A) Autoradiography
(B) Gel electrophoresis
(C) X-ray crystallography
(D) Microarray

251. What is primer?
(A) A short oligonucleotide strand
(B) 100-200bp length fragment
(C) Probe
(D) A short tandem repeat sequence

252. What is the function of DNA template in synthesis of daughter DNA?
(A) Identification of DNA polymerase
(B) Help in the separation of DNA strand
(C) Synthesis of cDNA
(D) Guide selection of each new nucleotide

253. At which position the primer reacts with an incoming deoxynucleoside triphosphate to form a new phosphodiester bond in cells?
(A) 3' hydroxyl
(B) 5' hydroxyl
(C) 2' hydroxyl
(D) 4' hydroxyl

254. What is the used in the Sanger sequencing procedure to interrupt DNA synthesis?
(A) Deoxynucleoside triphosphate (dNTP)
(B) Dideoxynucleoside triphosphate (ddNTP)
(C) dioxynucleoside triphosphate (doNTP)
(D) dinucleoside triphosphate (dnNTP)

255. Which of the following act as first messengers that produce second messengers by interacting with receptors on the cell surface to lead adaptive changes in the cell interior?
(A) Enzymes
(B) Cyclic AMP
(C) Extracellular chemical signals
(D) Nucleotides

256. What is also called Sanger sequencing method?
(A) Dideoxy sequencing method (ddNTP)
(B) Deoxynucleoside triphosphate (dNTP) sequencing method
(C) Deoxy sequencing method
(D) Dinucleoside triphosphate (dnNTP) sequencing method

257. Which is used to sequence DNA template in Sanger sequencing method?
(A) Radioactively or fluorescently labeled a short primer
(B) Radioactively or fluorescently labeled a short probe
(C) Only Radioactivly labeled a short primer
(D) Only fluorescently labeled a short primer

258. Which technique is used for separating short DNA molecules?
(A) Agarose electrophoresis
(B) Polyacrylamide gel electrophoresis
(C) Polymerase chain reaction
(D) Northern blotting

259. An oligonucleotide with sequence pAATCGACT, in which C residues labeled at the 5' en(D) How many fragments found in enzymatic sequencing?
(A) Two labeled fragments
(B) Three labeled fragments
(C) One labeled fragments
(D) Four labeled fragments

260. An oligonucleotide with sequence pAATCGACT in which C residues labeled at the 5' en(D) Which types of fragments found in enzymatic sequencing?
(A) A Three-nucleotide and a seven-nucleotide fragment
(B) A One-nucleotide, four-nucleotide and a seven-nucleotide fragment
(C) A one-nucleotide, three-nucleotide and a four-nucleotide fragment
(D) A four-nucleotide and a seven-nucleotide fragment

261. In automated DNA synthesis, the second nucleotide after attaching first nucleotide to solid support, is activated with .... and reacted with the bound nucleotide to form a 5', 3' linkage?
(A) Dipropylamino group
(B) Diisopropylamino group
(C) Isopropylamino group
(D) Diisopropyl group

262. In automated DNA synthesis, which chemical is used to oxidize a 5', 3' linkage that form of activation of second nucleotide to first nucleotide to produce a phosphotriester linkage?
(A) Acetic acid (B) DMT
(C) diisopropylamino group
(D) Iodine

263. Why the chemical synthesis of RNA is more complicated than DNA?
(A) Because ribose has 2' hydroxyl
(B) Because ribose has uracil
(C) Because 2' hydroxyl of ribose produces adversely affect
(D) Because 3' hydroxyl of ribose has no adversely affect

264. An oligonucleotide with sequence pAATCGACT in which G residue labeled at the 5' en(D) Which types of fragments found in enzymatic sequencing?
(A) A Three-nucleotide and a seven-nucleotide fragment
(B) A One-nucleotide, four-nucleotide and a seven-nucleotide fragment
(C) Only one labeled five-nucleotide fragment
(D) A four-nucleotide and a seven-nucleotide fragment

265. What is the length of human genome which has been sequenced?
(A) 2.9 billion base pairs
(B) 3.2 billion base pairs
(C) 5 billion base pairs
(D) Complete genome

266. First of all, who developed the chemical methods for synthesizing nucleic acids?
(A) H. Gobind Khorana
(B) Maxam-Gilbert
(C) F. Sanger
(D) Merrifield

267. First of all, who developed the chemical methods for synthesizing peptides?
(A) H. Gobind Khorana
(B) Maxam-Gilbert
(C) F. Sanger
(D) Merrifield

268. Which of the following base pairs have higher melting point in DNA?
(A) GC pairs (B) AT pairs
(C) AG pairs (D) CT pairs

269. Denatured single-stranded DNAs from two species can form a hybrid duplex. The degree of hybridization depending on the extent of ....
(A) Sequence variation
(B) Sequence similarity
(C) Base pairing capacity
(D) Identical sequence

270. Which are the following spontaneous reactions occur in DNA?
(A) Deamination of certain bases and oxidative damage
(B) Hydrolysis of base-sugar *N*-glycosyl bonds
(C) Radiation-induced formation of pyrimidine dimers
(D) All of the above

271. Which are building blocks of nucleic acids?
(A) Amino acids (B) Nucleotides
(C) Glucose (D) Nucleosides

272. Which is not the function of nucleotides?
(A) Energy carriers
(B) Intrien splicing
(C) Components of enzyme cofactors
(D) Chemical messengers

273. At which position the phosphate group covalently linked at in nucleic acids?
(A) 2' hydroxyl group
(B) 3' hydroxyl group
(C) 5' hydroxyl group
(D) 4' hydroxyl group

274. Which Greek letters are used for labeling the first phosphate groups at 5' end of nucleic acids?
(A) α, β and ω (B) a, b and c
(C) α, β and ϒ (D) ω, ψ and θ

275. Which of the following dNTP is most widely used as chemical energy carrier?
(A) Adenosine 5'-triphosphate (ATP)
(B) Guanosine 5'-triphosphate (GTP)
(C) Thymidine 5'-triphosphate (TTP)
(D) Cytosine 5'-triphosphate (CTP)

276. Which of the following phosphate form is most widely used as chemical energy carrier?
(A) Nucleoside monophosphates
(B) Nucleoside diphosphates
(C) Nucleoside triphosphates
(D) Nucleotide triphosphates

277. Which is the bond between the ribose and α phosphate in nucleoside triphosphates?
(A) Phosphoanhydrides
(B) An ester linkage
(C) β-Glycosidic bond
(D) Hydrogen bond

278. Which type of the bond form between α, β and β, γ phosphate in nucleoside triphosphates?
(A) Phosphoanhydrides
(B) An ester linkage
(C) β-Glycosidic bond
(D) Hydrogen bond

279. How many energy is released from hydrolysis of the ester linkage (ribose and α phosphate) in nucleoside triphosphates?
(A) About 4 kJ/mol
(B) About 5 kJ/mol
(C) About 24 kJ/mol
(D) About 14 kJ/mol

280. How many energy is released from hydrolysis of each anhydride bonds (α, β and β, γ phosphate) in nucleoside triphosphates?
(A) About 14 kJ/mol
(B) About 5 kJ/mol
(C) About 30 kJ/mol
(D) About 20 kJ/mol

281. Activation of MCM helicase activity is required to initiate cellular DNA replication and is regulated by specific protein kinases called ....
(A) S-phase cyclin-dependent kinases
(B) M-phase cyclin-dependent kinases
(C) G1-phase cyclin-dependent kinases
(D) G2-phase cyclin-dependent kinases

282. ATP-dependent resealing carried out by the enzyme ....
(A) DNA ligase (B) Primase
(C) SSBP (D) Polymerase

283. dnaK, dnaJ, and GrpE are .... in *E coli* that cooperate to remove the P protein and activate the dnaB helicase.
(A) Proline (B) Heat shock proteins
(C) LAE proteins (D) Dreb 1 protein

284. Eukaryotic chromosomal DNA molecules contain .... replication origins separated by tens to hundreds of kilobases.
(A) Ten
(B) Two at ends of DNA strand
(C) Multiple
(D) Only one

285. In *E coli,* the oriC is bound by the protein .....
....
(A) dnaB (B) dnaC
(C) dnaD (D) dnaA

286. In SV40 DNA, replication is initiated by binding of two large T-antigen hexameric .... to the single SV40 origin to form two replication forks.
(A) Helicases
(B) Primases
(C) DNA polymerases
(D) SSBP

287. Polynucleotide chain growth results from formation of a phosphodiester bond between the 3' oxygen of a growing strand and the .... of a dNTP.
(A) β phosphate (B) δ phosphate
(C) α phosphate (D) γ phosphate

288. .... is an expression of the number of nucleotides added to the nascent chain before the polymerase disengages from the template.
(A) Processivity (B) Polymerization
(C) Viscosity (D) Nuclease activity

289. A specific nuclease called .... degrades the RNA strand after synthesizing DNA-RNA hybrid molecule during replication in the RNA genome of the animal virus.
(A) RNase H
(B) RNase A
(C) DNase
(D) Restriction endonuclease

290. Unwinding of duplex DNA produces torsional stress, which is relieved by ....
(A) Topoisomerase II
(B) Topoisomerase IV
(C) Topoisomerase I
(D) Topoisomerase V

291. .... protein is a cell cycle regulator because it binds to and inactivates a transcription factor (E2F) necessary for the transcription of certain genes.
(A) Retinoblastoma (Rb)
(B) CDK proteins
(C) SSBP
(D) serine threonine protein kinase

292. .... remove the ribonucleotides at the 5' ends of Okazaki fragments.
(A) DNA polymerase δ and FEN I
(B) DNA polymerase γ and Ribonuclease A
(C) Ribonuclease H and FEN I
(D) DNA polymerase α and Ribonuclease H

293. Which enzyme adds deoxynucleotides to extend the upstream Okazaki fragment.
(A) DNA polymerase δ
(B) DNA polymerase γ
(C) Ribonuclease H
(D) DNA polymerase α

294. The .... oncogene linked with B cell lymphoma appears to be the cyclin D1 gene.
(A) *abl* (B) *erbb*
(C) *fms* (D) *bcl*

295. The D cyclins, CDK4 and CDK6 are nuclear proteins that assemble as a complex in late G1 phase called an active ....
(A) Non-functional psuedokinase
(B) Serine threonine protein kinase
(C) Histidine kinase
(D) Lipid kinases

296. The nucleotide sequences of origins from different organisms usually contain ....sequences.
(A) G-C rich (B) A-G rich
(C) G-T rich (D) A-T rich

297. The structure of the nucleosome proposed by
....
in 1997.
(A) Tim Richmond and coworkers
(B) François Jacob and Jacques Monod
(C) Linus Pauling and Dr Eze Benjamin
(D) Hugo de Vries, Carl Correns and Erich von Tschermak

298. The .... enzyme then reseals the DNA backbone to reform the DNA molecule after cutting in DNA molecules in either single or double-stranded breaks in the phosphodiester backbone.

(A) Ligase (B) Polymerase

(C) Topoisomerase (D) Primase

299. The oncoproteins .... reduced target the Rb transcription repressor for inactivation, inducing cell division inappropriately.

(A) *LYT-10/Cal*

(B) Proch36

(C) MAP Kinase Kinase

(D) *HER2/neu*

300. The origin recognition complex (ORC) in yeast cells is located adjacent to an approximately .... A+T rich sequence that is easy to unwin (D). This is called the DNA unwinding element (DUE).

(A) 80-bp (B) 180-bp

(C) 100-bp (D) 145-bp

Answer-Key

Q	Ans	Q	Ans	Q	Ans	Q	Ans	Q	Ans	Q	Ans	Q	Ans	Q	Ans
1.	(B)	41.	(A)	81.	(A)	121.	(C)	161.	(A)	201.	(A)	241.	(A)	281.	(A)
2.	(D)	42.	(D)	82.	(B)	122.	(D)	162.	(C)	202.	(C)	242.	(C)	282.	(A)
3.	(A)	43.	(B)	83.	(C)	123.	(A)	163.	(A)	203.	(B)	243.	(B)	283.	(B)
4.	(B)	44.	(A)	84.	(A)	124.	(D)	164.	(B)	204.	(C)	244.	(A)	284.	(C)
5.	(A)	45.	(C)	85.	(C)	125.	(A)	165.	(C)	205.	(D)	245.	(A)	285.	(D)
6.	(A)	46.	(B)	86.	(C)	126.	(C)	166.	(B)	206.	(B)	246.	(C)	286.	(A)
7.	(C)	47.	(D)	87.	(A)	127.	(A)	167.	(B)	207.	(B)	247.	(D)	287.	(C)
8.	(D)	48.	(A)	88.	(D)	128.	(C)	168.	(D)	208.	(A)	248.	(C)	288.	(A)
9.	(B)	49.	(A)	89.	(A)	129.	(B)	169.	(B)	209.	(C)	249.	(C)	289.	(A)
10.	(A)	50.	(C)	90.	(C)	130.	(A)	170.	(A)	210.	(D)	250.	(B)	290.	(C)
11.	(D)	51.	(B)	91.	(C)	131.	(C)	171.	(C)	211.	(A)	251.	(A)	291.	(A)
12.	(D)	52.	(D)	92.	(A)	132.	(B)	172.	(B)	212.	(D)	252.	(D)	292.	(C)
13.	(A)	53.	(B)	93.	(B)	133.	(D)	173.	(A)	213.	(C)	253.	(A)	293.	(A)
14.	(C)	54.	(A)	94.	(D)	134.	(A)	174.	(B)	214.	(B)	254.	(B)	294.	(D)
15.	(D)	55.	(C)	95.	(C)	135.	(B)	175.	(D)	215.	(D)	255.	(C)	295.	(B)
16.	(D)	56.	(A)	96.	(B)	136.	(A)	176.	(B)	216.	(B)	256.	(A)	296.	(D)
17.	(C)	57.	(A)	97.	(D)	137.	(A)	177.	(C)	217.	(A)	257.	(A)	297.	(A)
18.	(C)	58.	(B)	98.	(B)	138.	(C)	178.	(A)	218.	(D)	258.	(B)	298.	(C)
19.	(B)	59.	(A)	99.	(D)	139.	(D)	179.	(A)	219.	(A)	259.	(A)	299.	(B)
20.	(D)	60.	(D)	100.	(D)	140.	(B)	180.	(B)	220.	(A)	260.	(D)	300.	(A)
21.	(B)	61.	(A)	101.	(C)	141.	(D)	181.	(A)	221.	(C)	261.	(B)		
22.	(A)	62.	(D)	102.	(B)	142.	(A)	182.	(C)	222.	(A)	262.	(D)		
23.	(B)	63.	(C)	103.	(D)	143.	(B)	183.	(B)	223.	(C)	263.	(C)		
24.	(C)	64.	(C)	104.	(A)	144.	(C)	184.	(B)	224.	(D)	264.	(C)		
25.	(A)	65.	(C)	105.	(D)	145.	(B)	185.	(C)	225.	(C)	265.	(B)		
26.	(B)	66.	(C)	106.	(C)	146.	(D)	186.	(C)	226.	(B)	266.	(A)		
27.	(C)	67.	(A)	107.	(C)	147.	(A)	187.	(A)	227.	(D)	267.	(D)		
28.	(C)	68.	(B)	108.	(C)	148.	(C)	188.	(D)	228.	(A)	268.	(A)		
29.	(B)	69.	(B)	109.	(C)	149.	(A)	189.	(C)	229.	(C)	269.	(B)		
30.	(C)	70.	(A)	110.	(C)	150.	(B)	190.	(B)	230.	(D)	270.	(D)		
31.	(C)	71.	(D)	111.	(D)	151.	(C)	191.	(C)	231.	(D)	271.	(B)		
32.	(A)	72.	(B)	112.	(C)	152.	(D)	192.	(A)	232.	(C)	272.	(B)		
33.	(A)	73.	(B)	113.	(A)	153.	(A)	193.	(B)	233.	(B)	273.	(C)		
34.	(A)	74.	(A)	114.	(B)	154.	(A)	194.	(C)	234.	(B)	274.	(C)		
35.	(C)	75.	(C)	115.	(D)	155.	(C)	195.	(C)	235.	(B)	275.	(A)		
36.	(B)	76.	(B)	116.	(B)	156.	(B)	196.	(A)	236.	(C)	276.	(C)		
37.	(A)	77.	(D)	117.	(B)	157.	(A)	197.	(D)	237.	(A)	277.	(B)		
38.	(B)	78.	(C)	118.	(C)	158.	(D)	198.	(B)	238.	(D)	278.	(A)		
39.	(C)	79.	(D)	119.	(D)	159.	(A)	199.	(B)	239.	(A)	279.	(D)		
40.	(B)	80.	(C)	120.	(C)	160.	(D)	200.	(C)	240.	(B)	280.	(C)		

5

DNA Damage and Repair System

DNA is a highly stable and versatile biological molecule. But it damaged by different physical and chemical mechanisms. Accordingly, DNA repair systems require to overcome and correct the damages in DNA. Erroneousness in DNA replication and chemical changes in DNA lead to integration of wrong bases, which are mismatched with the complementary strand. The damage causing chemicals break the backbone of the strand and chemically alter the bases. Alkylation, oxidation and methylation cause damage to bases. X-rays and gamma radiations cause single or double stranded breaks in DNA. Damages in of DNA are responsible for various neurological and cancer disorders, such as Ataxia-telangiectasia, Fanconi anemia, and Xeroderma pigmentosum, that may be recovered by DNA repair systems.

Listing of damage types in DNA and DNA repair mechanisms

Damage types of DNA	
Simple mutations	Switching of one base for another base by point mutations like transition (one pyrimidine is substituted by another pyramidine and purine with another purine) and substitution (T by G or A and A by C or T).
Deamination	Deamination of cytosine (C) to form uracil (U) which base pairs with adenine (A) in next replication instead of guanine (G). Hypoxanthine results from adenine deamination
Missing bases	Cleavage of N-glycosidic bond between purine and sugar causes loss of purine base from DNA called depurination (non-coding lesion).
Chemical Modification of bases	Addition of methyl groups to various bases. Guanine forms 7-methylguanine, 3-methylguanine. Adenine forms 3-methyladenine. Cytosine forms 5- Methylcytosine. Replacement of amino group by a keto group converts 5-methylcytosine to thymine.
Formation of pyrimidine dimers (Thymine dimers)	A bacteria contains dimers after exposure to ultraviolet radiations.
Strand breaks	Peroxides, radiations and by enzymes like DNases break phosphodiester bonds in one strand
DNA repair mechanisms	
Direct Repair	Photoreactivation by enzyme DNA photolyase cleave the C-C bond of cyclobutyl ring of the thymine dimmers which are monomerized
Base excision repair or nucleotide excision repair	An enzyme N-glycosylase recognizes the abnormal base and hydrolyes glycosidic bond between it and sugar. DNA polymerase by its exonuclease activity removes the abnormal base.
Mismatch base repair	Wrong bases (G-T or C-A pairs) are repaired by methylation of adenine bases of the template strand using Mut" proteins (MutS, MutH and MutL)
Recombination repair or retrieval system (daughter strand gap repair mechanism)	Rec A protein retrieved a portion of the complementary strand from other side of the replication fork to fill the gap. Rec A is a strand exchange protein (homologous recombination).

SOS repair mechanism	Gene	Repair Function
	lexA	SOS repressor
	recA	SOS regulator; SOS mutagenesis; *recF*-dependent recombinational repair; *recB*-dependent repair of double-strand gaps; cross-link repair
	recN	*recF*-dependent recombinational repair; repair of double-strand gaps
	recQ	*recF*-dependent recombinational repair
	ruv	*recF*-dependent recombinational repair
	umuC	SOS mutagenesis (Error prone repair)
	umuD	SOS mutagenesis (Error prone repair)
	uvrA	Short-patch nucleotide-excision repair; long-patch nucleotide-excision repair; cross-link repair
	uvrB	Short-patch nucleotide-excision repair; long-patch nucleotide-excision repair; cross-link repair
	uvrD	Short-patch nucleotide-excision repair; *recF*-dependent recombinational repair; *recB*-dependent repair of double-strand gaps; cross-link repair; methylation-directed mismatch repair
	dinA	SOS mutagenesis
	sulA	Inhibitor of cell division

Human disease caused by DNA damage and lack of DNA repairing

Cells performed mechanisms DNA-damage response (DDR) to detect damaged DNA lesions and promote their repair. If cells fail to perform these mechanisms, defects in DNA caused many human diseases which may be hereditable. Several inheritable human diseases caused by lacking of DNA repair mechanism include Xeroderma pigmentosum, Ataxia-telangectasia, Bloom's and Werner's syndromes, Hereditary Non-Polyposis Colon Cancer, Li-Fraumeni-syndrome, and breast/ovarian cancer syndrome.

Gene therapy

Gene therapy is a gene repairing technique in which various approaches can be used to prevent inherited diseases caused by DNA damages and lacking DNA repairing mechanism.

- Replacing a mutated gene that causes disease with a healthy copy of the gene.
- Inactivating, or "knocking out," a mutated gene that is functioning improperly.
- Introducing a new gene into the body to help fight a disease.

Inherited DDR defects cause cancers and neurodegenerative diseases

Syndrome	Phenotypes	Mutated Genes
GGR/NER deficiency		
Xeroderma pigmentosum	Neurodegeneration and microcephaly, Photosensitivity, skin cancer	XPAXPG Pol η
Cockayne syndrome	Microcephaly and neuron demyelination	CSA, CSB, XPB, XPD, XPG
Trichothiodystrophy	Neurodevelopmental defects and dysmyelination (abnormal myelin)	XPD, XPB, TTDA

Syndrome	Phenotypes	Mutated Genes
Cerebrooculofacio skeletal (COFS) syndrome	Demyelination (loss of myelin), dysmyelination, brain calcification, microcephaly	XPD, XPG, CSB, ERCC1
DNA helicase deficiency		
Bloom's syndrome	Microcephaly, short stature, dysmorphic features, mild/moderate mental retardation, susceptible to infections, Elevated predisposition to all cancers	BLM
Werner's syndrome	Premature ageing, cancer predisposition	WRN
Rothmund–Thompson syndrome	Stunted growth, skeletal abnormalities, early cataracts, accelerated ageing, chromosomal instability and cancer predisposition	RECQL4
Ataxia with oculomotor apraxia 2	Ataxia, neurodegeneration and oculomotor apraxia	SETX
Exonuclease deficiency		
AicardiGoutières syndrome (AGS)	De/dysmyelination, brain calcification, microcephaly, elevated CSF (cerebrospinal fluid) IFNα, CSF lymphocytosis (CSF lymphocytes increased)	TREX1, RNASEH2
NHEJ/V(D)J recombination deficiency		
Inactivation of Ku70 or Ku80 in mouse models	Premature aging, cancer predisposition, lymphomas	Ku70 Ku80
Lig4 syndrome/Human immunodeficiency with microcephaly	Microcephaly, leukemia, immunodeficiency, and developmental and growth delay	DNA Ligase IV, XLF/Cerunnos
RSSCID (radiosensitive SCID)	Severecombined immunodeficiency, lymphomas and hypersensitivity to ionizing radiation	Artemis
Syndrome	Phenotypes	Mutated Genes
HR deficiency		
Breast cancer 1, early onset	Breast and ovarian cancer	BRCA1
Breast cancer 2, early onset	Breast and ovarian cancer; predisposition to pancreatic, prostate and gastric cancer and melanoma	BRCA2
DNA SSB repair deficiency		
Spinocerebellar ataxia with axonal neuropathy (SCAN1)	Ataxia, neurodegeneration, peripheral axonal motor and sensory neuropathy, muscle weakness	TDP1
Ataxia with oculomotor apraxia 1 (AOA1)	Ataxia, neurodegeneration, oculomotor apraxia and peripheral neuropathy	APTX
DNA crosslink repair deficiency		
Fanconi anaemia	Congenital abnormalities, progressive bone marrow failure, prone to AML, squamous carcinomas of head, neck or gynaecological system.	FANCAFANCL, BRCA2 (FANCD1)

Syndrome	Phenotypes	Mutated Genes
Mismatch repair deficiency		
Hereditary nonpolyposis colorectal cancer (HNPCC)	Colon and gynaecologic cancers	MSH2, MSH3, MSH6, MLH1, PMS2, APC
DNA DSB repair and signal transduction deficiency		
LiFraumeni syndrome	Soft tissue sarcomas, breast cancer, brain tumours	p53
Familial breast cancer (non BRCA1/2)	Predisposition to medium/lateonset breast cancer	Chk2, MRN, ATM BRIP1, PALB2,
Ataxia telangiectasia	Cerebellar ataxia, telangiectases, immune defects, predisposition to malignancy (mainly lymphomas but also breast cancer)	ATM
Ataxia telangiectasialike disorder	Mild AT like features, possibly cancer predisposed	MRE11
Seckel syndrome	Marked microcephaly, primordial dwarfism, dysmorphic facial features and mental retardation, possibly AML	ATR, SCKL2, SCKL3
Primary microcephaly 1	Microcephaly, mental retardation	MCPH1/BRIT1

Heritable syndromes caused by chromosomal aberrations

Syndrome	Phenotypes	Mutated Genes
DNA damage response impairment and defective DNA repair		
Down syndrome	Mental retardation, progeria	Trisomy of chromosome 21
Alzheimer's disease	Progressive neurodegeneration leading to dementia, memory loss and cognitive decline	Increased oxidative stress and damage
Parkinson's disease	Tremor, bradykinesia, posture rigidity and postural instability, degeneration of dopaminergic neurons in substantia nigra area	Mutations in α Synuclein and Parkin variants
Huntington's disease	Progressive chorea and dementia, severe neuronal loss in the striatum and cerebral cortex	CAG repeat expansion in *huntingtin (HD)*
Several spinocerebellar ataxias	Problems with bodily movements (similar to Huntington's disease), progressive neuron loss	Expanded CAG repeats in various genes
Friedreich's ataxia	Limb ataxia, cerebellar dysarthria, sensory loss, skeletal deformities	GAA expanded repeats in *frataxin (FXN)*
Myotonic dystrophy types 1 and 2	Muscle weakness and wasting, cataracts, testicular atrophy, cognitive decline	CTG expansion (type 1), CCTG expansion (type 2)
TripleA syndrome	Adrenal insufficiency, achalasia, alacrima, neurodegeneration, autonomic dysfunction	Mutation in *AAAS* gene
Amyotrophic lateral sclerosis	Progressive degeneration of motor neurons, muscle weakness and atrophy, leading to fatality	Defective CuZn superoxide dismutase (SODC, SOD1); mitochondrial DNA

Examples of DNA-damaging drugs used to treat cancer

Cancer Treatment	Types of DNA lesions induced
Radiotherapy and radiomimetics	
Ionizing radiation Bleomycin	Single-strand breaks, double-strand breaks, base damage
Monofunctional alkylators	
Alkylsulphonates Nitrosourea compounds Temozolomide	Base damage, replication lesions, bulky DNA adducts
Bifunctional alkylators	
Nitrogen mustard Mitomycin C Cisplatin	Double-strand breaks, DNA cross-links, replication lesions, bulky DNA adducts
Antimetabolites	
5-Fluorouracil (5FU) Thiopurines Folate analogs	Cytotoxic metabolite, inhibits base pairing leading to base damage and replication lesions
Topoisomerase inhibitors	
Camptothecins (Topo I); Etoposide (Topo II), anthracyclines (Doxorubicin, Epirubicin, Daunorubicin (Topo II)	Double-strand breaks, single-strand breaks, replication lesions; anthracyclines also generate free radicals
Replication inhibitors	
Aphidicolin Hydroxyurea	Double-strand breaks, replication lesions

Key notes

- DNA molecule can undertake a variety of chemical reactions however; changes in its structure are of much greater consequence than are alterations in other cell components.
- Mutation can results from the integration of incorrect bases during DNA replication.
- Various chemical changes arise in DNA either spontaneously or as a result of introduction to chemicals or radiation. Such damage to DNA can block replication or transcription, and can result in a high frequency of mutations.
- Mechanisms of DNA repair separated into two classes
  (1) Direct reversal of the chemical reaction accountable for DNA damage
  (2) Removal of the impaired bases and their replacement with freshly synthesized DNA.
- There are two main systems of spontaneous DNA damage
  (A) Deamination of adenine, cytosine, and guanine
  (B) Depurination (loss of purine bases) resultant from cleavage of the bond between the purine bases and deoxyribose
- UV light induces the development of pyrimidine dimers, in which two nearby pyrimidines (e.g., thymines) are joined by a cyclobutane ring structure
- Alkylation is the adding of methyl.

Direct reversal of DNA damage

- Most of the DNA damage is repaired by elimination of the damaged bases followed by re-synthesis of the deleted region.
- Some abrasions in DNA can be repaired by direct reversal of the damage. Examples: Particularly pyrimidine dimers resulting from exposure to ultraviolet (UV) light and alkylated guanine residues that have been modified by the addition of methyl or ethyl groups at the O^6 position of the purine ring.
- The main type of damage induced by UV light is the development of pyrimidine dimers.
- In both prokaryotic and eukaryotic cells, one mechanism of repairing UV-induced pyrimidine dimers is reversal of the dimerization reaction by photo-reactivation because energy derived from visible light is utilized to break the cyclobutane ring structure.
- Direct repair deals with damage between alkylating agents and DNA.
- Alkylating agents are responsive compounds that can transfer methyl or ethyl groups to a DNA base, and modify the base.
- A particularly important type of damage is methylation of the O^6 position of guanine, because the product, O^6-methylguanine, forms complementary base pairs with thymine instead of cytosine. This lesion can be repaired by an enzyme called O^6-methylguanine methyltransferase.
- O^6-methylguanine methyltransferase transfers the methyl group from O^6-methylguanine to a cysteine residue in the enzyme's active site.

Excision Repair

- Excision repair is a common process repairing a wide variety of chemical alterations to DNA in both prokaryotic and eukaryotic cells.
- In excision repair, the damaged DNA is recognized and deleted. The resulting gap is then filled in by synthesis of a new DNA strand, using the undamaged complementary strand as a template.
- Three types of excision repair found

Base-excision repair

- The repair of uracil-containing DNA is a good example of base-excision repair, in which single damaged bases are identified and deleted from the DNA molecule. Uracil can arise in DNA by two mechanisms: (1) Uracil is occasionally incorporated in place of thymine during DNA synthesis, and (2) uracil can be formed in DNA by the deamination of cytosine.
- The excision of uracil in DNA is catalyzed by DNA glycosylase that cleaves the bond connecting the base (uracil) to the deoxyribose of the DNA backbone.
- AP sites are formed as the result of the spontaneous loss of purine bases. These sites are repaired by AP endonuclease, which cleaves adjacent to the AP site and the resulting single-base gap is filled by DNA polymerase I and ligase.

Nucleotide-excision repair

- This repair system works to free DNA from UV-induced pyrimidine dimers and bulky groups added to DNA bases as a result of the reaction of many carcinogens with DNA.

- In *E. coli*, nucleotide-excision repair is catalyzed by the products of three genes (*uvrA*, *B*, and *C*) that were identified because mutations at these loci result in extreme sensitivity to UV light.
- The protein UvrA recognizes damaged DNA and recruits UvrB and UvrC to the site of the lesion.
- UvrB and UvrC then cleave on the 3′ and 5′ sides of the damaged site, respectively, thus excising an oligonucleotide consisting of 12 or 13 bases.
- The UvrABC complex is frequently called an excinuclease, a name that reflects its ability to directly excise an oligonucleotide and the resulting gap is filled by DNA polymerase I and sealed by ligase.
- In eukaryotes, several genes involved in DNA repair (called *RAD* genes for radiation sensitivity) have been identified by the isolation of mutants with increased sensitivity to UV light.
- The most extensively studied diseases is xeroderma pigmentosum (XP), a rare genetic disorder that affects approximately one in 250,000 people. Individuals with this disease are extremely sensitive to UV light and develop multiple skin cancers on the regions of their bodies that are exposed to sunlight.

Table 1. Enzymes Involved in Nucleotide-Excision Repair

Human	Yeast	Function
XPA	RAD14	Damage recognition
XPB	RAD25	Helicase
XPC	RAD4	DNA binding
XPD	RAD3	Helicase
XPF	RAD1	5′ nuclease
XPG	RAD2	3′ nuclease
ERCC1	RAD10	Dimer with XPF

Mismatch repair

- A third excision repair system recognizes mismatched bases that are incorporated during DNA replication. Many such mismatched bases are removed by the proofreading activity of DNA polymerase.
- In *E. coli*, the ability of the mismatch repair system to differentiate between parental DNA and newly synthesized DNA is based on the point that DNA of this bacterium is altered by the methylation of adenine residues inside the sequence GATC to form 6-methyladenine.
- Mismatch repair is started by the protein MutS, which distinguishes the mismatch and forms a complex with two other proteins called MutL and MutH.
- The MutH endonuclease cuts the unmethylated DNA strand at a GATC sequence. MutL and MutS then act together with an exonuclease and a helicase to remove the DNA between the strand break and the mismatch, with the resulting gap being filled by DNA polymerase and ligase.

Post-replication repair

- When direct reversal and excision repair systems fail, the cell has alternative mechanisms for dealing with damaged DNA at the replication fork.
- Pyrimidine dimers and several other types of lesions cannot be copied by the common action of DNA polymerases, so replication is choked at the sites of such damage.
- The result is a daughter strand that has a gap opposite the site of damage to the parental strand.
- One of two types of mechanisms may be used to repair such gaps in newly synthesized DNA: recombinational repair or error-prone repair.

Recombinational repair

- Recombinational repair based on the point that one strand of the parental DNA was undamaged and consequently copied during replication to yield an ordinary daughter molecule.
- The undamaged parental strand can be recycled to seal the gap reverse the site of damage in the other daughter molecule by recombination between homologous DNA sequences.
- Because the resulting gap in the formerly intact parental strand is opposite an undamaged strand, it can be occupied in by DNA polymerase.

Error-prone repair

- In error-prone repair, a break opposite a site of DNA damage is complete by newly synthesized DNA. Since the new DNA is synthesized from a damaged template strand, this form of DNA synthesis is very incorrect and leads to frequent mutations.
- It is used only in bacteria that have been exposed to possibly lethal disorders, such as wide UV irradiation. Such actions induce the SOS response, which may be observed as a mechanism for dealing with dangerous environmental stress.
- The SOS response comprises inhibition of cell division and initiation of repair systems to manage with a high level of DNA damage.
- Under these circumstances, error-prone repair mechanisms are applied, apparently as a way of dealing with damage so extensive that cell death is the only alternative.

Multiple Choice Questions

1. How many sets of genomic DNA is generally found in a cell?
   (A) Only one or two sets
   (B) Four sets (C) Only one set
   (D) Two sets

2. Which of the molecules cannot be quickly replaced by using information encoded in the DNA?
   (A) Damaged proteins
   (B) Damaged RNA molecules
   (C) Damaged DNA molecules
   (D) All of the above

3. Which of the following mechanism system is support maintaining the integrity of the information in DNA that is a cellular imperative?
   (A) Enzyme system
   (B) DNA replication systems
   (C) Genetic recombination systems
   (D) DNA repair systems

4. Which is the responsible for damaging in DNA?
   (A) Environmental agents
   (B) Polymerase enzymes
   (C) Topisomerase enzymes
   (D) Water

5. How can replication very occasionally damage the information content in DNA?
   (A) Cause of joining randome primer
   (B) Due to introduce mismatched base pairs
   (C) Due to tautomeric effect
   (D) Due to recombination system

6. Which of the following pairing is example of damaging in DNA?
   (A) G paired with C (B) A paired with T
   (C) G paired with T (D) All of the above

7. Which of the following is considered to illustrate the importance of DNA repair?
   (A) The effects of unrepaired DNA damage
   (B) Tautomeric effect
   (C) Hyperchromic effect
   (D) The integrity of the information in DNA

8. Which is the most serious outcome of a change in the base sequence of the DNA, if it replicated and transmitted to future cell generations?
   (A) A permanent change in the nucleotide sequence of DNA
   (B) A change in the nucleotide sequence of DNA reduce by environmental factors
   (C) A permanent change in the nucleotide sequence of transcript
   (D) A permanent change in the amino acid sequence of protein

9. What is called a permanent suddenly change in the nucleotide sequence of DNA?
   (A) Recomboination (B) Denaturation
   (C) Translation (D) Mutation

10. What is called the replacement of one base pair in DNA with another?
   (A) Insertion mutations
   (B) Substitution mutation
   (C) Deletion mutations
   (D) Silent mutation

11. What is called the addition of one or more base pairs in DNA?
   (A) Insertion mutations
   (B) Substitution mutation
   (C) Deletion mutations
   (D) Silent mutation

12. What is called the removal of one or more base pairs in DNA?
   (A) Insertion mutations
   (B) Substitution mutation
   (C) Deletion mutations
   (D) Silent mutation

13. If the mutation affects nonessential DNA or if it has a negligible effect on the function of a gene, it is known as a ....
   (A) Insertion mutations
   (B) Substitution mutation
   (C) Deletion mutations
   (D) Silent mutation

14. Which type of mutations are mostly nonsilences?
   (A) Harmless
   (B) Benificial
   (C) Deleterious
   (D) Sometimes deleterious

15. Who discovered a simple test for measuring the potential of a given chemical compound to promote certain easily detected mutations in a specialized bacterial strain?
(A) Bruce Ames
(B) G. Robert Greenberg
(C) Gertrude Elion
(D) Charles Best

16. Which of the following test is used to detect the potential of a given chemical compound to promote certain easily detected mutations in a specialized bacterial strain?
(A) Fehling's test
(B) Indole test
(C) Ames test
(D) LDL test

17. Which of the following test show strong correlation between mutagenesis and carcinogenesis in extensive animal trials?
(A) Ames test
(B) Indole test
(C) Fehling's test
(D) LDL test

18. Which of the following disease is caused by mutations in certain genes that regulate normal cell division fail to function that are activated at the wrong time or are altered?
(A) Sickel cell anemia (B) Heamophilia
(C) Hepatitis B (D) Human cancer

19. Which of the following does reflect both the importance of DNA repair to cell survival and the diverse sources of DNA damage in cells?
(A) The diversity of repair systems
(B) The number of repair systems
(C) both a and b
(D) The number and types of cells

20. How DNA damage in one strand can be removed and accurately replaced?
(A) By using the undamaged complementary RNA strand as a template
(B) By using the undamaged complementary DNA strand as a template
(C) By using the undamaged complementary primer for replication
(D) By repeating DNA replication from start point

21. In which DNA repair system does improve the overall fidelity of replication by an additional factor of 10^2 to 10^3 for the rare mismatches left after replication in E. coli?
(A) Mismatch Repair system
(B) Base-excision repair system
(C) Nucleotide-excision repair system
(D) Direct repair system

22. What is improved by correction of the rare mismatches left after replication in *E. coli* by an additional factor of 10^2 to 10^3?
(A) Activity of enzymes involved in replication
(B) Properly linking the primer to templare strand
(C) The overall fidelity of replication
(D) Rate of polymerization of template strand by primer

23. What is the importance of DNA repair system in replication?
(A) Discrimination between the template and the newly synthesized strand
(B) Improve the activity of enzymes involved in replication
(C) Increase the rate of polymerization of template strand
(D) Lost the tendency of primer link to template strand

24. How the cell distinguish template strand from newly synthesized strands in mismatch repair system?
(A) By tagging the template DNA with acyl groups
(B) By radioactive labeling of thymine in the template DNA
(C) By tagging the template DNA with methyl groups
(D) By tagging the newly synthesized DNA strand with methyl groups

25. How many proteins included in the mismatch repair system of *E. coli*?
(A) At least 100 protein components
(B) Proteins equal to number of primers used in lagging strand
(C) Only 1 protein component
(D) At least 12 protein components

26. For *E. coli* and some closely related bacteria, strand discrimination is based on .... in mismatch repair system.
(A) The action of DNA polymerase
(B) The action of DNA glycosylases
(C) The action of Dam methylase
(D) The action of ABC excinuclease

27. At which position in all adenine within (5') GATC sequences is done methylation for strand discrimination is based on the action of Dam methylase in mismatch repair system.
(A) *N9* position of all adenines
(B) *N6* position of all adenines
(C) *N1* position of all adenines
(D) *N7* position of all adenines

28. How many mismatches is efficiently repaired by the cell's methyl-directed mismatch repair system repairs mismatches from a hemimethylated GATC sequence.
(A) Less than 1,000 bp
(B) More than 1,00 bp
(C) up to 1,000 bp
(D) only 1 bp

29. Which of the following complex is involved in the mismatch correction directed by relatively distant GATC sequences?
(A) MutL-MutS complex
(B) Receptor-ligand complex
(C) PDH complex (D) Tus-Ter complex

30. Which of the following protein has a site-specific endonuclease activity that is inactive until the complex encounters a hemimethylated GATC sequence in mismatch repair system?
(A) MutL (B) MutH
(C) MutS (D) MSH2

31. Which of the sequences is binded to MutH by the MutL-MutS complex in mismatch repair system?
(A) CTAG sequences (B) CAAT sequences
(C) TATA sequences (D) GATC sequences

32. Which of the following structure is formed on both sides of the mismatch in DNA through the MutL-MutS complex in mismatch repair system?
(A) A dsDNA loop
(B) Degraded DNA loop
(C) ssDNA loop
(D) A triplex DNA loop

33. At which of the following sequence site MutH protein catalyzes cleavage of the unmethylated strand on the 5' side of the G in this site which marks the strand for repair in mismatch repair system?
(A) A hemimethylated CAAT sequence
(B) A hemimethylated GATC sequence
(C) A hemimethylated TATA sequence
(D) A hemimethylated CTAG sequence

34. Which of the following step is not involved in mismatch repair system, when the mismatch is on the 5' side of the GATC cleavage site?
(A) The unmethylated strand is unwound
(B) Degraded in the 5' to 3' direction from the cleavage site through the mismatch
(C) Mismatched segment is replaced with new DNA
(D) Degraded in the 3' to 5' direction from the cleavage site through the mismatch

35. When the mismatch is on the 5' side of the cleavage site then the unmethylated strand is unwound and degraded in the 3' to 5' direction from the cleavage site through the mismatch and this segment is replaced with new DNA. Which of the following enzemes and proteins are required in this process?
(A) DNA helicase II
(B) SSB, exonuclease I or exonuclease X
(C) DNA polymerase III and DNA ligase
(D) All of the above

36. Which of the following enzyme degrades single-stranded DNA in the 5' to 3' or 3' to 5' direction?
(A) Exonuclease II
(B) Exonuclease VII
(C) Endonuclease VII
(D) RecJ nuclease

37. Which of the following enzyme degrades single-stranded DNA in the 5' to 3' or 3' to 5' direction?
(A) Exonuclease II
(B) Exonuclease VII
(C) Endonuclease VII
(D) RecJ nuclease

38. Which of the following bacterial protein is not structurally and functionally analogous to proteins of all eukaryiotic cells except the bacterial MutS and MutL proteins?
(A) MutH (B) MSH2
(C) MSH3 (D) MSH6

39. Which of the following most common inherited disease is caused by MutS and MutL like proteins encoded by human genes?
(A) AIDS
(B) Hepatitis B
(C) Cancer-susceptibility syndromes
(D) Turnar's syndrome

40. Which of the following proteins in most eukaryotes (from yeast to humans) is not homolog to the main bacterial protein MutS that function in mismatch repair system?
(A) MSH6 (B) MutH
(C) MSH3 (D) MSH2

41. Which of the following heterodimer protein generally bind to single base-pair mismatches and bind less well to slightly longer mispaired loops during mismatch repar process of DNA?
(A) Heterodimers of MSH2 and MSH3
(B) Heterodimers of MSH3 and MSH6
(C) Heterodimers of MSH2 and MSH6
(D) Heterodimers of MutL and MutH

42. What is the size of the longer mismatches in many organisms may be bound instead by a heterodimer of MSH2 and MSH3 or are bound by both types of heterodimers in tandem.
(A) 6 bp (B) 15 to 20 bp
(C) 20 to 60 bp (D) 2 to 6 bp

43. What is the function of Homologs of MutL that is predominantly a heterodimer of MLH1 and PMS1 (*p*ost meiotic *s*egregation) bind to the MSH complexes in base-excision repair?
(A) Distabilize the MSH complexes
(B) Stabilize the MSH complexes
(C) Degrade the MSH complexes
(D) Debase the MSH complexes

44. Which of the following statement is not true?
(A) DNA damages can be repaired without removing a base or nucleotide
(B) Uracil glycosylase remove uracil from DNA after cytosine deamination
(C) Each DNA glycosylase is generally specific for one type of DNA lesion
(D) Bacteria and humans have 4 and 1 types of uracil DNA glycosylase, respectively

45. Which of the following enzyme promote direct photoreactivation of cyclobutane pyrimidine dimmers reaction?
(A) Uracil DNA glycosylase
(B) DNA photolyases
(C) MutL-MutS complex
(D) AP endonucleases

46. Which of the following reaction is promoted by DNA photolyases?
(A) Direct photoreactivation of cyclobutane pyrimidine dimmers
(B) Removing of uracil from DNA after cytosine deamination
(C) Methylation to distinguish parent from newly synthesized DNA strands in *E. coli*
(D) Removing the affected base by cleaving the *N*-glycosyl bond from DNA lesions

47. Which of the mechanism is based on DNA repairing without removing a base or nucleotide?
(A) Mismatch repair system
(B) Direct repair system
(C) Base-excision repair system
(D) Nucleotide-excision repair system

48. In which of the following DNA repair system involve the formation of pyrimidine dimers that result from an ultraviolet light–induced reaction with promoting by DNA photolyases?
(A) Mismatch repair system
(B) Nucleotide-excision repair system
(C) Base-excision repair system
(D) Direct repair system

49. Which of the following enzyme for its activity is used energy derived from absorbed light to reverse the damage?
(A) Exonuclease
(B) DNA polymerase
(C) DNA photolyase
(D) DNA glycosylase

50. Which of the following one of the chromophores as cofactor is always present in photolyases that serve as light-absorbing agents?
(A) $FADH^-$ (B) $NADH^-$
(C) $FADH^+$ (D) $NADH^+$

51. Which of the following class of enzymes that recognize particularly common DNA lesions and remove the affected base by cleaving the *N*-glycosyl bond in every cell?
(A) Exonucleases
(B) DNA polymerases
(C) DNA glycosylases
(D) Endonuclease

52. What result the removing of the affected base by cleaving the *N*-glycosyl bond trhough DNA glycosylases in every cell during base-excision repair?
(A) Creates an apyrimidinic site in the DNA
(B) Creates an apurinic site in the DNA
(C) Creates methylated purinic or pyrimidinic site in the DNA
(D) Both a and b

53. What is referred to an apurinic or apyrimidinic site in the DNA that created after removing the affected base by cleaving the *N*-glycosyl bond trhough DNA glycosylases in every cell during base-excision repair?
(A) An abasic site (B) A basic site
(C) A Dc site (D) An Ac site

54. What is referred to an apurinic or apyrimidinic site in the DNA that created after removing the affected base by cleaving the *N*-glycosyl bond through DNA glycosylases in every cell during base-excision repair?
(A) A basic site (B) An AP site
(C) A Dc site (D) An Ac site

55. Which of the enzyme is responsible for the absence of uracil in DNA?
(A) *O*6-methylguanine-DNA methyltransferase
(B) DNA polymerase
(C) DNA glycosylase
(D) Endonuclease

56. Which of the following enzyme remove uracil from DNA by spontaneous deamination of cytosine?
(A) Exonuclease
(B) Cytosine DNA glycosylase
(C) Uracil DNA glycosylase
(D) *O*6-methylguanine-DNA methyltransferase

57. What is the function of DNA glycosylase by spontaneously deaminase cytosine in DNA?
(A) Remove uracil from DNA
(B) Remove cytosine from DNA
(C) Remove thymine from RNA
(D) Remove uracil from RNA

58. Which of the following function is not of glycosylase enzyme?
(A) Remove uracil from DNA
(B) Remove cytosine from DNA
(C) Remove thymine from RNA
(D) Spontaneous deamination of cytosine

59. Which of the following production form by cytosine deamination?
(A) Thymine (B) Adenine
(C) Guanine (D) Uracil

60. How many types of uracil DNA glycosylases are found in bacteria and humans, respectively?
(A) One type and many types
(B) In both same four types
(C) One type and four types
(D) In both same one type

61. Which of the following cells does not contain enzyme DNA photolyases?
(A) The mammal RBCs
(B) The cells of placental mammals
(C) The cells of eye cornea
(D) The cells of large intestine

62. Which of the following is responsible the formation of the modified nucleotide *O*6-methylguanine and is a common and highly mutagenic lesion?
(A) UV radiation
(B) Nitrous acid
(C) Gamma radiation
(D) Alkylating agents

63. Which of the following mutation is caused by alkylating agents to form the modified nucleotide O^6-methylguanine in DNA?
(A) G≡C to A=T mutations
(B) A=T to G≡C mutations
(C) Purine to purine mutations
(D) Nonsense mutations

64. Which of the following change is caused by alkylating agents to form the modified nucleotide O^6-methylguanine in DNA?
(A) Cytosine replaced with uracil
(B) Cytosine replaced with adenine
(C) Cytosine replaced with thymine
(D) Cytosine replaced with guanine

65. Which of the following enzyme is carried out the direct repair of O^6-methylguanine in DNA?
(A) DNA polymerase
(B) DNA glycosylase
(C) Endonuclease
(D) O^6-methylguanine-DNA methyltransferase

66. Which of the following a protein that catalyzes transfer of the methyl group of O^6-methylguanine to one of its own Cys residues.
(A) DNA polymerase
(B) DNA glycosylase
(C) Endonuclease
(D) O^6-methylguanine-DNA methyltransferase

67. In which mechanism is used to repair 1-methyladenine and 3-methylcytosine y methylation of the amino groups of A and C residues when the DNA is single-stranded, and the methylation directly affects proper base pairing?
(A) Mismatch repair system
(B) Direct repair system
(C) Base-excision repair system
(D) Nucleotide-excision repair system

68. What is AlkB protein which mediates oxidative demethylation of these alkylated nucleotides in *E.* coli?
(A) Molecular chaperones
(B) Intermediate filament (IF) proteins
(C) A member of the α-ketoglutarate-Fe^{2+}–dependent dioxygenase superfamily
(D) Immunoglobulin-like Ca^{2+}-independent proteins

69. Which of the following reaction in *E. coli* is mediated by the AlkB protein that is a member of the α-ketoglutarate-Fe^{2+}–dependent dioxygenase superfamily?
(A) Homologous genetic recombination between any two DNA molecules
(B) DNA strand-exchange reaction
(C) Site-Specific recombination in DNA rearrangements
(D) Oxidative demethylation of alkylated nucleotides

70. Which of the following enzyme is associated with the human replisome where it eliminates the occasional U residue inserted in place of a T during replication?
(A) Human uracil glycosylase
(B) DNA polymerase
(C) Endonuclease
(D) O6-methylguanine-DNA methyltransferase

71. Which of the following enzyme remove only any U residues that occur in single-stranded DNA during replication or transcription in human?
(A) AP endonuclease (B) TDG
(C) hSMUG1 (D) MBD4

72. Which of the following reaction, that catalyzes by enzyme hSMUG1, is 100-fold faster in single stranded DNA than in double-stranded DNA?
(A) Deamination of C residues
(B) Direct photoreactivation of cyclobutane pyrimidine dimmers
(C) Methylation to distinguish parent from newly synthesized DNA strands in *E. coli*
(D) Removing the affected base by cleaving the *N*-glycosyl bond from DNA lesions

73. Which of the following does not belong to human DNA glycosylase?
(A) AP endonuclease (B) TDG
(C) hSMUG1 (D) MBD4

74. Which of the following a group of human DNA glycosylases that remove either U or T residues paired with G that generated by deamination of cytosine or 5-methylcytosine, respectively?
(A) AP endonuclease and TDG
(B) MBD4 and hSMUG1
(C) hSMUG1 and AP endonuclease
(D) TDG and MBD4

75. Which of the following damaged bases of DNA is recognized and removed by DNA glycosylases?
(A) Formamidopyrimidine
(B) 8-hydroxyguanine and Hypoxanthine
(C) 3-methyladenine and 7-methylguanine
(D) All of the above

76. DNA glycosylases recognize particularly common DNA lesions and remove the affected base by cleaving the *N*-glycosyl bond that creates an apurinic or apyrimidinic site in the DNA referred to as an AP site or abasic site. How repaired this AP site?
(A) By inserting a new base
(B) Re-forming the *N*-glycosyl bond
(C) Catalyses with AP endonucleases
(D) Replacing with complete DNA strand

77. Which of the following enzyme cut the DNA strand containing the AP site that contain an apurinic or apyrimidinic site in the DNA and formed by cleaving the *N*-glycosyl bond?
(A) Human uracil glycosylase
(B) DNA polymerase
(C) AP endonuclease
(D) O^6-methylguanine-DNA methyltransferase

78. What happen when AP endonucleaes cut the DNA strand containing the AP site that contain an apurinic or apyrimidinic site in the DNA and formed by cleaving the *N*-glycosyl bond?
(A) Removing deoxyribose 5'-phosphate
(B) Removing 3'-hydroxyl group
(C) Re-forming the *N*-glycosyl bond
(D) Replacing with complete DNA strand

79. Which of the following enzyme involve in nucleotide replacement in eukaryotic DNA?
(A) Human uracil glycosylase
(B) DNA polymerase
(C) AP endonuclease
(D) O^6-methylguanine-DNA methyltransferase

80. Which of the following mechanism is not involve in DNA repair system
(A) Nucleotide-Excision
(B) Base-excision
(C) Site-specific recombination
(D) Mismatch repair

81. In which of the following repair mechanism hydrolyzed two phosphodiester bonds by a multisubunit enzyme?
(A) Nucleotide-Excision
(B) Base-excision
(C) Site-specific recombination
(D) Mismatch repair

82. At which of the following position of phosphodiester bonds is hydrolyzed by a multisubunit enzyme in *E. coli* and other prokaryotes in nucleotide-excision repair?
(A) Fifth phosphodiester bond on the 3' side and the eighth on the 5' side
(B) Eighth phosphodiester bond on the 3' side and the fifth on the 5' side
(C) Fifth phosphodiester bond on the 5' side and the eighth on the 3' side
(D) Sixth phosphodiester bond on the 3' side and the twenty-second on the 5' side

83. How many nucleotides contained in the fragment that generated after hydrolyzing the fifth phosphodiester bond on the 3' side and the eighth phosphodiester bond on the 5' side by a multisubunit enzyme in *E. coli* and other prokaryotes in nucleotide-excision repair?
(A) 5 to 10 nucleotides
(B) 27 to 29 nucleotides
(C) 12 to 13 nucleotides
(D) 2 to 3 nucleotides

84. At which of the following position of phosphodiester bonds is hydrolyzed by a multisubunit enzyme in humans and eukaryotes in nucleotide-excision repair?
(A) Fifth phosphodiester bond on the 3' side and the eighth on the 5' side

(B) Eighth phosphodiester bond on the 3' side and the fifth on the 5' side
(C) Twenty-second phosphodiester bond on the 3' side and Sixth on the 5' side
(D) Sixth phosphodiester bond on the 3' side and the twenty-second on the 5' side

85. How many nucleotides contained in the fragment that generated after hydrolyzing Sixth phosphodiester bond on the 3' side and the twenty-second on the 5' side by a multisubunit enzyme in humans and eukaryotes in nucleotide-excision repair?
(A) 5 to 10 nucleotides
(B) 27 to 29 nucleotides
(C) 12 to 13 nucleotides
(D) 2 to 3 nucleotides

86. Which of the following enzymes involved in releasing the excised oligonucleotides from the duplex in *E. coli* and eukaryotes, respectively and the resulting gap is filled by sealing the nick with DNA ligase?
(A) DNA polymerase I in *E. coli* and DNA polymerase ε in humans
(B) DNA polymerase III in *E. coli* and DNA polymerase γ in humans
(C) DNA polymerase ε in *E. coli* and DNA polymerase I in humans
(D) DNA polymerase α in *E. coli* and DNA polymerase I in humans

87. In which organism the enzymatic complex present the ABC excinuclease which has three subunits UvrA, UvrB, and UvrC that are used to catalyze two specific endonucleolytic cleavages.
(A) Human
(B) *Saccharomyces cerevisiae*
(C) *Escherichia coli*
(D) *Arabidopsis thaliana*

88. What is meaning of the term "excinuclease" is used to describe the unique capacity of ABC enzyme complex?
(A) To catalyze only one specific endonucleolytic cleavage
(B) To catalyze two specific endonucleolytic cleavages
(C) To catalyze two specific endonucleolytic recombination
(D) To catalyze two specific endonucleolytic recombination

89. In which of the following repair system a complex of the UvrA and UvrB proteins (A_2B) scans the DNA and binds to the site of a DNA lesion?
(A) Nucleotide-Excision
(B) Base-excision
(C) Site-specific recombination
(D) Mismatch repair

90. What the sequence for completion of nucleotide-excision repairing after a complex of the UvrA and UvrB proteins (A_2B) scans the DNA and binds to the site of a DNA lesion?
i. A UvrC-mediated incision at the eighth phosphodiester bond on the 5' side
ii. The UvrA dimer dissociates with leaving a tight UvrB-DNA complex
iii. UvrB makes an incision at fifth phosphodiester bond on the 3' side of the lesion
iv. 12 to 13 nucleotide fragment is removed by UvrD helicase
(A) ii, iii, i and iv (B) ii, iii, iv and i
(C) iii, i, ii and iv (D) ii, iv, iii and i

91. Which of the following factor that is responsible for the rate of DNA repair?
(A) Cell type
(B) Age of the cell
(C) Extracellular environment
(D) All of the above

92. Which of the following enzyme remove the 12 to 13 nucleotide fragment after UvrC-mediated incision at the eighth phosphodiester bond on the 5' side?
(A) UvrA helicase (B) UvrB helicase
(C) UvrC helicase (D) UvrD helicase

93. Which of the following enzymes are involved in filling the short gap created by removing the 12 to 13 nucleotide fragment after UvrC-mediated incision at the eighth phosphodiester bond on the 5' side?
(A) UvrD helicase and DNA ligase
(B) DNA polymerase I and DNA ligase
(C) Exonuclease and UvrD helicase
(D) DNA polymerase I and UvrD helicase

94. Which of the following lesions is not primarily repaired by nucleotide-excision repair pathway?
(A) Cyclobutane pyrimidine dimmers
(B) 6-4 photoproducts
(C) O^6-Methylguanine
(D) Benzo[*a*]pyrene-guanine

95. Which of the following lesion is primarily repaired by nucleotide-excision repair pathway and formed in DNA by exposure to cigarette smoke?
(A) Cyclobutane pyrimidine dimmers
(B) 6-4 photoproducts
(C) O^6-Methylguanine
(D) Benzo[*a*]pyrene-guanine

96. Which of the following enzyme has the nucleolytic activity that two cuts are made in the DNA during nucleotide-excision repair?
(A) Human uracil glycosylase
(B) ABC excinuclease
(C) AP endonuclease
(D) O^6-methylguanine-DNA methyltransferase

97. Which of the following enzyme is required for the dual incision that two cuts are made in the DNA during nucleotide-excision repair?
(A) Human uracil glycosylase
(B) ABC excinuclease
(C) AP endonuclease
(D) O^6-methylguanine-DNA methyltransferase

98. Which of the following the types of lesions in DNA most often arise when a replication fork encounters an unrepaired DNA lesion?
(A) Double strand breaks
(B) Double-strand cross-links
(C) Lesions in a single-stranded DNA
(D) both (A) and (C)

99. What are major reasons of forming double strand breaks, lesions in a single-stranded DNA lesions and DNA cross-links?
(A) Ionizing radiation and oxidative reactions
(B) Nitrous acid and EMS
(C) Inactive DNA polymerase and ligase
(D) x- and γ- rays

100. What is called the DNA repair in which the information is required for accurate repair must come from a separate homologous chromosome in the absence of a second strand.
(A) Nucleotide-Excision
(B) Base-excision
(C) Recombinational DNA repair
(D) Mismatch repair

101. After the activation of which DNA repair pathway, DNA repair becomes significantly less accurate and a high mutation rate can result?
(A) Nucleotide-Excision
(B) Base-excision
(C) Error-prone translesion DNA synthesis (TLS)
(D) Mismatch repair

102. In bacteria, error-prone translesion DNA synthesis is part of a cellular stress response to extensive DNA damage known as ....
(A) SOS response
(B) Nucleotide-Excision
(C) Base-excision
(D) Mismatch repair

103. Which of the following proteins are normally present in the cell but are induced to higher levels as part of the SOS response and participate in the pathway for error-prone repair?
(A) UvrA and UvrB proteins
(B) UmuC and UmuD proteins
(C) UvrC helicase and UvrD helicase
(D) Both (A) and (B)

104. What is the function of the UmuC and UmuD proteins in DNA repair pathway?
(A) Required for the dual incision that two cuts are made in the DNA
(B) Eliminate error-prone repair with the umutable gene function
(C) Form double strand breaks, lesions in a single-stranded DNA lesions
(D) Lack of the umutable gene function eliminates error-prone repair

105. In which process the UmuD protein is cleaved into a shorter form called UmuD' that forms a complex with UmuC to create a

specialized DNA polymerase V that can replicate of the DNA lesions and replication normally blocked.
(A) Nucleotide-Excision repair process
(B) Base-excision repair process
(C) Mismatch repair process
(D) SOS-regulated process

106. Which of the following enzyme created by UmuD'-UmuC complex and the enzyme can replicate past many of the DNA lesions that would normally block replication through producing mutations?
(A) DNA polymerase IV
(B) DNA polymerase III
(C) DNA polymerase I
(D) DNA polymerase V

107. Which of the following statement id true about DNA polymerase V?
(A) Replicate DNA lesions that block replication
(B) Act as DNA polymerase III
(C) Separate daughter DNA strands from parental DNA after replication
(D) Participate in DNA repair

108. Which of the following enzyme is induced during SOS response?
(A) DNA polymerase I
(B) DNA polymerase III
(C) DNA polymerase III
(D) DNA polymerase IV

109. Which of the following is a product of the *dinB* gene?
(A) DNA polymerase IV
(B) DNA polymerase III
(C) DNA polymerase III
(D) DNA polymerase I

110. Which of the following enzymes are part of a family of TLS polymerases (error-prone translesion DNA synthesis) found in all organisms?
(A) bacterial DNA polymerases III and V
(B) bacterial DNA polymerases I and III
(C) bacterial DNA polymerases IV and V
(D) bacterial DNA polymerases I, III, IV and V

111. Which of the following enzymes participate in error-pron DNA replication?
(A) bacterial DNA polymerases III and V
(B) bacterial DNA polymerases I and III
(C) bacterial DNA polymerases IV and V
(D) bacterial DNA polymerases I, III, IV and V

112. Which of the following activity lack in DNA polymerases IV and V that participate in error-prone DNA synthesis?
(A) Helicase activity
(B) Proofreading exonuclease activity
(C) 5' to 3' exonuclease activity
(D) 5'-deoxyribose phosphate lyase activity

113. Which of the following enzyme is a TLS polymerase found in all eukaryotes and promotes translesion synthesis primarily across cyclobutane T–T dimmers?
(A) DNA polymerases ί (iota)
(B) DNA polymerases β (beta)
(C) DNA polymerases λ (lemda)
(D) DNA polymerase η (eta)

114. DNA polymerase η (eta) is a TLS polymerase found in all eukaryotes and promotes translesion synthesis primarily across cyclobutane T–T dimers. Why?
(A) Because the enzyme preferentially inserts two C residues
(B) Because the enzyme preferentially inserts two T residues
(C) Because the enzyme preferentially inserts two A residues
(D) Because the enzyme preferentially inserts two U residues

115. Which of the following enzyme is not involved in in eukaryotic base-excision repair?
(A) DNA polymerases ί (iota)
(B) DNA polymerases β (beta)
(C) DNA polymerases λ (lemda)
(D) DNA polymerase III

116. Which of the following activity present in DNA polymerases ί (iota), β (beta) and λ (lemda) of eukaryotes?
(A) Helicase activity
(B) Proofreading exonuclease activity
(C) 3' to 5' exonuclease activity
(D) 5'-deoxyribose phosphate lyase activity and polymerase activity

117. Which of the following activity of DNA polymerases ί (iota), β (beta) and λ (lemda) of eukaryotes fill the very short gap that craet by removing the abasic site (a 5'-deoxyribose phosphate) after base removal by a glycosylase and backbone cleavage by an AP endonuclease?

(A) Polymerase activity
(B) Proofreading exonuclease activity
(C) 3' to 5' exonuclease activity
(D) 5'-deoxyribose phosphate lyase activity

118. How minimized the the frequency of mutations due to DNA polymerase η activity?

(A) 5'-deoxyribose phosphate lyase activity
(B) Proofreading exonuclease activity
(C) To backbone cleavage by an AP endonuclease
(D) To synthesize the very short lengths of DNA

119. How many genes encode proteins dedicated to the repair of DNA in the human genome?

(A) More than 130 genes
(B) 100 Genes
(C) Less than 130 genes
(D) Only five genes

120. What are the results the loss of function of genes encode proteins dedicated to the repair of DNA in the human genome?

(A) Increase genomic stability
(B) Genomic instability and an occurrence of oncogenesis
(C) Increase fidelity of replication
(D) Increase polyrization activity of TLS polymerase

121. Abnormal regions of DNA either from copying errors or DNA damage are replaced by four mechanisms that are ....

(A) Mismatch repair
(B) Base excision repair
(C) Nucleotide excision-repair and double strand break repair
(D) All of the above

122. Which of the following figure shows deaminates to uracil?

(A)

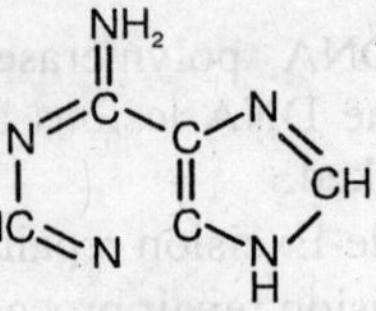

(B)

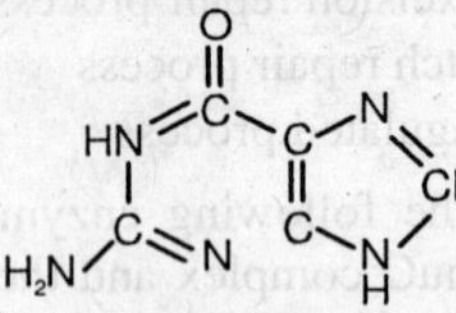

(C)

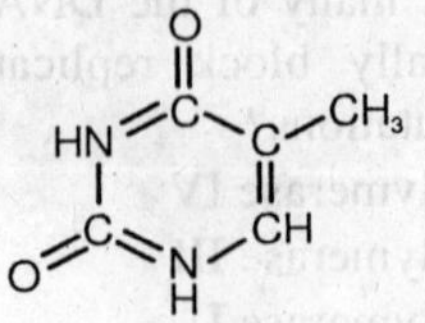

(D)

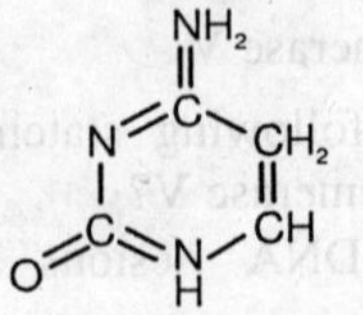

123. Replication in *E. coli* is initiated by the generation of short RNA primers using

(A) DNA ligase
(B) Primase
(C) DNA polymerase II
(D) reverse transcriptase

124. The RNA primers used to initiate replication in *E. coli*.

(A) Result in Okazaki fragments on the leading strand
(B) Are removed by Pol I "nick translation"
(C) Are joined together by DNA ligase
(D) Are removed by helicase + ATP

125. The two exonuclease activities of DNA polymerase I.

(A) Occur at two different active sites
(B) Are coupled during the formation of "gapped DNA"
(C) Degrade DNA in a 5' --> 3' direction
(D) Degrade DNA in a 3' --> 5' direction.

126. The *E. coli* chromosome has $5x10^6$ bp; a replication fork progresses at about 1000 nucleotides/se(C) Therefore, the minimum time required to complete replication is about.

(A) 10 min (B) 20 min
(C) 40 min (D) 80 min

127. Which feature of DNA replication in eukaryotes is different from that in prokaryotes?
(A) Not inhibited by Actinomycin D
(B) More than two DNA polymerases are found
(C) RNA primers are not required
(D) Multiple origins of replication are used

128. The melting of 40-45 bp at the oriC site of *E. coli* requires
(A) DnaA + ATP
(B) DnaA + ATP + HU
(C) DnaA + ATP + HU + DnaB
(D) DnaA + ATP + HU + DnaB + DnaC

129. The b subunit of *E. coli* DNA polymerase III holoenzyme
(A) Unwinds DNA ahead of the replication fork
(B) Does the 3' --> 5' editing
(C) Repairs single strand nicks in DNA
(D) Is the essential processivity factor

130. The replication error frequency in *E. coli* ($\sim 1\times10^{-9}$/bp replicated) does not depend on
(A) The presence of dUTP in cells
(B) Template-directed binding of dNTPs on DNA polymerase
(C) The 3' --> 5' exonuclease functions of Pol I and Pol III
(D) Post-replication repair activities

131. The rapid appearance of HIV-1 strains that are resistant to AZT is due in part to this property of its reverse transcriptase
(A) The RNase H domain of p66 causes error prone synthesis
(B) It lacks a 5' --> 3' exonuclease
(C) AZT is too expensive to be taken in effective doses
(D) It lacks a proofreading exonuclease.

132. Which of the following processes is not catalyzed by a DNA repair enzyme?
(A) Removal of a uracil base to form an apyrimidinic (AP) site
(B) Removal of a methyl group from O^6-methylguanine
(C) Methylation of dU to form T using S-adenosylmethionine (SAM)
(D) Removal of 10-12 nucleotides around a pyrimidine dimer

133. Which of the following processes produce variations in genomes?
(A) Mutation and recombination
(B) Recombination and replication
(C) Replication and transcription
(D) Replication, mutation and recombination

134. .... is a rearrangement of a DNA molecule that involves the adding or deletion and reunion of DNA.
(A) Mutation (B) Replication
(C) Transcription (D) Recombination

135. Recombination occurs during .... phase in meiosis of eukaryotic cells.
(A) Prophase (B) Metaphase
(C) Anaphase (D) Telophase

136. What is called the exchange of homologous segments between any two homologous DNA molecules?
(A) Homologous recombination
(B) Site-specific recombination
(C) Reverse recombination
(D) Heterogous recombination

137. Which process is a type of recombination?
(A) Transformation
(B) Transduction
(C) Transposition
(D) Transcription

138. Homologous recombination occur during .... in eukaryotes.
(A) Cell plate formation
(B) Telophase
(C) Formation of equator line
(D) Crossing-over

139. Which model presents the mechanism of homologous recombination?
(A) Nucleosome model
(B) Holliday model
(C) Hoogsteen model
(D) Heteroduplex model

140. Which of the following process is proposed by Robin Holliday in 1964?
(A) DNA denaturation
(B) DNA damage
(C) DNA recombination
(D) Mutation

141. Who proposed the holliday junction model for homologous recombination in 1964?
(A) Robin Holliday
(B) Robert Brown
(C) Robert Hook
(D) Robinson

142. Holliday model is also known as ....
(A) Homoduplex model
(B) heteroduplex model
(C) Homo-heteroduplex model
(D) Heterologous model

143. A enzyme .... is primarily responsible for correcting apurinic or apyrimidinic sites during DNA repair mechanism.
(A) AP endonuclease
(B) DNase
(C) Restriction endonuclease
(D) Ligase

144. A human has 10^{14} nucleated cells each with .... base pairs of DNA.
(A) 5×10^9 (B) 2785
(C) 3.2×10^9 (D) 3×10^{10}

145. An irreversible state of dormancy is known as ....
(A) Senescence (B) Fluorenscence
(C) Photoperod (D) Vernalization

146. AP repair acts on nucleotides that lost their ... .
(A) Thymine (B) Adenine
(C) Cytosine (D) Any base

147. .... is a collection of processes by which a cell identifies and corrects damage to the DNA molecules that encode its genome.
(A) DNA replication (B) DNA repair
(C) DNA damage (D) Mutation

148. During a normal cell's life, the SOS system is turned off, because .... represses expression of all the critical proteins.
(A) *lexA* (B) *lacA*
(C) *lacZ* (D) *lacY*

149. If about 10^{16} cell divisions occur in a lifetime and 10 mutations/base pair per cell generation escape repair, there may eventually be .... per 10^6 bp in the genome.
(A) Two mutations (B) Ten mutations
(C) One mutation (D) 10^6 mutations

150. In human beings the inherited autosomal recessive disease .... is the result of a defect in the repair of ultraviolet light damaged DNA.
(A) Xeroderma pigmentosum
(B) AIDS
(C) Sickle cell anemia
(D) Hemophilia

151. In mammalian cells, enzyme .... is responsible for the replication of mitochondrial DNA.
(A) DNA polymerase δ
(B) DNA polymerase β
(C) DNA polymerase γ
(D) DNA polymerase α

152. Increased risk of cancer is one clinical feature shared by disorders caused by defects in ....
(A) DNA repair mechanisms
(B) DNA replication
(C) DNA damaging mechanisms
(D) DNA transfer mechanisms

153. .... is being copied in the direction of the advancing replication fork and is synthesized almost continuously.
(A) Sense strand
(B) Antisense strand
(C) Lagging strand
(D) Leading strand

154. *Lex*A cleaves itself in the presence of bound .... , which causes cellular levels of LexA to drop that causes coordinate depression of the ....
(A) RecB and SOS regulon genes
(B) RecA and SOS regulon genes
(C) RecC and SOS regulon genes
(D) Only SOS regulon genes

155. Methylation of cytosine followed by .... can cause a cytosine to become a thymine in the genome.
(A) Deamination
(B) Nitrification
(C) Denitrification
(D) Amination

156. Mutations in homologs of bacterial MutS and MutL in humans cause hereditary ....

(A) Non-polyposis colon cancer
(B) Xeroderma pigmentosum
(C) Sickle cell anemia
(D) Polyposis colon cancer

157. .... prevents tautomer-induced misincorporation from occurring more frequently than once every 10^8–10^{10} base pairs of DNA synthesized.

(A) Proofreading activity
(B) Endonuclease activity
(C) Exonuclease activity
(D) Repairing mechanism

158. Proofreading is performed by the .... during replication to correct mismatch bases.

(A) DNA polymerase II
(B) DNA polymerase III
(C) DNA polymerase I
(D) DNA polymerase I and II

159. RecA stimulates the .... of *lexA*'s gene product.

(A) Autoglucolysis
(B) Autophagouse
(C) Autolysis
(D) Autoproteolysis

160. The DNA sequence 5'-GATTCAAGTAGTC-3' as a template used to form RNA during transcription. RNA sequence after synthesis is 5'.................... -3'.

(A) UUGAAUCGACUAC
(B) CUAAGUUCAUCAG
(C) GACUACUUGAAUC
(D) ACUUGAAUCGACU

161. The enzyme ....is necessary in order to remove a uracil base from DNA.

(A) Thymine glycosylase
(B) Uracil glycosylase
(C) Glycosylase
(D) All of the above

162. The primary control for the SOS regulon is the gene product of *lexA*, which serves as a repressor for.

(A) *recA, lexA* (B) *recA, lexB*
(C) *recB, lexA* (D) *recB, lexB*

163. The primary function of primosomes creates .... for DNA replication.

(A) DNA primers
(B) RNA primers
(C) DNA-RNA hybrid primers
(D) DNA or RNA primers changed on the condition and process

164. The .... is a state of high-activity DNA repair and is activated by bacteria that have been exposed to heavy doses of DNA-damaging agents.

(A) Chemo response
(B) SOS response
(C) Phyto response
(D) Hypersensitive response

165. The SOS system is a .... that is, it controls expression of several genes distributed throughout the genome simultaneously.

(A) Regulon (B) Cistron
(C) Mutaton (D) Recon

166. When DNA damage occurs in SOS system, RecA binds to .... when a lesion creates a gap in daughter DNA to repair the damage.

(A) Single-stranded DNA
(B) Double-stranded DNA
(C) Hybrid DNA
(D) DNA-RNA hybrid

167. Which of the following DNA repair mechanism is known as the 'cut and patch mechanism'?

(A) Photoreactivation
(B) Nucleotide excision repair
(C) Base excision repair
(D) Mismatch repair

168. DNA helicase enzyme involved in base excision repair mechanism is ....

(A) DNA helicase I
(B) DNA helicase II
(C) DNA helicase III
(D) DNA helicase IV

169. In mismatch repair mechanism, which of the following protein recognize DNA mismatches in the genomic DNA?

(A) MutH (B) MutS
(C) MutL (D) UvrD

170. The main difference between nucleotide excision repair (NER) and base excision repair (BER) is
   (A) In NER double strand breaks are repaired where as in BER single strand breaks repaired
   (B) NER is a light dependent reaction whereas BER is light independent process
   (C) In NER phosphodiester backbone is first cleaved where as in BER phosphodiester backbone is cleaved later
   (D) All of these

171. Which enzyme is responsible for photoreactivation of DNA
   (A) Photoligase
   (B) Photoreductase
   (C) Photo-oxidase
   (D) Photolyase

172. Which of the following enzyme(s) is involved base excision repair?
   (A) DNA glycosylase
   (B) AP endonuclease
   (C) AP exonuclease
   (D) Both (a) and (b)

173. *Xeroderma pigmentosum* in human is associated with a mutation in
   (A) Photoreactivation
   (B) Nucleotide excision repair
   (C) Base excision repair
   (D) Mismatch repair

174. Which of the following gene is NOT involved in mismatch repair system in bacteria?
   (A) MutS (B) MutA
   (C) MutL (D) UvrD

175. In eukaryotes, the mismatch repair mechanism is initiated and directed by ....
   (A) Methylated DNA strand
   (B) Acetylated DNA strand
   (C) Strand specific nicks
   (D) Double strand breaks

176. In nucleotide excision repair mechanism which of the following proteins first recognize DNA lesion?
   (A) UvrA (B) UvrB
   (C) UvrC (D) UvrD

177. The two chromophores present in photolyase enzyme are
   (A) NAD and FAD (B) FAD and Folate
   (C) FADH and Folate (D) FAD- and Folate

178. Which of the following statement is true regarding the nucleotide excision repair (NER) mechanism?
   (A) NER mechanism is present only in eukaryotes
   (B) NER present only in prokaryotes
   (C) NER Present in both prokaryotes and eukaryotes but the exact mechanisms varies
   (D) NER Present in both prokaryotes and eukaryotes with similar molecular mechanism in both

179. Photolyase enzyme involved in the photoreactivation can remove ....
   (A) UV induced double strand breaks
   (B) UV induced thymine dimers
   (C) Mismatches in DNA
   (D) Single strand breaks in DNA

180. DNA polymerase enzyme involved in base excision repair mechanism is ....
   (A) Pol-α (B) Pol-β
   (C) Pol-γ (D) Pol-δ

181. DNA polymerase enzyme involved in mismatch repair system of E. coli is ....
   (A) DNA Pol-I (B) DNA Pol-II
   (C) DNA Pol-III (D) DNA Pol-β

182. In NER machinery, which of the following protein processes the helicase activity?
   (A) UvrA (B) UvrB
   (C) UvrC (D) UvrD

183. The catalytic activity of photolyase enzyme in the bacterial cells is activated by ....
....
   (A) UV light
   (B) IR light
   (C) Visible light near blue region
   (D) Visible light near red region

184. In NER (nucleotide excision repair) mechanism, the incision at the 3' position of the lesion is done by ....
   (A) Uvr A (B) UvrB
   (C) UvrC (D) UvrD

185. In mismatch repair mechanism, the endonuclease activity which cut the nascent DNA strand is done by:
(A) MutH (B) MutL
(C) MutS (D) UvrD

186. The activity of AP endonuclease is involved in ....
(A) Base excision repair
(B) Nucleotide excision repair
(C) Mismatch repair
(D) Double strand break repair

187. Which of the following statement is correct?
(A) Photoreactivation is absent in human
(B) Photoreactivation is present in all organism
(C) Photoreactivation present only in plants and animals
(D) Photoreactivation is present only in bacteria

188. The enzyme photolyase repair the thymine dimer in the DNA by .... mechanism.
(A) Oxidation mechanism
(B) Free radical mechanism
(C) Direct bond breaking
(D) Adduct formation

189. In *E. coli*, inactivation of cellular DNA methylase enzyme causes severe mutation in the genomic DNA. Which of the following DNA repair mechanism would be most probably inhibited?
(A) Double strand break repair
(B) Mismatch repair
(C) Base Excision Repair
(D) Nucleotide Excision Repair

190. The function of UvrC in NER mechanism is ....
(A) To identify the lesion in the DNA
(B) Helicase activity to unwind the DNA
(C) Catalyze the incision at the 3' end of the lesion
(D) Catalyze the incision at the 5' side of the lesion

191. First DNA glycosylase enzyme discovered is ....
(A) Uracil DNA glycosylase
(B) Thymine DNA glycosylase
(C) Adenine DNA glycosylase
(D) Methyl Adenine Glycosylase

192. Which of the following is not an effect of a mutation?
(A) Are permanent changes in the DNA sequence or structure
(B) Produce allelic variation
(C) Are more likely to be harmful than beneficial
(D) All of the above

193. Mutations that cause loss of a chromosome would be termed
(A) Structural mutations
(B) Chromosome mutations
(C) Genome mutations
(D) Single-gene mutations

194. A frameshift mutation could be caused by
(A) Transition
(B) Transversion
(C) Deletion of 3 bp
(D) None of the above

195. Which of the following point mutations would be most likely to affect protein function?
(A) TAA to TGA
(B) CAA to TAA
(C) AGG to AGA
(D) CTT to CTC

196. A wild type allele
(A) Is the allele that is the most common in the population
(B) Is the allele with the highest level of functional protein
(C) Is the most favorable allele in the population
(D) All of the above

197. The mutation which causes sickle cell anemia in humans
(A) Is a base substitution
(B) Is a deleterious mutation
(C) Is a missense mutation
(D) All of the above

198. The gene that is mutated in cases of cystic fibrosis encodes an ion transporter protein normally expressed in the respiratory and digestive tracts along with other locations. Individuals homozygous for the most common mutation in this gene (ΔF508) died at very early ages prior to the advent of advanced medical intervention. Despite this, the mutation exists at relatively high frequency in populations of European descent. One potential explanation is that heterozygotes for such mutations may be relatively immune to the lethal effects of diseases such as typhoid fever. Which statement below is most accurate?

(A) The ΔF508 cystic fibrosis mutation is a beneficial mutation
(B) The ΔF508 cystic fibrosis mutation is a lethal mutation
(C) The ΔF508 cystic fibrosis mutation is a conditional mutation
(D) The ΔF508 cystic fibrosis mutation is a neutral mutation

199. In Himalayan rabbits, the ears, feet, and muzzle are pigmented while the rest of the body is white. This is because they express an enzyme variant that is ineffective at body temperature but functions at the lower temperatures experienced by cells of the extremities. This is an example of ....

(A) A wild type allele
(B) A lethal mutation
(C) A suppressor mutation
(D) A conditional mutation

200. A woman with mild symptoms of myotonic muscular dystrophy has a daughter with a very severe form of the disease. Which of the following statements is most correct?

(A) The daughter probably has a mutation with the same number of repeats as her mother
(B) The mother probably has a mutation with a larger number of repeats than the daughter
(C) This must be an X-linked dominant condition
(D) This is an example of anticipation

201. Which of these is NOT a mechanism by which a chromosome rearrangement could affect gene function?

(A) The breakpoint is in the middle of the coding region
(B) The breakpoint is between the coding region and the regulatory elements
(C) The new arrangement puts a gene near new regulatory elements
(D) None of the above

202. Which of the following would be an example of a germline mutation?

(A) Exposure to excessive UV radiation causes changes in the DNA of a skin cell, leading to basal cell carcinoma (skin cancer
(B) In a very early human embryo, a mistake in mitosis causes loss of a Y chromosome in one daughter cell. The resulting child is a fertile male
(C) A man with normal chromosomes has a hip X-ray without a protective shield. Ten months later, his wife gives birth to a child with a chromosomal deletion
(D) An embryo missing one copy of the third chromosome is miscarried very early in pregnancy

203. Mitochondrial DNA polymerase does not have an error checking mechanism such as that of nuclear DNA polymerase. This would be expected to lead to a higher rate of which type of mutation in mitochondrial DNA?

(A) Spontaneous mutations
(B) Induced mutations
(C) Deletions
(D) Chemical mutation

204. Based on the physiological adaptation theory, bacterial populations that had not been exposed to T1 infection.

(A) Should have a high level of resistance to T1 phage
(B) Should be completely susceptible to infection by T1 phage
(C) Should replicate more rapidly in the presence of T1 phage
(D) None of the above

205. The main difference between directed mutation theory and the random mutation theory is

(A) The chemical nature of the mutagens
(B) The effect of the mutation on the phenotype
(C) The cause of the mutation
(D) The heritability of the trait

206. In a haploid organism, lethal alleles would be expected to
(A) Have a high mutation frequency
(B) Have a high mutation rate
(C) Be caused only by induced mutations
(D) None of the above

207. In which situation would you expect the mutation frequency to increase over time?
(A) The deleterious effect of the mutation is balanced with the mutation rate
(B) A lethal allele is created at a high mutation rate
(C) A neutral allele is created at low mutation rate
(D) All of the above

208. Examples of environmental mutagens include
(A) Alkylating agents, which chemically modify bases
(B) Nucleotide base analogues
(C) Ionizing radiation
(D) All of these

209. A bacterial cell experiences a mutation as a result of exposure to nitrogen mustard and then divides several times to produce a total of eight cells. How many of the resulting cells would you expect to contain the mutation?
(A) One (B) Two
(C) Four (D) Eight

210. In the Ames test
(A) Mutagens cause lethal mutations, reducing the number of colonies
(B) Mutagens cause mutations that disrupt the ability of the cell to produce histidine
(C) Mutagens will cause an increase in the number of revertants
(D) Only mutagens that cause transitions can be identified

211. Which of the following forms of DNA repair does NOT require DNA polymerase?
(A) Direct DNA repair
(B) Base excision repair
(C) Nucleotide excision repair
(D) Mismatch repair

212. Why is alkyltransferase only able to be used once?
(A) The protein is hydrolyzed in the process of repairing the DNa
(B) The protein is chemically changed by the addition of a methyl group
(C) The protein becomes attached to the DNA strand
(D) All of these

213. How does recombinational repair differ from nucleotide excision repair (NER)?
(A) Unlike NER, recombinational repair is unable to repair damage caused by thymine dimers
(B) NER replaces the thymine dimmer while recombinational repair leaves it in place
(C) NER requires DNA polymerase and ligase function, while recombinational repair does not
(D) NER involves nick translation, which recombinational repair does not

214. Photo reactivation
(A) Repairs dimers in DNA using an end nuclease
(B) Uses light to activate repair enzymes
(C) Removes alkylating agents from bases
(D) Delaminates bases

215. Photo reactivation in DNA repair requires P. an end nuclease, Q. pyrimidine dimmers R activation by ultraviolet light, S. an enzyme that cleaves - C-C - bonds
(A) P,Q (B) Q,R
(C) R,S (D) Q,S

216. Enzyme DNA photolyase which is involved in direct repair of damaged DNA like thymine dimer
(A) Can work only in the presence of light
(B) Can work in dark
(C) Is independent of life
(D) None of the above

217. Nucleotide excision repair of DNA in *E. Coli*
(A) replaces both strands of the DNA in the damaged region
(B) Uses high energy phosphate bonds
(C) utilizes RNA polymerase to make a primer
(D) Requires uvrABC excinuclease

218. Xeroderma pigment sum is an autosomal recessive mutation that affects the mechanism for repair of ultraviolet light-induced thymine dimer formation in DNA.

Which of the following enzymes is most likely to be defective?
(A) DNA polymerase I
(B) DNA polymerase II
(C) DNA ligase
(D) enzymes involved in nucleotide excision repair

219. Which of the following abnormalities in DNA can be corrected by excision repair systems?
P. intrastrand pyrimidine dimmers
Q. alkylated bases
R. cytosine deamination
S. missing base-pairs
(A) P,Q (B) Q,R
(C) P,R (D) P,Q,R,S

220. How does nucleotide-excision repair differ from base-excision repair?
(A) Base-excision repair recognizes and removes single damaged bases, whereas nucleotide-excision repair is more general, recognizing many different kinds of lesions that distort the DNA molecule
(B) Nucleotide-excision repair reverses the chemical reaction that caused the lesion, whereas base excision repair removes the damaged bases and replaces them with normal ones
(C) Only the base is removed in base-excision repair, whereas the entire nucleotide is removed in nucleotide-excision repair
(D) Base-excision repair requires no protein components, and can occur by simple absorption of UV light, whereas nucleotide-excision repair requires several enzymes

221. the most common modification of DNA is methylation. In prokaryotes, this serves to P. identify damaged DNA for repair
P. Identify damaged DNA for repair
Q. Distinguish the strands after replication and Allow/disallow continued replication
R. Identify methylated foreign DNA for recombination into the genome
S. protect its *own* DNA but not Foreign DNA from restriction by endonuclease
(A) P, Q (B) Q, R
(C) R, S (D) P, Q, S

222. In *Escherichia coli,* which of the following repair systems is most error-prone?
(A) Photo activation
(B) excision repair
(C) recombination repair
(D) SOS repair

223. The SOS repair mechanism is activated by : ....
(A) 5-Bromouracil
(B) Acridine Orange
(C) Hydroxylamine
(D) Thymidine dimmers

224. Old and new strands of DNA in bacteria can be distinguished by
(A) DNA glycosylases
(B) Methylation patterns
(C) 3' → 5' exonuclease activity
(D) AP endonucleases

225. If a mutation results in multiple contiguous amino acid changes in a protein, the mutation is likely to be a
(A) Frameshift (B) Base analogue
(C) Transversion (D) Transition

226. Which statement is true?
(A) DNA damage and mutation are the same thing
(B) DNA damage is defined as permanent change in the nucleotide sequence
(C) Although DNA damage doesn't necessarily lead to mutation, when a mutation does occur it is always caused by DNA damage
(D) DNA damage can lead to mutation, but doesn't always

227. A tautomeric shift
(A) Is Typically Induced By High-Energy Radiation
(B) Occurs Only In Prokaryotes
(C) May Lead To A Transversion Mutation.
(D) May Produce a transition mutation

228. In mammals, DNA double-strand breaks may be repaired by
(A) Mismatch Repair
(B) Base Excision Repair

(C) Nucleotide Excision Repair
(D) Nonhomologous end-joining

229. Thymine dimers may be repaired by
(A) Photoreactivation In Humans
(B) Excision Repair In Humans
(C) Oxidative Damage
(D) Telomerase

230. In the first step of base excision repair
(A) A Phosphodiester Bond Is Broken
(B) Two Nicks Are Made Surrounding The Dna Damage
(C) A Break In Dna Is Sealed By Dna Ligase
(D) A Base is removed by the action of a DNA glycosylase

231. When the base cytosine undergoes deamination, it becomes
(A) Uracil (B) Adenine
(C) Guanine (D) Phosphorylated

232. In a physical map of a chromosome, distances are measured in units of
(A) Percent recombination
(B) RFLPs
(C) Centimorgans
(D) Base pairs

233. Which technique would be used to make a *genetic map* of the genome of Mendel's pea plants?
(A) Dihybrid cross with two linked genes
(B) Map-Based Sequencing
(C) Restriction Mapping
(D) Shotgun sequencing

234. What is the function of dideoxynucleotides in Sanger DNA sequencing?
(A) They act as primers for DNA polymerase.
(B) They act as primers for reverse transcriptase.
(C) They cut the sequenced DNA at specific sites.
(D) They stop synthesis at a specific site, so the base at that site can be determined.

235. The human genome contains about
(A) 300 billion base pairs
(B) 30 billion base pairs
(C) 3 billion base pairs
(D) 30 million base pairs

236. Based on the most recent analysis of the sequence of the human genome, which is closest to the estimated number of human genes?
(A) 3,000 (B) 30,000
(C) 100,000 (D) 500,000

237. Regarding the sequencing of genomes, which statement is true?
(A) Most of the human genome has been sequenced
(B) No eukaryotic genome has yet been sequenced
(C) Dna sequencing has revealed a complete lack of polycistronic transcription units in eukaryotic genomes
(D) Eukaryotic genomes display a higher gene density than do prokaryotic genomes

238. Analysis of eukaryotic genomes has revealed that
(A) Large segments of genomes have undergone duplication during evolution
(B) Genes are never found within the introns of other genes
(C) Genome size always correlates well with organism complexity
(D) Transposons are rarely found in genomes

239. Imagine a world in which proteins are made from 65 different amino acids. Assume that in this imaginary world the genetic code is nonoverlapping and utilizes the smallest possible size for a codon that accommodates all amino acids unambiguously. In this particular world, which of the following mutations in the coding region of a gene would not cause a frame-shift?
(A) An insertion of 2 nucleotides
(B) An insertion of 3 nucleotides
(C) A deletion of 8 nucleotides
(D) A deletion of 5 nucleotides

240. Which of the following is true in our world?
(A) Every amino acid is coded for by a single codon
(B) There are more amino acids than there are codons
(C) Every codon codes for an amino acid
(D) Each codon in a gene codes for no more than one single amino acid

241. What is the term which refers to a contiguous set of bacterial genes which are under coordinate control?
(A) Lysogen (B) Prototroph
(C) Operon (D) Allosteric

242. The *lac* operon
(A) Is under negative and positive control
(B) Is under positive control *only*
(C) Is normally expressed constitutively
(D) Is an example of tissue-specific expression

243. What is the function of cAMP in regulation of the *lac* operon?
(A) Activates a repressor protein
(B) Activates an activator protein
(C) Inactivates a repressor protein
(D) Inactivates an activator protein

244. One form of post-translational modification of a protein might be
(A) Removal of introns
(B) Shuffling of exons
(C) Removal or modification of amino acids.
(D) Removal of exons

245. In eukaryotes, actively transcribed chromatin is
(A) DNase insensitive
(B) In a highly condensed form
(C) Usually dnase sensitive
(D) Sometimes referred to as heterochromatin

246. The term "chromatin remodeling" refers to
(A) Alteration of chromatin structure in association with transcription
(B) A process that only bacteria perform since they contain no nucleus
(C) A process that is exclusively associated with transcription by RNA polymerase iii in eukaryotes
(D) Alteration in chromatin structure to facilitate loading and translation by ribosomes and thus, enhance gene expression

247. DNA methylation may be a significant mode of genetic regulation in eukaryotes. Methylation refers to
(A) Altering RNA polymerase activity by methylation of RNA polymerase
(B) Altering translational activity especially of highly methylated trnas.
(C) Alteration of DNA polymerase activity by addition of methyl groups to glycine residues.
(D) Addition of methyl groups to the cytosine of CG doublets

248. Those mutations that occur by environmental damage or mistakes during DNA replications are
(A) Acquired mutations
(B) Inherited mutations
(C) (A) and (B)
(D) Non-of them

249. .... is a genetic changes that occur in more than 1 percent of the population.
(A) Polymorphisms
(B) Monotheism
(C) Frameshift mutation
(D) Monomorphisms

250. Which of the following base changes in DNA is an example of a transition?
(A) A-to-C (B) G-to-C
(C) C-to-A (D) A-to-G

251. Eukaryotic influenza and Sindbis viruses have ....
(A) ssRNA genome (B) ssDNA genome
(C) dsRNA genome (D) dsDNA genome

252. Which of the following enzyme catalyze the RNA-dependent RNA synthesis in single stranded RNA chromosomes of ssRNA viruses?
(A) DNA polymerase
(B) Transposase
(C) Reverse transcriptase
(D) RNA replicase

253. Which of the following enzyme catalyze the RNA synthesis in single stranded RNA chromosomes of ssRNA viruses?
(A) RNA-dependent RNA polymerase
(B) DNA-dependent RNA polymerase
(C) RNA-dependent DNA polymerase
(D) DNA-dependent DNA polymerase

254. Why all RNA viruses encode a protein with RNA-dependent RNA polymerase activity in host cells?

(A) Host cells do not possess this enzyme
(B) This enzyme use duplex DNA of host cells for RNA replication
(C) Host cells provide specific enzymes for viral protein synthesis
(D) All of the above

255. Which of the following enzyme of most RNA bacteriophages has a molecular weight of ~210,000 and consists of four subunits?
(A) DNA polymerase
(B) Transposase
(C) Reverse transcriptase
(D) RNA replicase

256. Which of the following function is related to the RNA replcase enzyme?
(A) The product of the replicase gene encoded by the viral RNA and has the active site for replication
(B) One subunit of *Mw* 65,000 has the active site for replication
(C) Three subunits of the enzyme are host proteins involved in host-cell protein synthesis
(D) All of the above

257. Which of the following *E. coli* elongation factors that synthesized by three subunits of RNA replicase enzyme?
(A) Tu (B) Ts
(C) S1 (D) All of the above

258. What is the function of *E. coli* elongation factor Ts that synthesized by subunits of RNA replicase enzyme of RNA bacteriophage?
(A) Carry amino acyl–tRNAs to the ribosomes
(B) It is a part of 30S ribosomal subunit
(C) Encoded by the viral RNA
(D) It has the active site for replication

259. Which of the following *E. coli* elongation factor an integral part of the 30S ribosomal subunit?
(A) Tu protein (B) Ts protein
(C) S1 protein (D) GAG protein

260. Cells respond to their environment by taking signal from hormones or other external chemical signals called .... and the interaction of these extracellular chemical signals with receptors on the cell surface leads to the production of .... inside the cell.
(A) First messengers and second messengers
(B) Enzymes and Isozymes
(C) Factors and elements
(D) Acids and bases

261. What is the role of secondary messengers that produced by the interaction of extracellular chemical signals called first messengers, with receptors on the cell surface?
(A) Act as active site of enzymes
(B) Help in protoplasm assimilation
(C) Lead to adaptive changes in the cell interior
(D) Provide signals for replication and transcription

262. Which of the following act as second messenger into cells formed from ATP in a reaction catalyzed by adenylyl cyclase present on inner face of the plasma membrane?
(A) Enzymes (B) Cyclic AMP
(C) Harmones and other extracellular chemical
(D) Proteins

263. Which of the following enzyme associated with the inner face of the plasma membrane catalyze cyclic AMP formation from ATP?
(A) β-ketoacyl-CoA transferase
(B) Adenylyl cyclase
(C) Superoxide dismutase
(D) Succinyl-CoA transferase

264. What are the function of Cyclic AMP and cyclic GMP in virtually every cell outside the plant kingdom?
(A) Regulatory functions
(B) Signal transduction
(C) Acetylation
(D) Lipid synthesis

265. Which of the following regulatory nucleotide is produced in a bacterium which inhibits the synthesis of the rRNA and tRNA molecules needed for protein synthesis by preventing the unnecessary production of nucleic acids?
(A) ppApp (B) ppTpp
(C) ppCpp (D) ppGpp

266. Which of the biomolecules is not act as regulatory nucleotides?
(A) ppGpp
(B) Cyclic AMP
(C) *N*6-methyladenosine
(D) Cyclic GMP

267. Which is the active form vitamin B_2 that functions in electron transfer?
(A) NAD^+
(B) FAD
(C) Adenosine is 5'-deoxyadenosylcobalamin
(D) Ribofavin

268. Which functions in hydride transfers?
(A) NAD^+ (B) FAD
(C) Adenosine (D) Ribofavin

269. Which group is attached to the Coenzyme A (CoA) through a thioester linkage to the β-mercaptoethylamine moiety?
(A) Acyl group (B) Sulphur group
(C) Nicotinamide group
(D) Succinyl group

270. Find the group of positive and negative amino acids.
(A) Aspartic acid and glutamic acid
(B) Lysine, arginine and histidine
(C) Aspartic acid, arginine and histidine
(D) Lysine, arginine and glycine

271. Humans can only synthesise .... of the 20 standard amino acids called essential amino acids.
(A) 5 (B) 11
(C) 10 (D) 15

272. Most amino acids are synthesized from α-ketoacids and later transaminated from another amino acid by using the enzyme ....
(A) Aminotransfrase (B) Ketotransfrase
(C) Aldotransfrase (D) Aminoisomerase

273. Lysine is synthesized from aspartate via ....
(A) De novo synthesis pathway
(B) Glycolysis pathway
(C) β-oxidation pathway
(D) Diaminopimelate (DAP) pathway

274. Glycine is synthesized from serine using the enzyme…………………
(A) SHMT (B) ALDC
(C) MSUD (D) DHPS

275. Tyrosine is synthesized by the hydroxylation of ....
(A) Glutamine (B) Histidine
(C) Phenylalanine (D) Methionine

276. The biosynthesis of threonine is regulated via .... of homoserine.
(A) Competitive inhibition
(B) Allosteric regulation
(C) Noncompetitive inhibition
(D) Amphibolic Pathway

277. .... is a collection of databases dealing with genomes, biological pathways, diseases, drugs, and chemical substances.
(A) BRENDA (B) ExPASy
(C) IntEnz (D) KEGG

278. Which is not a protein database?
(A) SwissProt (B) PDBTM
(C) LMSD (D) SWISS-MODEL

279. .... is repressible negative regulation of gene expression.
(A) *Trp* operon (B) *Lac* operon
(C) *Phase* operon (D) *Sugar* operon

280. What is hydrophobic effect by proteins?
(A) The tendency of hydrophobic groups to come together
(B) The tendency of hydrophobic groups to go away from each other
(C) Influences protein architecture
(D) Strongly absorb ultraviolet light

281. Which of the following is not a aromatic amino acid?
(A) Tryptophan (B) Phenylalanine
(C) Methionine (D) Tyrosine

282. .... has an indole ring joined to a methylene ($-CH_2^-$) group.
(A) Tryptophan (B) Phenylalanine
(C) Methionine (D) Tyrosine

283. Why the aromatic rings of tryptophan and tyrosine strongly absorb ultraviolet light?
(A) They have imino groups
(B) These contain delocalized p electrons
(C) They have additional chiral center
(D) They contain high molecular weight

284. RNA replicase was isolated from .... that catalyzes the formation of an RNA complementary to the viral RNA?
(A) Qβ-infected *E. coli* cells
(B) MS2-infected *E. coli* cells
(C) R17-infected *E. coli* cells
(D) f2-infected *E. coli* cells

285. Which of the following enzyme reuire RNA as template to form an RNA complementary to the viral RNA?
(A) DNA polymerase (B) Transposase
(C) Reverse transcriptase
(D) RNA replicase

286. Which of the following activity is not found in RNA replicase?
(A) Polymerisation activity
(B) Proofreading endonuclease activity
(C) Mutation activity
(D) All of the above

287. Who proposed the unveiling of the structural and functional complexity of RNA in the 1960s and stated that this macromolecule might serve as both information carrier and catalyst?
(A) Howard Temin and David Baltimore
(B) Marianne Grunberg-Manago and Severo Ochoa,
(C) Carl Woese, Francis Crick, and Leslie Orgel
(D) Thomas Cech

288. Which of the following chemical solution generates adenine in yields of up to 0.5% after refluxing for a few days?
(A) Concentrated ammonium cyanide
(B) Concentrated HCN
(C) Concentrated ethidium bromide
(D) Concentrated silver nitrate

289. Which of the following nitrogen base may have the first and most abundant constituent to appear on Earth?
(A) Thymine (B) Uracil
(C) Guanine (D) Adenine

290. What is the main role of the self-splicing rRNA intron of *Tetrahymena*?
(A) Makes the transition to the modern DNA-based system
(B) Catalyzes the reversible attack of a 151uanosine residue on the 5' splice junction
(C) Provides energy for translocation of the protein
(D) Forms a closed polyhedral structures

291. What happen if the 5' splice site and the internal guide sequence are removed from the intron of *Tetrahymena* that catalyzes the reversible attack of a guanosine residue on the 5' splice junction?
(A) The rest of the intron can bind RNA strands paired with short oligonucleotides
(B) The rest of the intron can from hairpic structure
(C) The rest of the intron can bind Host DNA strands
(D) The rest of the intron can move from one position to another

292. SELEX (*s*ystematic *e*volution of *l*igands by *ex*ponential enrichment) is used
(A) To select a DNA species that binds tightly to ATP
(B) To select an RNA species that binds tightly to ATP
(C) To select a protein species that binds tightly to ATP
(D) To select a glucose species that binds tightly to ATP

293. Which of the following activity or technique is used to to generate RNA molecules that bind to amino acids, organic dyes, nucleotides, cyanocobalamin, and other molecules?
(A) SELEX (*s*ystematic *e*volution of *l*igands by *ex*ponential enrichment)
(B) NMR (nuclear magnetic resonance)
(C) SDS-PAGE (sodium Dodecyl sulfate-polyacrylamide gel electrophoresis)
(D) Microarray

294. What is called the combine binding of RNA molecule and ATP?
(A) Octamer
(B) Polymer
(C) Aptamer
(D) Nucleosome

295. The processes of DNA and RNA synthesis are similar in
(A) The general steps of initiation, elongation, and termination with 5′ to 3′ polarity
(B) Large, multicomponent initiation complexes
(C) Adherence to Watson-Crick base-pairing rules
(D) All of the above

296. RNA and DNA synthesis processes differ in several important ways including
(A) Ribonucleotides are used in RNA synthesis rather than deoxyribonucleotides
(B) U replaces T as the complementary base pair for A in RNA
(C) A primer is not involved in RNA synthesis
(D) All of the above

297. Which of the following process is not performed during RNA synthesis?
(A) Proofreading
(B) Polymerasation
(C) 5’ capping
(D) Poly (A) tail formation

298. The appearance of different sigma (σ) factors can be correlated temporally with which various programs of gene expression in prokaryotic systems?
(A) Bacteriophage development
(B) Sporulation
(C) Response to heat shock
(D) All of the above

299. Which of the following largest subunit of RNA polymerase II in eukarytic cells?
(A) RBP11 (B) RBP12
(C) RBP1 (D) RBP3

300. How the expression of many of the genes encoding DNA repair enzymes is induced?
(A) By high levels of DNA damage
(B) By high rate of DNA replication
(C) By high rate of mutation
(D) By high rate of protein modification

Answer-Key

Q	Ans	Q	Ans	Q	Ans	Q	Ans	Q	Ans	Q	Ans	Q	Ans	Q	Ans
1.	(A)	41.	(C)	81.	(A)	121.	(D)	161.	(B)	201.	(D)	241.	(C)	281.	(C)
2.	(C)	42.	(D)	82.	(A)	122.	(C)	162.	(A)	202.	(C)	242.	(A)	282.	(A)
3.	(D)	43.	(B)	83.	(C)	123.	(B)	163.	(B)	203.	(A)	243.	(B)	283.	(B)
4.	(A)	44.	(D)	84.	(D)	124.	(B)	164.	(B)	204.	(D)	244.	(C)	284.	(A)
5.	(B)	45.	(B)	85.	(B)	125.	(A)	165.	(A)	205.	(C)	245.	(C)	285.	(D)
6.	(C)	46.	(A)	86.	(A)	126.	(C)	166.	(A)	206.	(D)	246.	(A)	286.	(B)
7.	(A)	47.	(B)	87.	(C)	127.	(D)	167.	(B)	207.	(C)	247.	(D)	287.	(C)
8.	(A)	48.	(D)	88.	(B)	128.	(B)	168.	(C)	208.	(D)	248.	(A)	288.	(A)
9.	(D)	49.	(C)	89.	(A)	129.	(D)	169.	(B)	209.	(C)	249.	(A)	289.	(D)
10.	(B)	50.	(A)	90.	(A)	130.	(A)	170.	(C)	210.	(C)	250.	(D)	290.	(B)
11.	(A)	51.	(C)	91.	(D)	131.	(D)	171.	(D)	211.	(A)	251.	(A)	291.	(A)
12.	(C)	52.	(D)	92.	(D)	132.	(C)	172.	(D)	212.	(B)	252.	(D)	292.	(B)
13.	(D)	53.	(A)	93.	(B)	133.	(A)	173.	(B)	213.	(B)	253.	(A)	293.	(A)
14.	(C)	54.	(B)	94.	(C)	134.	(D)	174.	(B)	214.	(B)	254.	(A)	294.	(C)
15.	(A)	55.	(C)	95.	(D)	135.	(C)	175.	(C)	215.	(D)	255.	(D)	295.	(D)
16.	(C)	56.	(C)	96.	(B)	136.	(A)	176.	(A)	216.	(A)	256.	(D)	296.	(D)
17.	(A)	57.	(A)	97.	(B)	137.	(C)	177.	(D)	217.	(D)	257.	(D)	297.	(A)
18.	(D)	58.	(B)	98.	(D)	138.	(D)	178.	(C)	218.	(D)	258.	(A)	298.	(C)
19.	(C)	59.	(D)	99.	(A)	139.	(B)	179.	(B)	219.	(C)	259.	(C)	299.	(C)
20.	(B)	60.	(C)	100.	(C)	140.	(C)	180.	(B)	220.	(A)	260.	(A)	300.	(A)
21.	(A)	61.	(B)	101.	(C)	141.	(A)	181.	(C)	221.	(D)	261.	(C)		
22.	(C)	62.	(D)	102.	(A)	142.	(A)	182.	(D)	222.	(D)	262.	(B)		
23.	(A)	63.	(A)	103.	(D)	143.	(A)	183.	(C)	223.	(D)	263.	(B)		
24.	(C)	64.	(C)	104.	(B)	144.	(C)	184.	(B)	224.	(B)	264.	(A)		
25.	(D)	65.	(D)	105.	(D)	145.	(A)	185.	(A)	225.	(A)	265.	(D)		
26.	(C)	66.	(D)	106.	(D)	146.	(D)	186.	(A)	226.	(D)	266.	(C)		
27.	(B)	67.	(B)	107.	(A)	147.	(B)	187.	(A)	227.	(D)	267.	(B)		
28.	(C)	68.	(C)	108.	(D)	148.	(A)	188.	(B)	228.	(D)	268.	(A)		
29.	(A)	69.	(D)	109.	(A)	149.	(C)	189.	(B)	229.	(B)	269.	(A)		
30.	(B)	70.	(A)	110.	(C)	150.	(A)	190.	(D)	230.	(D)	270.	(C)		
31.	(D)	71.	(C)	111.	(C)	151.	(C)	191.	(A)	231.	(A)	271.	(B)		
32.	(A)	72.	(A)	112.	(B)	152.	(A)	192.	(D)	232.	(D)	272.	(A)		
33.	(B)	73.	(A)	113.	(D)	153.	(D)	193.	(C)	233.	(A)	273.	(D)		
34.	(B)	74.	(D)	114.	(C)	154.	(B)	194.	(D)	234.	(D)	274.	(A)		
35.	(D)	75.	(D)	115.	(D)	155.	(A)	195.	(B)	235.	(C)	275.	(C)		
36.	(B)	76.	(C)	116.	(D)	156.	(A)	196.	(A)	236.	(B)	276.	(B)		
37.	(D)	77.	(C)	117.	(A)	157.	(A)	197.	(D)	237.	(A)	277.	(D)		
38.	(A)	78.	(A)	118.	(D)	158.	(C)	198.	(B)	238.	(A)	278.	(C)		
39.	(C)	79.	(B)	119.	(A)	159.	(D)	199.	(D)	239.	(C)	279.	(A)		
40.	(B)	80.	(C)	120.	(B)	160.	(C)	200.	(D)	240.	(D)	280.	(A)		

6

Transcription and RNA Processing

DNA is transcribed into RNA, which is translated into protein. First, DNA converted into primary transcript that is processed into a mature RNA. Many mRNAs, especially in eukarytoes, are derived from pre mRNAs by splicing and other processing events. In prokaryotes, RNA polymerase is main enzyme which catalyzes the DNA dependent RNA synthesis (Transcription). RNA polymerase requires DNA as a template. RNA polymerase does not require a primer to initiate transcription. One RNA polymerase synthesizes all classes of RNA (mRNA, rRNA, tRNA). RNA polymerase holoenzyme is composed of two parts: a. Core ($\alpha2\beta\beta'$) and b. sigma factor (σ). RNA polymerase holoenzyme binds to the promoter using sigma factor ($\sigma70$) to form a closed complex; at this stage there is no unwinding of DNA. Promoter site has a sequence of 6 base pairs (TATAAT) in DNA is called as Pribnow box in prokaryotes. While in eukaryotes, promoter site contains 7 base pairs sequence (TATAAAA) in DNA called as TATA box or Hogness box. The polymerase-promoter complex undergoes the closed to open transition which is a melting or unwinding of about 12 bp. The initiating nucleotides can bind to the enzyme, as directed by their complementary nucleotides in the DNA template strand, and the enzyme will catalyze formation of a phosphodiester bond between them. This polymerase-DNA-RNA complex is referred to as the ternary complex. When transcription starts and enzyme now translocates to an new position on the template, sigma factor leaves the complex. RNA polymerase enzyme forms phosphodiester bonds between adjacent ribonucleotides. Sigma factor released and core enzyme move along the sense strand till it reaches termination site. In prokaryotes there are two ways in which transcription is terminated. In Rho-dependent termination, a protein factor called "Rho" is responsible for disturbing the complex composed of the template strand, RNA polymerase and RNA molecule. In Rho-independent termination, a loop forms at the end of the RNA molecule which causes it to detach itself. Termination in eukaryotes is more complicated, involving the addition of additional adenine nucleotides at the 3' of the RNA transcript (a process referred to as polyadenylation).

Inhibitors to block transcription

1. Rifampicin binds the β subunit of RNA polymerase to block initiation. The drug prevents addition of the 3rd or 4th nucleotide, hence the initiation process cannot be completed.
2. Streptomycin binds to the β subunit to inhibit chain elongation.
3. Heparin, a polyanion, binds to the β' subunit to prevent binding to DNA *in vitro*.

Different RNA polymerases in eukaryotes

RNA polymerase	Localization	Synthesizes	Effect of α-amanitin
RNA polymerase I	Nucleolus	Pre-rRNA	None
RNA polymerase II	Nucleoplasm	Pre-mRNA some snRNAs	Inhibited by low concentrations

RNA polymerase	Localization	Synthesizes	Effect of α-amanitin
RNA polymerase III	Nucleoplasm	Pre-tRNA, other small RNAs some snRNAs	Inhibited by high concentrations

Key notes

- François Jacob and Jacques Monod first hypothesized that DNA allow to code protein.
- Severo Ochoa won a Nobel Prize in Physiology or Medicine in 1959 for developing a process for synthesizing RNA *in vitro* with polynucleotide phosphorylase, which was useful for cracking the genetic code.
- In 1972, Walter Fiers became the first person to actually prove the existence of the terminating enzyme.
- Roger D. Kornberg won Nobel Prize in Chemistry (in 2006) "for his studies of the molecular basis of eukaryotic transcription".
- Gene expression involves first process called transcription in which a certain piece of DNA is converted into RNA by the enzyme RNA polymerase (DNA dependent RNA polymerase).
- During transcription, RNA polymerase identify a DNA sequence which produces a complementary and antiparallel RNA strand called primary transcript.
- The segment of DNA that transcribed into an RNA molecule is called a transcription unit and contains at least one gene.
- If the gene encodes a protein, the transcription produces messenger RNA (mRNA) by enzyme RNA polymerase II.
- Further, the mRNA serves as a template for the protein's synthesis during translation. Alternatively, the transcribed gene may encode for either non-coding RNA (such as microRNA), ribosomal RNA (rRNA), transfer RNA (tRNA), or other enzymatic RNA molecules called ribozymes.
- In retroviruses, mRNA synthesis from an RNA molecule called RNA replication by enzyme viral RNA replicase (RNA dependent RNA polymerase). Retroviral genome is a negative-sense single-stranded RNA (-ssRNA) serves as a template for a positive-sense single-stranded RNA (+ ssRNA). This is because the positive-sense strand contains the information needed to translate the viral proteins for viral replication.
- A DNA transcription unit contains a coding sequence which translated into the protein and other region is regulatory sequences which direct and regulate the synthesis of that protein.
- The regulatory sequence at upstream site before the coding sequence is called the five prime untranslated region (5' UTR) and the sequence at downstream site after the coding sequence is called the three prime untranslated region (3' UTR).
- Transcription always performed from the 5' end to the 3' end.
- RNA is complementary to antisense or template DNA strand.
- Non-template sense strand of DNA is called the coding strand, because its sequence is the same as the newly created RNA transcript (except for the substitution of uracil for thymine).
- Transcription has some proofreading mechanisms, but they are fewer and less effective than the controls for copying DNA; therefore, transcription has a lower copying fidelity than DNA replication.

Initiation of Transcription

- One strand of the DNA, the template strand (noncoding strand) is used as a template for mRNA synthesis.
- As transcription proceeds, RNA polymerase II pass through the template strand and uses base pairing complementarity with the DNA template to create an RNA copy.
- Although RNA polymerase pass through the template strand from 3' → 5', the coding (non-template) strand and newly formed RNA can also be used as reference points, so transcription can be described as occurring 5' → 3'.
- This produces an RNA molecule from 5' → 3', an exact copy of the coding strand (except that thymines are replaced with uracils, and the nucleotides are composed of a ribose (5-carbon) sugar where DNA has deoxyribose (one fewer oxygen atom) in its sugar-phosphate backbone).
- In eukaryotes, promoter DNA is fully double-stranded DNA sequence which participates in transcription initiation after binding of RNA polymerase II with one or more transcription factors and form an RNA polymerase-promoter "closed complex".
- RNA polymerase II unwinds approximately 14 base pairs of DNA to form an mRNA polymerase-promoter "open complex" (called an "open complex" or "transcription bubble" because the promoter DNA is partly unwound and single-stranded).
- RNA polymerase II identifies a transcription start site in the transcription bubble and binds to an initiating nucleotide triphosphate (NTP) and an extending NTP which complementary to the transcription start site sequence.
- In bacteria, RNA polymerase core enzyme consists of five subunits: 2 α subunits, 1 β subunit, 1 β' subunit, and 1 ω subunit.
- In bacteria, there is one general RNA transcription factor: sigma70 (σ_{70}).
- RNA polymerase core enzyme binds to the bacterial general transcription factor σ_{70} to form RNA polymerase holoenzyme and then binds to a promoter.
- In archaea and eukaryotes, RNA polymerase contains subunits homologous to each of the five RNA polymerase subunits in bacteria and also contains additional subunits.
- In archaea, there are three general transcription factors: TBP, TFB, and TFE.
- In eukaryotes, RNA polymerase II-dependent transcription involves six general transcription factors: TFIIA, TFIIB (an ortholog of archaeal TFB), TFIID (a multisubunit factor and an ortholog of archaeal TBP), TFIIE (an ortholog of archaeal TFE), TFIIF, and TFIIH.
- In archaea and eukaryotes, the RNA polymerase-promoter closed complex usually is referred to as the "preinitiation complex."
- Transcription initiation is regulated by activator and repressor proteins and in some cases, associated coactivators or corepressors which modulate formation and function of the transcription initiation complex.

Promoter clearance or escape

- During escape RNA polymerase from the promoter site, truncated RNA transcript released. This is called *abortive initiation* and is common for both eukaryotes and prokaryotes.
- Abortive initiation continues to occur until an RNA product of a threshold length of approximately 10 nucleotides is synthesized, at which point promoter escape occurs and a transcription elongation complex is formed.

- In bacteria, upon and following promoter clearance, the σ factor is released according to a stochastic model.
- In eukaryotes, at an RNA polymerase II-dependent promoter, upon promoter clearance, TFIIH phosphorylates serine 5 on the carboxy terminal domain of RNA polymerase II, leading to the recruitment of capping enzyme (CE).

Elongation

- As transcription proceeds, RNA polymerase pass through the template strand and uses base pairing complementarity with the DNA template to create an RNA copy.
- Although, RNA polymerase crosses the template strand from 3' → 5', the coding (non-template) strand and newly formed RNA (except that thymines are replaced with uracils), so transcription described as opposite direction occurring 5' → 3'.
- mRNA transcription can involve multiple RNA polymerases on a single DNA template and multiple rounds of transcription, so many mRNA molecules can be rapidly produced from a single copy of a gene.
- Elongation also involves proofreading mechanism that replaces incorrectly incorporated bases.

Termination

- Bacteria use two different strategies for transcription termination - *Rho-independent termination* and *Rho-dependent termination.*
- In Rho-independent transcription termination (intrinsic termination), RNA transcription stops when the newly synthesized RNA molecule forms a G-C-rich hairpin loop followed by a run of Us.
- When the hairpin forms, the mechanical stress breaks the weak rU-dA bonds, now filling the DNA–RNA hybrid.
- This pulls the poly-U transcript out of the active site of the RNA polymerase.
- In the "Rho-dependent" type of termination, a protein factor called "Rho" destabilizes the interaction between the template and the mRNA, thus releasing the newly synthesized mRNA from the elongation complex.
- Transcription termination in eukaryotes is less understood but involves cleavage of the new transcript followed by template-independent addition of adenines at its new 3' end, in a process called polyadenylation.

Transcription inhibitors

- Many antibiotics works as transcription inhibitors
- An antibacterial is rifampicin, which inhibits prokaryotic DNA transcription into mRNA by inhibiting DNA-dependent RNA polymerase by binding its beta-subunit.
- 8-Hydroxyquinoline is an antifungal transcription inhibitor.
- The effects of histone methylation also work to inhibit the action of transcription.
- α-Amanitin binds to the largest subunits of RNA polymerase II (RNAP II) and RNAP III, with RNAP II being the most sensitive. As a consequence, the incorporation of new ribonucleotides into the nascent RNA chains is blocked.

- Actinomycin D is generally thought to intercalate into DNA thereby preventing the progression of RNA polymerases, with RNAP I being the most sensitive.

Measurement and detection

- G-Less Cassette transcription assay: measures promoter strength.
- Run-off Transcription assay: identifies transcription start sites (TSS).
- Nuclear Run-on assay: measures the relative abundance of newly formed transcripts.
- RNase protection assay and ChIP-Chip of RNAP: detect active transcription sites.
- RT-PCR: measures the absolute abundance of total or nuclear RNA levels, which may however differ from transcription rates.
- DNA microarrays: measures the relative abundance of the global total or nuclear RNA levels; however, these may differ from transcription rates.
- In situ hybridization: detects the presence of a transcript.
- MS2 tagging: by incorporating RNA stem loops, these become incorporated into newly synthesized RNA. The stem loops can then be detected using a fusion of GFP and the MS2 coat protein, which has a high affinity, sequence-specific interaction with the MS2 stem loops.
- Northern blot: the traditional method, and until the advent of RNA-Seq, the most quantitative.
- RNA-Seq: applies next-generation sequencing techniques to sequence whole transcriptomes, which allows the measurement of relative abundance of RNA, as well as the detection of additional variations such as fusion genes, post-transcriptional edits and novel splice sites.

Reverse transcription

- Retroviruses (HIV contains ssRNA genome), have the ability to transcribe RNA into DNA called reverse transcription.
- The main enzyme responsible for reverse transcription is called reverse transcriptase (RNA dependent DNA polymerse).
- Reverse transcriptase synthesizes a complementary DNA strand (cDNA) to the viral RNA genome and form RNA-DNA hybris.
- The enzyme ribonuclease H then digests the RNA strand and reverse transcriptase synthesizes a complementary strand of DNA to form a double helix DNA structure ("cDNA").
- The cDNA is integrated into the host cell's genome by the enzyme integrase, which causes the host cell to generate viral proteins that reassemble into new viral particles.
- In HIV, the host cell undergoes programmed cell death or apoptosis of T cells.
- Some eukaryotic cells contain an enzyme with reverse transcription activity called telomerase.
- Telomerase is a reverse transcriptase that lengthens the ends of linear chromosomes. Telomerase carries an RNA template from which it synthesizes a repeating sequence of DNA (telomere) or "junk" DNA.

Post-transcriptional modification

- Post-transcriptional modification or Co-transcriptional modification is a process involves the conversion of primary transcript RNA into mature RNA.
- Post-transcriptional modification involves RNA splicing (remove of introns), 5' capping and 3' Poly(A) tailing.
- Primary RNA transcript contains both exons (coding sequences of the transcript) and introns (non-coding sequences of the primary RNA transcript).
- The pre-mRNA molecule undergoes three main modifications which are 5' capping, 3' polyadenylation, and RNA splicing, which occur in the cell nucleus before the RNA is translated.

5' cap processing

- Capping of the pre-mRNA involves the addition of 7-methylguanosine (m^7G) to the 5' end.
- The terminal 5' phosphate requires removal which is done with the aid of a phosphatase enzyme.
- The enzyme guanosyl transferase then catalyses the reaction which produces the diphosphate 5' end.
- The diphosphate 5' end then attacks the alpha phosphorus atom of a GTP molecule in order to add the guanine residue in a 5'5' triphosphate link.
- The enzyme (guanine-N^7-)-methyltransferase ("cap MTase") transfers a methyl group from S-adenosyl methionine to the guanine ring.
- This type of cap (m^7G) in position is called a cap 0 structure.

Cleavage and 3' polyadenylation processing

- The pre-mRNA processing at the 3' end of the RNA molecule involves cleavage of its 3' end and then the addition of about 250 adenine residues to form a poly(A) tail.
- The cleavage and adenylation reactions occur if a polyadenylation signal sequence (5'-AAUAAA-3') is located near the 3' end of the pre-mRNA molecule which is followed by another sequence, which is usually (5'-CA-3') and is the site of cleavage.
- A GU-rich sequence is also usually present further downstream on the pre-mRNA molecule. After the synthesis of the sequence elements, two multisubunit proteins called cleavage and polyadenylation specificity factor (CPSF) and cleavage stimulation factor (CStF) are transferred from RNA Polymerase II to the RNA molecule.
- The two factors bind to the sequence elements. A protein complex forms that contains additional cleavage factors and the enzyme Polyadenylate Polymerase (PAP).
- This complex cleaves the RNA between the polyadenylation sequence and the GU-rich sequence at the cleavage site marked by the (5'-CA-3') sequences.
- Poly(A) polymerase then adds about 200 adenine units to the new 3' end of the RNA molecule using ATP as a precursor.
- As the poly(A) tail is synthesised, it binds multiple copies of poly(A) binding protein, which protects the 3'end from ribonuclease digestion.

RNA Splicing

- RNA splicing is the process of removal of introns from pre-RNA that do not code for protein and the remaining exons connected to re-form a single continuous molecule.
- RNA splicing occurs after the complete synthesis and end-capping of the pre-mRNA.
- The splicing reaction is catalyzed by a large protein complex called the spliceosome assembled from proteins and small nuclear RNA molecules that recognize splice sites in the pre-mRNA sequence.
- Many pre-mRNAs including those encoding antibodies, can be spliced in multiple ways to produce different mature mRNAs that encode different protein sequences. This process is known as alternative splicing and allows production of a large variety of proteins from a limited amount of DNA.

Histone mRNA processing

- Histones H2A, H2B, H3 and H4 form the core of a nucleosome and thus are called core histones.
- Processing of core histones is done differently because typical histone mRNA lacks several features of other eukaryotic mRNAs, such as poly(A) tail and introns.
- Thus such mRNAs do not undergo splicing and their 3'-prime processing is done independent of most cleavage and polyadenylation factors.
- Core histone mRNAs have a special stem-loop structure at 3-prime end that is recognized by a stem–loop binding protein and a downstream sequence, called histone downstream element (HDE) that recruits U7 snRNA.
- Cleavage and polyadenylation specificity factor 73 cuts mRNA between stem-loop and HDE.
- Histone variants, such as H2A.Z or H3.3, however, have introns and are processed as normal mRNAs including splicing and polyadenylation, in mammals.

Multiple Choice Questions

1. Which of the following the first process of the expression of the information in a gene generally involves production of an RNA molecule from a DNA template.
(A) Transcription (B) Translation
(C) Replication (D) Mutation

2. What are the differences between strands of RNA and DNA?
(A) RNA has a hydrogen group at the 2' carbon of the ketopentose and uracil
(B) RNA has a hydroxyl group at the 2' carbon of the aldopentose and thymine
(C) RNA has a hydroxyl group at the 2' carbon of the aldopentose and uracil
(D) RNA has a hydrogen group at the 2' carbon of the aldopentose and thymine

3. Which of the following one only macro-molecule has a role in both the storage and transmission of information and in catalysis?
(A) DNA (B) RNA
(C) Protein (D) Enzymes

4. What are ribozymes?
(A) RNAs withot enzymatic activity
(B) A cluster of RNAs with nuclease activity
(C) A cluster of RNAs with Proofreading activity
(D) RNAs with enzymatic activity

5. Which of the macromolecule has both structural and catalytic roles in nucleic acids?
(A) DNA (B) RNA
(C) Protein (D) Enzymes

6. All RNA molecules except .... are derived from information permanently stored in DNA?
(A) RNA genomes of bacteria
(B) RNA genomes of Mycoplasma
(C) RNA genomes of certain viruses
(D) RNA genomes of fungus

7. What is called the conversion of the genetic information in a segment of double-stranded DNA into an RNA strand with a base sequence complementary to one of the DNA strands?
(A) Translation (B) Replication
(C) Transcription (D) Mutation

8. Which of the following is not a major kind of RNA in all organisms?
(A) mRNA (B) rRNA
(C) snRNA (D) tRNA

9. Which kind of RNA encodes the amino acid sequence of one or more polypeptides specified by a gene or set of genes?
(A) mRNA (B) rRNA
(C) snRNA (D) tRNA

10. Which kind of RNA read the information encoded in the mRNA and transfer the appropriate amino acid to a growing polypeptide chain during protein synthesis?
(A) mRNA (B) rRNA
(C) snRNA (D) tRNA

11. Which kind of RNA is constituents of ribosomes that synthesize proteins?
(A) mRNA (B) rRNA
(C) snRNA (D) tRNA

12. Which of the portions of the DNA genome are transcribed at any one time?
(A) Entire chromosome
(B) Only particular genes
(C) Groups of genes
(D) Both (B) and (C)

13. What is called the specific regulatory sequences that mark the beginning the DNA segments to be transcribed in the DNA template?
(A) Terminator sequence
(B) Promoter sequence
(C) Operator sequence
(D) Structural gene sequence

14. What is not similarity between transcription and replication of DNA?
(A) Fundamental chemical mechanism
(B) Polarity (direction of synthesis)
(C) Use of a template
(D) Require primer

15. What are the difference between transcription and replication of DNA?
(A) Fundamental chemical mechanism
(B) Polarity (direction of synthesis)
(C) Use of a template (D) Require primer

16. What is called the transcribed segments of only one DNA strand?
(A) Template (B) Gene
(C) Promoter (D) Primer

17. Which of the following enzyme synthesize RNA complementary to a DNA strand from ribonucleoside 5'-triphosphates?
(A) RNA-dependent RNA polymerase
(B) DNA-dependent DNA polymerase
(C) DNA-dependent RNA polymerase
(D) RNA-dependent DNA polymerase

18. Which of the following correct match between enzymes and biological processes?
I. RNA-dependent RNA polymerase = Translation
II. DNA-dependent DNA polymerase = Replication
III. DNA-dependent RNA polymerase = Transcription
IV. RNA-dependent DNA polymerase = Reverse transcription
(A) I, II and III (B) II and III
(C) II, III and IV (D) I, II, III and IV

19. Which of the following chemical is not used in transcription?
(A) DNA-dependent RNA polymerase
(B) Mg^{2+}
(C) NTPs
(D) dNTPs

20. Which of the following ribonucleoside 5'-triphosphate is not used in RNA transcription?
(A) ATP (B) TTP
(C) UTP (D) CTP

21. Which of the following divalent metal ion act as precursors of the nucleotide units of RNA during transcription?
(A) Mg^{2+} (B) Zn^{2+}
(C) Ca^{2+} (D) Mn^{2+}

22. Which of the following divalent ion is bind to the protein?
(A) Mg^{2+} (B) Zn^{2+}
(C) Ca^{2+} (D) Mn^{2+}

23. At which position in growing RNA strand, RNA polymerase adds new ribonucleotide units during transcription.
(A) 2'-hydroxyl group
(B) 5'-phosphate group
(C) 3'-hydroxyl group
(D) 2'- and 3'-hydroxyl group

24. In which direction, RNA is synthesized by RNA polymerase from DNA template?
(A) 5' to 3' direction
(B) 5' to 2' direction
(C) 2' to 5' direction
(D) 3' to 5' direction

25. What is the nature of the 3'-hydroxyl group in RNA strand that synthesized from DNA template?
(A) A Hydrophobic (B) A positive group
(C) A nucleophobic (D) A nucleophile

26. Which of the following phosphate group of the incoming ribonucleoside triphosphate is involved in the formation of phosphodiester bond during transcription?
(A) β phosphate (B) α phosphate
(C) γ phosphate (D) λ phosphate

27. Which of the phosphate derivative is released after making phosphodiester bond in RNA strand with attacking α phosphate of the incoming ribonucleoside triphosphate?
(A) Hydrogen phosphate
(B) Deoxyribo phosphate
(C) Pyrophosphate
(D) Glucophosphate

28. What is pyrophosphate that is released after making phosphodiester bond in RNA strand with attacking α phosphate of the incoming ribonucleoside triphosphate?
(A) Combination of α phosphate and β phosphate
(B) Combination of α phosphate, β phosphate and γ phosphate
(C) Combination of β phosphate and γ phosphate
(D) Combination of α phosphate and γ phosphate

29. Which of the following strand of DNA serve as template for RNA synthesis?
(A) Both strands
(B) Only one strand
(C) None of the DNA strand
(D) Short DNA segment serves as template

30. What is direction of template DNA that antiparallel to the new RNA strand?
(A) 5' to 3' direction (B) 5' to 2' direction
(C) 2' to 5' direction (D) 3' to 5' direction

31. Which of the following overall reaction show adding ribonucleotide units to the 3'-hydroxyl end to build RNA in the 5' to 3' direction and and releasing pyrophosphate?
(A) $(NMP)_n + NTP \rightarrow (NMP)_{n+1} + PPi$
RNA Lengthened RNA
(B) $(NMP)_n + NTP \rightarrow (NMP)_{n+1} + PPi$
DNA Lengthened RNA
(C) $(NMP)_n + dNTP \rightarrow (NMP)_{n+1} + PPi$
RNA Lengthened RNA
(D) $(NMP)_n + NTP \rightarrow (NMP)_{n+1} + PPi$
RNA Lengthened DNA

32. Which type of base-pairing selected by each nucleotide in the newly formed RNA from template DNA?
(A) Chargaff base-pairing interactions
(B) Franklin-Wilkins base-pairing interactions
(C) Measelson-Stahl base-pairing interactions
(D) Watson-Crick base-pairing interactions

33. Which of the following base residue is inserted in the RNA to pair with adenine (A) residues in the DNA template?
(A) Thymine (T) (B) Guanine (G)
(C) Uracil (U) (D) Cytosine (C)

34. Which of the following base pairing is correct in RNA?
(A) A=T (B) C≡G
(C) A=U (D) No base pairing

35. What is the role of base-pair geometry in DNA or RNA synthesis?
(A) In base selection
(B) In base deletion
(C) In hydrogen bonding
(D) In glycosidic bonding

36. What does not require to RNA polymerase, Unlike DNA polymerase, to initiate the synthesis of RNA?
(A) Template DNA (B) NTPs
(C) Mg^{2+} (D) Primer

37. RNA synthesis is initiated when RNA polymerase binds at specific DNA sequences called ....?
(A) Promoter
(B) Operator
(C) Primer
(D) Regulator

38. Promoter sequence is located on the .... for RNA synthesis?
(A) 5' region of template DNA
(B) 3' region of RNA primer
(C) 3' region of template DNA
(D) 3' region of RNA primer

39. Which of the group in a newly formed RNA molecule is not cleaved to release PPi but instead remains intact throughout the transcription process?
(A) 3'-hydroxyl group of the first residue
(B) 5'-triphosphate group of the first residue
(C) 5'-triphosphate group of the last residue
(D) 5'-triphosphate group of the ninth residue

40. What is the size a short hybrid RNA-DNA double helix that formed by joining the growing end of the new RNA strand base-pairs temporarily with the DNA template during the elongation phase of transcription,?
(A) 18 bp long
(B) 100 bp long
(C) 8 bp long
(D) Less than 8 bp long

41. Which of the following structure does RNA polymerase form after unwinding of DNA to synthesize an RNA strand complementary to one of the DNA strands?
(A) Holliday junction
(B) Transcription bubble
(C) Klenow fragment
(D) Single strand nick

42. What is the size of unwound DNA during transcription for forming transcription bubble in *E. coli*?
(A) About 8 bp
(B) About 50 bp
(C) About 17 bp
(D) About 3.2 bp

43. Which of the following structure is 8 bp in size that formed by *E. coli* RNA polymerase to synthesize an RNA strand complementary to one of the DNA strands through unwinding the DNA duplex over a short distance?
(A) RNA-DNA hybrid
(B) Transcription bubble
(C) Klenow fragment
(D) Single strand nick

44. What is the rate of elongation of a transcript by *E. coli* RNA polymerase?
(A) 15 to 19 nucleotides/s
(B) 5 to 9 nucleotides/s
(C) 500 to 900 nucleotides/s
(D) 50 to 90 nucleotides/s

45. Which of the following participate to restrict DNA strand rotation that require for movement of a transcription bubble?
(A) DNA-binding proteins
(B) Hydrogen bonding between bases
(C) Minor and major grooves
(D) Topoisomerases

46. What is the result occurs after movement of a transcription bubble that requires considerable strand rotation of the nucleic acid molecules.
(A) It generates waves of positive supercoils behind of the transcription bubble
(B) It generates waves of only negative supercoils ahead of the transcription bubble
(C) It generates waves of positive supercoils ahead of the transcription bubble and negative supercoils behind
(D) It generates waves of negative supercoils behind of the transcription bubble and negative supercoils also in same direction

47. How relived the topological problems caused by transcription in which a moving RNA polymerase generates waves of positive supercoils ahead of the transcription bubble and negative supercoils behind?
(A) DNA-binding proteins
(B) The action of topoisomerases
(C) Breaking of hydrogen bonds between bases
(D) Winding of DNA template

48. What is called the DNA strand that is used for RNA synthesis?
(A) Template strand (B) Coding strand
(C) Sense strand (D) Nontemplate strand

49. What is called the DNA strand that is identical in base sequence to the RNA transcribed from the gene with U in the RNA in place of T in the DNA?
(A) Template strand
(B) Noncoding strand
(C) Antisense strand
(D) Nontemplate strand

50. What is the main difference between DNA template and RNA transcript?
(A) T in the RNA in place of U in the DNA
(B) U in the Both RNA and DNA
(C) U bind to A In RNA
(D) U in the RNA in place of T in the DNA

51. Which of the following enzyme play a important role in transcription in *E. coli*?
(A) DNA-dependent DNA polymerase
(B) RNA-dependent RNA polymerase
(C) RNA-dependent DNA polymerase
(D) DNA-dependent RNA polymerase

52. Which of the following subunits present in the core complex of DNA-dependent RNA polymerase of *E. coli*?
(A) α_2 (B) ββ'
(C) ω (D) All of the above

53. A ribozyme made up of .... is a segment of mRNA in the ciliated protozoan *Tetrahymena thermophila*?
(A) Only intron (B) Only exon
(C) Both (A) and (B) (D) None of above

54. Which of the following drug is used to treat African sleeping sickness (African trypanosomiasis) that caused by parasitic protists (trypanosomes) in which lack pathways for de novo nucleotide biosynthesis?
(A) Trimethoprim (B) Fluorouracil
(C) Allopurinol (D) Aminopterin

55. Which of the following most common σ subunit present in the RNA polymerase holoenzyme with core complex in *E. coli*?
(A) σ^{70} (B) σ^{32}
(C) σ^{60} (D) $\sigma^{18.2}$

56. Which of the following activity lack in RNA polymerases?
(A) 3' to 5' exonuclease activity
(B) 5' to 3' exonuclease activity
(C) 5' to 3' polymerase activity
(D) All of the above

57. What is the error rate for transcription that is higher than chromosomal DNA replication due to lack proofreading activity in RNA polymerases?
(A) Approximately ten error for every 104 to 105 ribonucleotides
(B) Approximately one error for every 100 to 200 ribonucleotides
(C) Approximately one error for every 10 to 50 ribonucleotides
(D) Approximately one error for every 10^5 to 10^9 ribonucleotides

58. Why the error rate for transcription is higher than chromosomal DNA replication
(A) Due to lack 5' to 3' exonuclease activity in RNA polymerases
(B) Due to inactivation of RNA polymerases
(C) Due to lack polymerase activity in RNA polymerases
(D) Due to lack proofreading activity in RNA polymerases

59. How can remove mismatched nucleotides from the 3' end of a transcript when a mispaired base is added during transcription?
(A) By direct reversal of the polymerase reaction
(B) By 5' to 3' exonuclease activity in RNA polymerases
(C) By polymerase activity in RNA polymerases
(D) By proofreading activity in RNA polymerases

60. Which of the following specific sequence in DNA to direct the transcription of adjacent segments of DNA (genes) and used to bind for initiation of RNA synthesis RNA polymerase?
(A) Promoter (B) Operator
(C) Primer (D) Regulator

61. In *E. coli,* RNA polymerase binding occurs within a region stretching from about .... before the transcription start site to .... beyond it.
(A) About -70 bp and +30 bp
(B) About -10 bp and +30 bp
(C) About +10 bp and -35 bp
(D) About -70 bp and -30 bp

62. Which of the following region extends between positions -70 and +30 and recognized by an RNA polymerase holoenzyme containing σ^{70}?
(A) Promoter region (B) Operator region
(C) Primer region (D) Regulator region

63. Which of the following enzyme recognize region extends between positions -70 and +30 during transcription in *E. coli*?
(A) RNA polymerase core enzyme
(B) RNA-dependent RNA polymerase
(C) RNA-dependent DNA polymerase
(D) RNA polymerase holoenzyme containing σ^{70}

64. At which of the following position similarities but not identical have revealed in two short sequences in the most common class of bacterial promoters found after analysis and comparisons?
(A) About -70bp and +30 bp
(B) About -10bp and +35bp
(C) About -10bp and -35bp
(D) About -70bp and -30bp

65. Why important about -10bp and -35bp sequence positions that have similarities but not identical in the most common class of bacterial promoters found after analysis and comparisons?
(A) These are important interaction sites for the ρ factor
(B) These are important interaction sites for the σ^{70} subunit
(C) These are important termination sites
(D) These are important binding sites for ribosome

66. Which of the following nucleotide sequences are common at -10 and -35 position of promoter in *E. coli oriC,* respectively?
(A) (5')TATAAT(3') and (5')TATAAT(3')
(B) (5')TTGACA(3') and (5')TATAAT(3')
(C) (5')TATAAT(3') and (5')TTGACA(3')
(D) (5')TTGACA(3') and (5')TTGACA(3')

67. What are called the nucleotide sequences that are common at -10 and -35 position of promoter in *E. coli oriC*?
(A) Consensus sequence
(B) Jumping sequence
(C) Terminator sequence
(D) Transposon sequence

68. Which of the following AT-rich recognition element occurs between positions -40 and -60 in the promoters of certain highly expressed genes in *E. coli*?
(A) Downstream elements
(B) Upstream elements
(C)Transcription factors
(D) Downstream promoter elements

69. What is the main feature of upstream promoter element occurs between positions -40 and -60 in the promoters of certain highly expressed genes in *E. coli*?
(A) GC-rich recognition element
(B) AU -rich recognition element
(C) Both GC and AT-rich recognition element
(D) AT-rich recognition element

70. Which of the subunit of RNA polymerase bind to AT-rich upstream element that occurs between positions -40 and -60 in the promoters of certain highly expressed genes in *E. coli*?
(A) β (B) α (C) β' (D) δ

71. Which measures do determine the efficiency with which an RNA polymerase binds to a promoter and initiates transcription in *E. coli*?
(A) *E. coli-oriC* consensus sequences
(B) Spacing between consensus sequences
(C) Distance of consensus sequences from the transcription start site
(D) All of the above

72. .... affect the function of a given promoter often involve a base pair in these regions and .... in the consensus sequence also affect the efficiency of RNA polymerase binding and transcription initiation.
(A) Genome complexity and C-values
(B) Chromosome number and variations
(C) Mutations and variations
(D) Repressors and Terminators

73. The pathway of transcription initiation consists of two major parts namely,
i ....
ii ....
(A) Binding and initiation
(B) Termination and elongation
(C) Initiation and elongation
(D) Binding and initiation

74. What is a closed complex that formed after binding of RNA polymerase to promoter?
(A) Bound DNA is intact
(B) Bound DNA is intact and partially unwound near the -10 sequence
(C) Totally unbound DNA
(D) bound DNA is intact and partially unwound near the -30 sequence

75. What is promoter clearance?
(A) RNA polymerase binds to the promoter forminga closed complex and an open complex
(B) A conformational change in DNA
(C) Movement of the transcription complex away from the promoter
(D) Dissociation of σ subunit from promoter

76. Which of the following step is not involved in transcription process?
(A) RNA polymerase binds to the promoter
(B) ATP is hydrolyzed by σ protein during the termination process
(C) Dissociation of σ subunit from promoter
(D) Movement of the transcription complex away from the promoter

77. What are the ribozymes?
(A) RNAs with exonuclease activity
(B) Ribosomes with enzymatic activity
(C) RNAs without enzymatic activity
(D) RNAs with enzymatic activity

78. What is called a newly synthesized RNA molecule from DNA template through RNA polymerase?
(A) Primary transcript
(B) Progressive transcript
(C) Repetitive transcript
(D) Transposable transcript

79. The sequences for at least how many genes are contained in the primary transcript for a eukaryotic mRNA that encode the polypeptide?
(A) Two genes
(B) One gene
(C) Four genes
(D) Many genes

80. What is called the noncoding tracts that break up the coding region of the transcript?
(A) Exons (B) Cistrons
(C) Introns (D) Recons

81. What is called the coding region that breaked up by noncoding tracts of the transcript?
(A) Exons (B) Cistrons
(C) Introns (D) Recons

82. What is called the process of removing of introns from the primary transcript and the exons are joined to form a continuous sequence that specifies a functional polypeptide.
(A) Trans-splicing (B) RNA splicing
(C) Remodeling (D) Proof reading

83. What are the characteristics of modification of eukaryotic mRNAs?
(A) 3' cap and 5' poly(A) "tail
(B) Methylation of adenine residues in mRNAs
(C) 5' cap and 3' poly(A) "tail
(D) Removing of exons from premRNAs transcript

84. How many adenine residues are added to create a poly(A) "tail" at 3' end of eukaryotic mRNAs?
(A) 50 to 100 adenine residues
(B) 80 to 250 adenine residues
(C) 9 to 11 adenine residues
(D) 1000 to 2000 adenine residues

85. Unusual bases are found in only ....
(A) tRNAs (B) mRNAs
(C) rRNAs (D) miRNAs

86. Which of the following type of eukaryotic RNA has 5' cap and 3' poly 'A' tail?
(A) tRNA (B) mRNA
(C) rRNA (D) miRNA

87. Which of the following residue is capped at 5' end of most eukaryotic mRNAs and linked to the 5'-terminal residue of the mRNA through an unusual 5', 5'-triphosphate linkage?
(A) 7-methylguanosine
(B) 5-bromo uracil
(C) 5-Hydroxymethylcytosine
(D) 2-deoxyuridine

88. By which linkage a residue of 7-methylguanosine at 5' cap on most eukaryotic mRNAs linked to the 5'-terminal residue of the mRNA?
(A) Phosphodieter bond
(B) 5',5'-triphosphate linkage
(C) Hydrogen bonds
(D) Covalent bond

89. What is the function of 5' capping of a residue of 7-methylguanosine on most eukaryotic mRNAs linked to the 5'-terminal residue of the mRNA through an unusual 5', 5'-triphosphate linkage?
(A) Removing of exons from premRNAs transcript
(B) Methylation of adenine residues in mRNAs
(C) Protection of mRNA from ribonucleases
(D) blocking the activity of the polymerase

90. 5' cap of most eukaryotic mRNAs binds to a specific cap binding complex of proteins and participates in binding of the mRNA to the ribosome to initiate ....
(A) Transcription
(B) Translation
(C) Replication
(D) RNA splicing

91. Which of the molecule help in formation of the 5' cap at the 5' end of the transcript in eukaryotes?
(A) ATP (B) CTP
(C) UTP (D) GTP

92. AT which position on guanine is methylated to form the 5' cap by condensation of a molecule of GTP with the triphosphate at the 5' end of the transcript?
(A) N-9 (B) N-7
(C) N-1 (D) N-3

93. Methyl group is derived from .... to form the Methylated 5' cap by condensation of a molecule of GTP with the triphosphate at the 5' end of the transcript in eukaryotes?
(A) *S*-adenosylmethionine
(B) 5-Methylcytosine
(C) α-Methyldopa
(D) Methylhistidine

94. When occur all the 5' capping reactions after methylation by condensation of a molecule of GTP with the triphosphate at the 5' end of the transcript in eukaryotes?
(A) RNA catalyzes the splicing of introns
(B) After blocking the activity of the polymerase
(C) After adding the first 20 to 30 nucleotides of the transcript
(D) After forming a transcription "bubble.

95. Which of the following enzyme is involved in the 5' capping of the transcript in eukaryotes?
(A) Ligase
(B) Ribonuclease
(C) Isomerase
(D) RNA polymerase II CTD

96. Who independently discovered that many genes for polypeptides in eukaryotes are interrupted by noncoding sequences (introns) in 1977?
(A) Phillip Sharp and Richard Roberts
(B) Stanley Cohen and Francis S. Collins
(C) Gerty and Carl Cori
(D) Paul Berg and W.J. Beal

97. Genes encoding which proteins do not contain introns in eukaryotes?
(A) Trypsin
(B) Histones
(C) Hemoglobins
(D) Myoglobins

98. What is the size of most exons in genes of eukaryotic mRNAs?
(A) 10 to 20 nucleotides
(B) 30 to 60 nucleotides
(C) Less than 1,000 nucleotides
(D) 50 to 20,000 nucleotides

99. What is the size range of exons in genes of eukaryotic mRNAs?
(A) 80 to 250 nucleotides
(B) 30 to 60 nucleotides
(C) 100 to 200 nucleotides
(D) 50 to 20,000 nucleotides

100. Mostly how many amino acids are encoded by eukaryotic gene that contains exons in the 100 to 200 nucleotide size range and introns in size from 50 to 20,000 nucleotides?
(A) less than 1,000 amino acids
(B) 30 to 60 amino acids
(C) 100 to 200 amino acids
(D) 50 to 20,000 amino acids

101. How many classes of introns found in eukaryotic cells?
(A) Four classes (B) Three classes
(C) Two classes (D) Ten classes

102. How occurs self-splicing of introns in eukaryotic cells?
(A) by RNA-protein complexes
(B) By spliceosome
(C) No protein enzymes are involved in splicing
(D) Only two enzymes help in self splicing

103. In which process not require a high energy cofactor ATP?
(A) Protein synthesis
(B) Formation of phosphodiester linkage
(C) DNA replication
(D) Intron splicing

104. Which of the following reaction is involved in the intron splicing mechanisms in both group I and group II of introns?
(A) Pentose phosphate pathway
(B) Trans-esterification reaction
(C) Homopolymer tailing
(D) Protein sorting

105. Which of the following nucleoside is required as a nucleotide cofactor for the group I splicing reaction but the cofactor is not used as a source of energy?
(A) Adenine nucleoside
(B) Cytosine nucleoside
(C) Uracil nucleoside
(D) Guanine nucleoside

106. Which of the following is used as a nucleophile in the first step of the splicing pathway of the group I intron splicing reaction?

(A) 3'-hydroxyl group of guanosine
(B) 2'-hydroxyl group of adenine
(C) 3'-hydroxyl group of guanosine
(D) 2'-hydroxyl group of adenin

107. Which of the following is used as a nucleophile in the first step of the splicing pathway of the group II intron splicing reaction?

(A) 3'-hydroxyl group of guanosine
(B) 2'-hydroxyl group of adenine
(C) 3'-hydroxyl group of guanosine
(D) 2'-hydroxyl group of adenin

108. Who first revealed self-splicing of introns in 1982 in studies of the splicing mechanism of the group I rRNA intron from the ciliated protozoan *Tetrahymena thermophila* conducted?

(A) Phillip Sharp and Richard Roberts
(B) Stanley Cohen and Francis S. Collins
(C) Gerty and Carl Cori
(D) Thomas Cech and colleagues

109. In which organism first revealed self-splicing of introns in 1982 in studies of the splicing mechanism of the group I rRNA intron by Thomas Cech and colleagues?

(A) Backer yeast (*Saccharomyces cerevisiae)*
(B) Bacterium *Nocardia coralline*
(C) Ciliated protozoan (*Tetrahymena thermophila)*
(D) Bacterium *Escherichia coli*

110. Which of the following enzyme is used in the isolation of ciliated protozoan (*Tetrahymena thermophila)* DNA (including the intron) *in vitro* for revealing the self- splicing mechanism of the group I rRNA intron?

(A) Purified bacterial RNA polymerase
(B) Purified eukaryotic DNA polymerase
(C) Purified eukaryotic RNA polymerase
(D) Purified bacterial DNA polymerase

111. What is called introns that are not self-spliced found in nuclear mRNA primary transcripts?

(A) Group I rRNA intron
(B) Spliceosomal introns
(C) Group II rRNA intron
(D) All of the above

112. What is called the large protein complex that catalyzes the removal of introns that are not self-spliced found in nuclear mRNA primary transcripts?

(A) Nucleosome (B) Spherosome
(C) Spliceosome (D) Peroxisome

113. What is called specialized RNA-protein complexes that make up spliceosome for catalyzing the removal of introns that are not self-spliced found in nuclear mRNA primary transcripts?

(A) Small nuclear ribonucleoproteins
(B) miRNA proteins
(C) hnRNA proteins
(D) All of the above

114. How many nucleotides contained in small nuclear RNAs (snRNAs) that make up spliceosome for catalyzing the removal of introns?

(A) less than 1,000 nucleotides
(B) 30 to 60 nucleotides
(C) 100 to 200 nucleotides
(D) 50 to 20,000 nucleotides

115. Which types of five small nuclear RNAs (snRNAs) involved in splicing reactions are generally found in eukaryotic nuclei?

(A) A, T, G, C and U
(B) TFIIA, TFIIB, TFIIE, TFIIF and TFIIH
(C) U1, U2, U4, U5, and U6
(D) ATP, GTP, CTP, UTP and TTP

116. What are dinucleotide sequences of spliceosomal introns generally found at the 5' and 3' ends, respectively and these sequences mark the sites where splicing occurs.

(A) AT and GC
(B) GU and AG
(C) AU and GC
(D) GG and AU

117. What are U1, U2, U4, U5, and U6 that involved in intron removing reactions are generally found in abundance in eukaryotic nuclei?

(A) small nuclear RNAs (snRNAs)
(B) nuclear mRNA introns
(C) Heterogenous RNA (hnRNA)
(D) Types of uracil nucleotide

118. The U1 snRNA contains a sequence complementary to sequences near .... of nuclear mRNA introns (Fig. 26–16à), and the U1 snRNP binds to this region in the primary transcript.

(A) 3' splice site
(B) Both 3' and 5' splice site
(C) 5' splice site
(D) None splice site

119. How many proteins contributed by the snRNPs together with five RNAs to the spliceosome?

(A) About 50 proteins
(B) About 30 proteins
(C) More than 100 proteins
(D) 50 to 20,000 proteins

120. Which of the following energy molecule is required for assembly of the spliceosome?

(A) GTP (B) CTP
(C) UTP (D) ATP

121. Which of the following does not require ATP?

(A) Assembly of the spliceosome
(B) Protein synthesis
(C) Formation of phosphodiester linkage
(D) RNA cleavage-ligation reactions

122. What is a open complex that formed after binding of RNA polymerase to promoter?

(A) Bound DNA is intact
(B) Bound DNA is intact and partially unwound near the -10 sequence
(C) Totally unbound DNA
(D) bound DNA is intact and partially unwound near the -30 sequence

123. Which of the following introns requires ATP and an endonuclease in splicing reaction?

(A) Introns of snRNAs
(B) Introns of rRNAs
(C) Introns of tRNAs
(D) Introns of mRNAs

124. How work the splicing endonuclease?

(A) Cleaves the phosphodiester bonds at both ends of the intron
(B) Two exons are joined
(C) Two introns are joined
(D) Cleaves the glycosidic bonds at both ends of the intron

125. How many adinine residues found at the 3' end of most eukaryotic mRNAs to make up the poly(A) tail.

(A) 80 to 250 nucleotides
(B) 30 to 60 nucleotides
(C) 100 to 200 nucleotides
(D) 50 to 20,000 nucleotides

126. The 3' end of most eukaryotic mRNAs contains poly(A) tail that serves as a binding site for one or more specific proteins. What is the role of The poly(A) tail and its associated proteins?

(A) Assembly of the spliceosome
(B) Intron splicing
(C) Protect mRNA from enzymatic destruction
(D) Methylation of adenine residues in mRNAs

127. Which of the following enzyme cleave transcript at the poly(A) addition site that again associated with the CTD of RNA polymerase II?

(A) Endonuclease
(B) Exonuclease
(C) RNA polymerase I
(D) Polyadenylate polymerase

128. Cleave at the poly(A) addition site in the mRNA by an endonuclease component of a large enzyme complex site where is marked by two sequence elements. Which of the following sequence are highly conserved at their position?

(A) The sequence of G and U residues on upstream and (5') AAUAAA (3') on downstream
(B) (5') TTGACA (3') on upstream and (5') TATAAT (3') on downstream
(C) (5') AAUAAA (3') on upstream and sequence rich in G and U residues on downstream
(D) (5') TATAAT (3') on upstream and (5') TTGACA (3') on downstream

129. The mRNA site where cleavage occurs is marked by two sequence elements which contain a highly conserved sequence (5')AAUAAA(3') .... on the 5' side (upstream) of the cleavage site and a less well-defined sequence rich in G and U

residues .... downstream of the cleavage site.
(A) 10 to 30 nucleotides and 20 to 40 nucleotides
(B) 100 to 300 nucleotides and 20 to 40 nucleotides
(C) 10 to 30 nucleotides and 200 to 400 nucleotides
(D) 1 to 3 nucleotides and 2 to 4 nucleotides

130. Cleavage at the poly(A) addition site in the mRNA by an endonuclease component of a large enzyme complex site where generates the free 3'-hydroxyl group that defines the end of the mRNA, to which A residues are immediately added by ....
(A) RNA polymerase II
(B) Endonuclease
(C) Polyadenylate polymerase
(D) Ribonuclease

131. Which of the following reaction is catalyzed by enzyme polyadenylate polymerase that adds 'A' residues at the free 3'-hydroxyl group that defines the end of the mRNA generated by cleavage at the poly(A) addition site by an endonuclease component of a large enzyme complex?
(A) RNA + nATP →RNA–(ADP)n + nPPi
(B) RNA + nATP →RNA–(AMP)n + nPPi
(C) RNA + nATP →RNA–(AMP)n
(D) RNA + nAMP →RNA–(nATP)+ nPPi

132. Which of the following enzyme does not require a template but does require the cleaved mRNA as a primer?
(A) RNA polymerase
(B) DNA polymerase
(C) Reverse transcriptase
(D) Polyadenylate polymerase

133. What happen in RNA editing?
(A) Modification in polypeptide- coding region of the mRNA
(B) Help in RNA splicing
(C) Termination of primary transcript
(D) Protect mRNA from enzymatic destruction

134. Which of the following is not involved in RNA editing?
(A) Modification in polypeptide- coding region of the mRNA
(B) add or delete bases in the coding regions of primary transcripts
(C) enzymatic deamination of a C residue to create a U residue
(D) Protect mRNA from enzymatic destruction

135. Which process play a important role in synthesizing of large regions of an mRNA without any uridylate, and the U residues are inserted later in parasitic protozoa trypanosomes?
(A) RNA splicing (B) RNA editing
(C) RNA sequencing
(D) RNA modification

136. In eukaryotic mRNAs, Introns vary in size from .... nucleotides.
(A) 80 to 250
(B) 30 to 60
(C) 100 to 200
(D) 50 to 20,000

137. Which of the following may be considered as the vestiges of a molecular parasite?
(A) Exons (B) Cistrons
(C) Recons (D) Introns

138. If there are two or more sites for cleavage and polyadenylation, use of the one closest to the 5' end that will remove more of the primary transcript sequence. What is called this mechanism?
(A) poly(A) tailing
(B) poly(A) site choice
(C) Intron splicing
(D) RNA silencing

139. What is the function of poly(A) site choice mechanism in which If there are two or more sites for cleavage and polyadenylation, use of the one closest to the 5' end that will remove more of the primary transcript sequence.
(A) To generate diversity in immunoglobulin heavy chains
(B) Alternative splicing
(C) Act as precursor of pre-mRNA
(D) Act as transfer RNA precursors for post-transcriptional processing

140. In which immunoglobulin, three different forms from a common primary transcript of the heavy chain produced by alternative splicing patterns at different stages of fruit fly development.

(A) Chitin (B) Heamoglobin
(C) Myosin (D) Myoglobulin

141. Which of the following different hormones are produced from a single RNA transcript by alternative splicing patterns and poly(A) site choice mechanisms?

(A) Calcium-regulating hormone calcitonin in rat thyroid
(B) Calcitonin-gene-related peptide (CGRP) in rat brain
(C) Calcium-gene-related to degrading in rat thyroid and brain
(D) Both (A) and (B)

142. Which of the following act as longer precursor for making ribosomal RNAs of both prokaryotic and eukaryotic cells?

(A) Pre-tRNAs (B) pre-mRNAs
(C) pre-rRNAs
(D) Both pre-mRNAs and pre-rRNA

143. Which of the following enzyme synthesize rRNA molecules?

(A) RNA polymerase II
(B) RNA polymerase I
(C) RNA polymerase III
(D) Reverse transcriptase

144. Which of the following act as precursor of about 6,500 nucleotides to arise 16S, 23S, and 5S rRNAs in bacteria?

(A) Endonuclease RNase P
(B) A 45S pre-rRNA transcript
(C) Exonuclease RNase D
(D) A single 30S RNA precursor

145. How many pre-rRNA molecules are encoded by the genome of *E. coli*?

(A) Five (B) One
(C) Seven (D) Ten

146. Which of the following act as precursor to form the 18S, 28S, and 5.8S rRNAs characteristic of eukaryotic ribosomes in eukaryotes?

(A) Endonuclease RNase P
(B) A 45S pre-rRNA transcript
(C) Exonuclease RNase D
(D) A single 30S RNA precursor

147. Which of the following rRNA subunit of most eukaryotes is made as a completely separate transcript by a different polymerase (Pol III instead of Pol I)?

(A) 16S rRNA (B) 23S rRNA
(C) 5S rRNA (D) 45S rRNA

148. What is the range of number of tRNA contained in most cells?

(A) 100 to 200 tRNAs
(B) 1 to 10 tRNAs
(C) 4 to 5 tRNAs
(D) 40 to 50 tRNAs

149. Which of the following endonuclease enzyme removes intron sequences of RNA at the 5' end of tRNAs in all eukaryotic organisms by enzymatic cleavage?

(A) RNase P (B) RNase A
(C) RNase H (D) RNase D

150. RNase P is made up of ....

(A) RNA
(B) Protein
(C) Both protein and RNA
(D) both histone protein and RNA

151. Which of the following endonuclease enzyme removes intron sequences of RNA at the 3' end of tRNAs in all eukaryotic organisms by enzymatic cleavage?

(A) RNase P (B) RNase A
(C) RNase H (D) RNase D

152. Which of the following trinucleotide sequence at 3'-terminal of tRNA to which an amino acid will be attached during protein synthesis is absent from some bacterial and all eukaryotic tRNA precursors and is added during processing?

(A) AUG(3') (B) CCA(3')
(C) GUG(3') (D) UUG(3')

153. Which of the following enzyme binds the three ribonucleoside triphosphate precursors in separate active sites and catalyzes formation of the phosphodiester bonds to produce the 3' terminal trinucleotide CCA(3') sequence in tRNA to which an amino acid will be attached during protein synthesis is absent from some bacterial and all eukaryotic tRNA precursors?

(A) tRNA nucleotidyltransferase
(B) DNA ligase
(C) Endonuclease RNase D
(D) Polynucleotide phosphorylase

154. Which types of modifications of some of the bases in tRNA are done during processing?

(A) Methylation
(B) Deamination
(C) Reduction
(D) All of the above

155. Which of the following characteristic in all tRNAs that not found in other types of RNA?

(A) Presence of modified bases
(B) Presence of pseudouridine (ψ)
(C) Presence of introns
(D) Both a and b

156. Modified bases are found in which type of RNA?

(A) mRNA (B) tRNA
(C) rRNA (D) All of the above

157. Which of the following base is removed from tRNA and reatachhed pseudouridine (ψ) to sugar through C-5?

(A) Thymine (B) Adenine
(C) Uracil (D) Guanine

158. Which of the structure is shown pseudouridine (ψ) found in tRNA?

(A) (B)

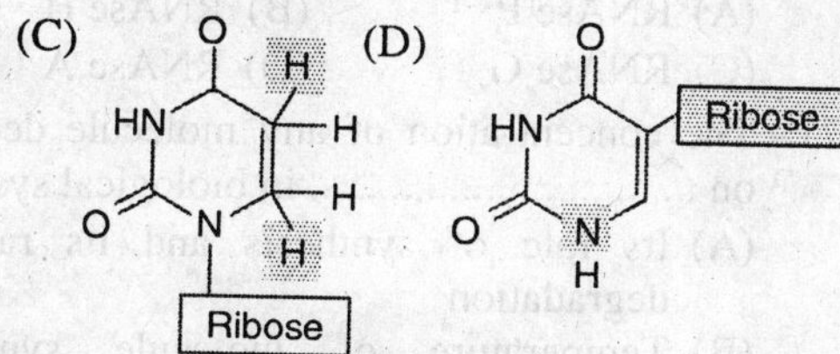

159. Which of the structure is shown Dihydrouridine (D) found in tRNA?

(A) (B)

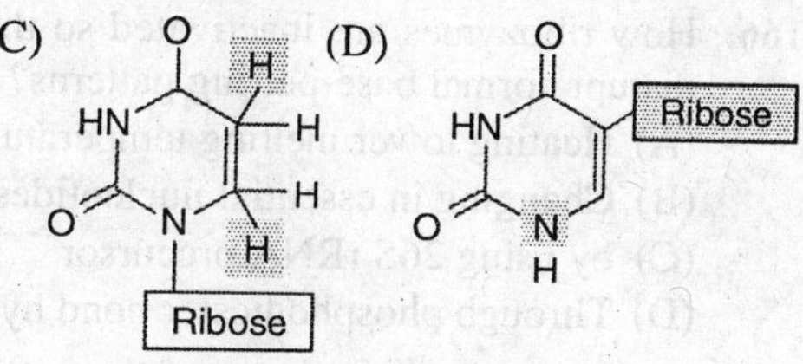

160. What is shape of tRNA?

(A) Domb shape structure
(B) Lotus leaf structure
(C) Cone shape structure
(D) Clover leaf structure

161. Which of the following best-characterized ribozymes are the self-splicing group I introns and the hammer head ribozyme?

(A) RNase H (B) RNase A
(C) RNase P (D) RNase B

162. What are the activities of ribozymes RNase P are based on two fundamental reactions?

(A) Transesterification and phosphodiester bond hydrolysis
(B) Proofreading and exonucleae activity
(C) Polymerase and ligation activity
(D) Cleavage and Splicing activity

163. What are the size of ribozymes that belongs to a self-splicing group I intron?

(A) 40 to 50 nucleotides
(B) Only 41 nucleotides
(C) More than 400 nucleotides
(D) About 6,500 nucleotides

164. The best-characterized ribozymes are the self-splicing group I introns and the hammer head consists of two RNA strands with .... in all.

(A) 40 to 50 nucleotides
(B) Only 41 nucleotides
(C) More than 400 nucleotides
(D) About 6,500 nucleotides

165. Ribozymes are inactivated by which processes that it can disrupt normal base-pairing patterns?
(A) Heating above their melting temperature
(B) Addition of denaturing agents
(C) Complementary oligonucleotides
(D) All of the above

166. How ribozymes are inactivated so that it can disrupt normal base-pairing patterns?
(A) Heating lower melting temperature
(B) Changing in essential nucleotides
(C) by using 26S rRNA precursor
(D) Through phosphodiester bond hydrolysis

167. Which of the following completely inhibit the binding of the guanosine cofactor to the *Tetrahymena* group I rRNA intron that is saturable ($Km \approx 30\ \mu M$)?
(A) 3'-deoxyguanosine
(B) 3'-deoxyadenosine
(C) 3'-deoxyuridine
(D) 3'-guanosine

168. What is called the large segments of excised introns that can base-pair with exon sequences near the 5' splice site?
(A) Shine-delgarno sequence
(B) Kozac sequence
(C) Internal guide sequence
(D) Promoter sequence

169. What happen when the large intron segment called the internal guide sequence that can base-pair with exon sequences near the 5' splice site?
(A) This pairing promotes the alignment of specific bonds to be cleaved and rejoined
(B) Internal guide sequence undergo further posttranscriptional processing
(C) Modified bases occur at characteristic positions in all tRNAs
(D) Ribozymes inactivats if essential nucleotides are changed after base pairing

170. How many nucleotides introns from *Tetrahymena* rRNA can act as a true enzyme after excision *in vitro* but *in vivo*, it is quickly degraded?
(A) 500 (B) 200
(C) 414 (D) 150

171. What is called the remaining 395 nucleotode sequence of linear RNA that obtained from cleaving of *Tetrahymena* rRNA and promotes nucleotidyl transfer reactions in which some oligonucleotides are lengthened at the expense of others?
(A) L-19 IVS (B) SINEs
(C) LINEs (D) snRNA

172. Why is considered intron as a catalyst in biological reaction?
(A) Due to dependent on temperature
(B) Because intron is non-coding sequence
(C) Because intron found only in mRNA
(D) Because intro is not altered in the reaction

173. Intron act as a catalyst due to not alter itself in biological reaction with following
(A) Michaelis-Menten kinetics
(B) Thermodynamic rules
(C) Electromagnetic induction mechanism
(D) Nuclear magnetic resonance mechanism

174. How many size of M1 RNA in RNase P?
(A) 100 nucleotides (B) 377 nucleotides
(C) 395 nucleotides (D) 414 nucleotides

175. Who discovered in 1983 that the M1 RNA in RNase P alone is capable of catalysis with cleaving tRNA precursors at the correct position?
(A) Stanley Cohen and Francis S. Collins
(B) Phillip Sharp and Richard Roberts
(C) Sidney Altman and Norman Pace
(D) Gerty and Carl Cori

176. Which of the following RNAse recognizes the three dimensional shape of its pre-tRNA substrate, along with the CCA sequence, and thus can cleave the 5' leaders from diverse tRNAs?
(A) RNAse P (B) RNAse H
(C) RNAse G (D) RNAse A

177. The concentration of any molecule depends on .... in biological system
(A) Its rate of synthesis and its rate of degradation
(B) Temperature of molecule synthesis reaction
(C) pH of molecule synthesis reaction
(D) Temperature and pH of molecule synthesis reaction

178. What may be the half-life of mRNA generally in organisms?
(A) 24 hours
(B) 5-10days
(C) Only minutes or even seconds
(D) Within one hour

179. What is the average half-life of a vertebrate cell mRNA?
(A) About one hours
(B) About 5-10 hours
(C) Only minutes or even seconds
(D) About 3 hours

180. What is the average half-life of a bacterial mRNAs?
(A) About one hours
(B) About 1.5 min
(C) With in 10 seconds
(D) About 3 hours

181. Which help in the degradation of messenger RNA in all cells?
(A) Polymerases (B) Helicases
(C) Ribonucleases (D) Topoisomerases

182. Which of the following enzymes are involved in beginning of the mRNA degradation in *E. coli*?
(A) Endoribonuclease
(B) Exoribonuclease
(C) Restriction endonuclease
(D) Both a and b

183. Which of the following enzyme degrades mRNA at 3' to 5' in *E. coli*?
(A) Endoribonuclease
(B) Exoribonuclease
(C) Restriction endonuclease
(D) Both a and b

184. Which of the following process is done during mRNA degradation in lower eukaryotes?
(A) Shortening the poly(A) tail
(B) Decapping the 5' end
(C) Degrading the mRNA in the 5' to 3' direction
(D) Cutting of mRNA by exoribonuclease in 3' to 5' direction

185. What is the sequential steps of mRNA degradation process in lower eukaryotes?
i. Decapping the 5' end
ii. Shortening the poly (A) tail
iii. Degrading the mRNA in the 5' to 3' direction
(A) i, ii and iii (B) ii, i, iii
(C) iii, ii, i (D) iii, i, ii

186. Which of the following steps is not involved in mRNA degradation in lower eukaryotes?
(A) Shortening the poly(A) tail
(B) Decapping the 5' end
(C) Degrading the mRNA in the 3' to 5' direction
(D) Degrading the mRNA in the 5' to 3' direction

187. What is called the complex contains up to ten 3' to 5' exoribonucleases in higher eukaryotes?
(A) Nucleosome (B) Ribozyme
(C) Exosome (D) Polysome

188. Which of the following step of mRNA degradation found in higher eukaryotes, not in lower eukaryotes?
(A) Shortening the poly(A) tail
(B) Decapping the 5' end
(C) Degrading the mRNA in the 5' to 3' direction
(D) Degrading the mRNA in the 3' to 5' direction

189. Which of the following structure is involved in the processing of the 3' end of rRNAs and tRNAs as well as the degradation of mRNAs?
(A) A hairpin mRNA (B) Ribozyme
(C) Exosome (D) Polysome

190. Which of the following structure with a ρ-independent terminator confer stability against mRNA degradation in bacteria?
(A) A hairpin structure in mRNAs
(B) Ribozyme
(C) Exosome (D) Polysome

191. What is the function of both the 3' poly(A) tail and the 5' cap in eukaryotic cells?
(A) Stability of mRNAs
(B) Stability of tRNAs
(C) Stability of rRNAs
(D) Stability of DNAs

192. Who discovered the bacterial enzyme polynucleotide phosphorylase which in vitro catalyzes the reaction of lengthened polynucleotide?
(A) Marianne Grunberg-Manago and Severo Ochoa (1955)
(B) Howard Temin and David Baltimore (1962)
(C) Carl Woese, Francis Crick and Leslie Orgel (1960s)
(D) Arthur Kornberg (1957)

193. What is called the movement of the transcription complex away from the promoter?
(A) Promoter activation
(B) Promoter checking
(C) Proofreading
(D) Promoter clearance

194. Which of the following σ subunit of the promoters of the heat-shock genes is bound by RNA polymerase holoenzymes in *E. coli* transcription?
(A) σ^{70} (B) σ^{32}
(C) $\sigma^{18.9}$ (D) σ^{60}

195. Which of the following protein does activate transcription in *E. coli* and increase the transcription of genes coding for enzymes that metabolize sugars other than glucose when cells are grown in the absence of glucose?
(A) Chymotrypsinogen
(B) Ribisco
(C) cAMP receptor protein
(D) Endoproteinase Lys C

196. How activates transcription by the cAMP receptor protein (CRP) in *E. coli*?
(A) It forms the hairpin structure in the RNA that disrupts several AUU base pairs in the RNA-DNA hybrid segment
(B) It blocks the synthesis of RNA of specific genes
(C) Promotes the unwinding of DNA near the RNA start site thereby creating an open complex
(D) It increases the transcription of genes coding for enzymes that metabolize sugars other than glucose when cells are grown in the absence of glucose

197. What is the function of repressor proteins during transcription of *E. coli*?
(A) It forms the hairpin structure in the RNA that disrupts several AUU base pairs in the RNA-DNA hybrid segment
(B) It blocks the synthesis of RNA of specific genes
(C) Promotes the unwinding of DNA near the RNA start site thereby creating an open complex
(D) It increases the transcription of genes coding for enzymes that metabolize sugars other than glucose when cells are grown in the absence of glucose

198. When transcription of the genes for the enzymes of lactose metabolism is blocked in lac repressor of *E. coli*?
(A) Lactose is unavailable
(B) Lactose is available
(C) Glucose is unavailable
(D) Galactose is unavailable

199. Which of the following protein factor provides terminal signals to stop transcription in *E. coli*?
(A) σ (sigma) factor (B) ρ (rho) factor
(C) Transcription factor IIB
(D) Transcription factor IIH

200. *E. coli* has at least two classes of termination signals for transcription that are
(A) σ (sigma) dependent and σ-independent
(B) THIIB dependent and THIIB-independent
(C) *ρ* (rho) dependent and *ρ*-independent
(D) THIIH-dependent and THIIH-independent

201. Which of the following feature is not related to most ρ-independent terminators for transcription in *E. coli*?
(A) A region that produces an RNA transcript with self-complementary sequences
(B) Above sequences permit the formation of a hairpin structure centered 15 to 20 nucleotides before the projected end of the RNA strand
(C) ρ-independent terminators lack the sequence of repeated A residues in the template strand but usually include a CA-rich sequence

(D) A highly conserved string of three A residues in the template strand that are transcribed into U residues near the 3' end of the hairpin.

202. In which of the following class of termination signals, a region that produces an RNA transcript with self-complementary sequences that permit the formation of a hairpin structure centered 15 to 20 nucleotides before the projected end of the RNA strand for transcription in *E. coli*?

(A) σ (sigma) dependent
(B) σ-independent
(C) *ρ* (rho) dependent
(D) *ρ*-independent

203. What happen when a RNA polymerase arrives at a termination site with hairpin RNA structure?

(A) Transcription is continued
(B) Transcription is paused
(C) Transcription is initiated
(D) Translation started

204. What happen when the hairpin structure in the RNA is formed?

(A) Disrupt several A=U base pairs in the RNA-DNA hybrid segment
(B) Disrupt important interactions between RNA and DNA that facilitate dissociation of the transcript
(C) Release transcript through ρ-dependent termination of transcription
(D) All of the above

205. Which of the following feature includes in ρ-independent termination of transcription?

(A) Formation of the hairpin structure in the RNA
(B) Lack the sequence of repeated A residues in the template strand
(C) A CA-rich sequence
(D) ρ-independent terminators includes a *rut* (*r*ho *ut*ilization) element

206. Which of the following statements regarding termination of transcription in prokaryotes is correct?

(A) Termination often involves a stem-loop structure forming in the RNA transcript
(B) Rho factor has topoisomerase activity for relieving supercoiling
(C) In Rho dependent termination the Rho factor moves along the DNA template ahead of the RNA polymerase
(D) Termination often involves a stem-loop structure forming in the DNA template

207. Which of the following sequence is coorect for the process of ρ-dependent termination of transcription?

i. Transcription is paused after reaching transcription complex at a termination site
ii. RNA transcript is release
iii. ρ protein migrates in the 5' to 3' direction until it reaches the transcription complex
iv. The ρ protein associates with the RNA at specific binding sites

(A) i, ii, iii and iv (B) iv, iii, i and ii
(C) ii, iii, iv and i (D) i, iv, iii and ii

208. Which of the following protein play a role to release a transcript during transcription?

(A) Sigma protein (σ)
(B) Rho protein (ρ)
(C) Omega protein (ω)
(D) TFII protein

209. Which of the following protein has an ATP-dependent RNA-DNA helicase activity that promotes translocation of the protein along the RNA and ATP is hydrolyzed by ρ protein during the termination process of transcription?

(A) Sigma protein (σ)
(B) TFII protein
(C) Omega protein (ω)
(D) Rho protein (ρ)

210. Which of the following activity is found in rho (ρ) protein that promotes translocation of the protein along the RNA and ATP is hydrolyzed by ρ protein during the termination process of transcription?

(A) ATP-dependent RNA-DNA helicase activity
(B) ATP-dependent polymerase activity
(C) ATP-dependent proofreading activity
(D) ATP-dependent endonuclease activity

211. Which process is performed during the termination process of transcription?
(A) ATP methylation
(B) ATP glycosylation
(C) ATP hydrolyzed
(D) ATP phosphorylation

212. How many kinds of nuclear RNA polymerases contained eukaryotic cells?
(A) Four (B) One
(C) Three (D) Two

213. Where transcription process is performed in eukaryotic cells?
(A) Cytoplasm
(B) Nucleus
(C) Golgi body
(D) Endoplasmic reticulum

214. Which of the following process is completed in the nucleus of eukaryotic cells?
(A) Translation
(B) Photosynthesis
(C) Transcription
(D) Protein synthesis

215. Which of the following process is completed in the cytoplasm of eukaryotic cells?
(A) DNA replication
(B) Transcription and relication
(C) Transcription (D) Protein synthesis

216. Which of the following RNA polymerase is not found in eukaryotic cells?
(A) RNA polymerase I
(B) RNA polymerase II
(C) RNA polymerase III
(D) RNA polymerase IV

217. Which of the following enzyme is responsible for the synthesis of preribosomal RNA (pre-rRNA) in eukaryotic cells except 5S rRNA?
(A) RNA polymerase II
(B) RNA polymerase I
(C) RNA polymerase III
(D) RNA polymerase α

218. Which of the following type of RNA is synthesized by RNA polymerase I in eukaryotic cells?
(A) mRNA (B) tRNA
(C) rRNA (D) siRNA

219. What is difference in RNA polymerase I and II from one species to another in eukaruotes?
(A) Nuclotoide sequences
(B) RNA structure
(C) RNA function
(D) RNA modification

220. Which of the following RNA contain the precursor for the 18S, 5.8S, and 28S subunits of RNAs?
(A) mRNA (B) tRNA
(C) rRNA (D) siRNA

221. What is the principal function of RNA polymerase II (Pol II) in eukaryotic cells?
(A) mRNA synthesis
(B) tRNA synthesis
(C) rRNA synthesis
(D) siRNA synthesis

222. Which of the following enzyme is responsible for the synthesis of messenger RNA in eukaryotic cells?
(A) RNA polymerase II
(B) RNA polymerase I
(C) RNA polymerase III
(D) RNA polymerase β

223. What are the common features in many RNA Polymerase II promoters of eukarytic cells?
(A) TATA box
(B) Complete nucleotide sequences
(C) Initiator sequence (Inr) near the RNA start site at +1
(D) Both (A) and (C)

224. Which of the following enzyme is responsible for the synthesis of transfer RNA in eukaryotic cells?
(A) RNA polymerase II
(B) RNA polymerase I
(C) RNA polymerase III
(D) RNA polymerase γ

225. What is the principal function of RNA polymerase II in eukaryotic cells?
(A) mRNA synthesis
(B) tRNA synthesis
(C) rRNA synthesis
(D) siRNA synthesis

226. What is the function of RNA polymerase II in eukaryotic cells?

(A) mRNA synthesis
(B) 5S rRNA synthesis
(C) hnRNA synthesis
(D) siRNA synthesis

227. Which of the central enzyme to eukaryotic gene expression?
(A) RNA polymerase II
(B) RNA polymerase I
(C) RNA polymerase III
(D) RNA polymerase γ

228. How many subunits contained by the structure of RNA polymerase II in eukarytic cells?
(A) 10 (B) 5
(C) 12 (D) 100

229. Which of the following subunits has a high degree of homology for RNA polymerase?
(A) RBP2 of eukaryotic RNA polymerase II and α subunit of bacteria RNA polymerase
(B) RBP12 of eukaryotic RNA polymerase II and β subunit of bacteria RNA polymerase
(C) RBP10 of eukaryotic RNA polymerase II and γ subunit of bacterial RNA polymerase
(D) RBP1 of eukaryotic RNA polymeraseII and β'subunit of bacterial RNA polymerase

230. Which of the following best describes the 'cap' modification of eukaryotic mRNA?
(A) A modified guanine nucleotide added to the 5' end of the transcript
(B) A modified guanine nucleotide added to the 3' end of the transcript
(C) A string of adenine nucleotides added to the 3' end of the transcript
(D) A string of adenine nucleotides added to the 5' end of the transcript

231. Which of the following the largest subunit of RNA polymerase II in eukaryotic cell has a long carboxyl-terminal tail consisting of many repeats of a consensus heptad amino acid sequence –YSPTSPS–?
(A) RBP11 (B) RBP12
(C) RBP1 (D) RBP3

232. How many repeats of a consensus heptad amino acid sequence –YSPTSPS– in a long carboxyl-terminal tail the largest subunit of RNA polymerase II in yeast?
(A) 10 (B) 27
(C) 52 (D) 72

233. How many repeats of a consensus heptad amino acid sequence –YSPTSPS– in a long carboxyl-terminal tail the largest subunit of RNA polymerase II in mouse and human?
(A) 10 (B) 27
(C) 52 (D) 72

234. How many exactly matching in the repeats of a consensus heptad amino acid sequence –YSPTSPS– in a long carboxyl-terminal tail the largest subunit of RNA polymerase II in yeast, mouse and human?
(A) Yeast = 18, Mouse = 52 and Human = 52
(B) Yeast = 52, Mouse = 18 and Human = 52
(C) Yeast = 52, Mouse = 52 and Human = 18
(D) Yeast = 18, Mouse = 18 and Human = 52

235. Which of the following domain is separated from the main body of the RNA polymerase II enzyme by an unstructured linker sequence and causes a conformational change in the overall complex, initiating transcription?
(A) Carboxylterminal domain (CTD)
(B) TATA-binding protein (TBP) domain
(C) cAMP receptor protein (CRP) domain
(D) Zinc finger protein domain

236. Which of the following proteins require RNA polymerase II in order to form the active transcription complex?
(A) Transcription factors
(B) Elongin (SIII)
(C) cAMP receptor protein (CRP)
(D) Zinc finger proteins

237. Which of the following protein first bind to the TATA box to begin the formation of a closed complex by the transcription factor TFIIB which binds to DNA during transcription?
(A) Transcription factors
(B) TATA-binding protein (TBP)
(C) cAMP receptor protein (CRP)
(D) Zinc finger proteins

238. Which of the following transcription factor can stabilize the TFIIB-TBP complex on the DNA and can be important at nonconsensus promoters where TBP binding is relatively weak?
(A) TFIIA (B) TFIIH
(C) TFIIF (D) TFIIE

239. What is the role of transcription factor TFIIF when binds to the TFIIB-TBP complex after stabilizing?
(A) Causes a conformational change in the overall TFIIB-TBP complex
(B) Promotes the unwinding of DNA near the RNA start site
(C) Helps target Pol II to its promoters
(D) Helps in the elongation phase of transcription

240. What is sequence of binding of different transcription factors to TATA-binding protein (TBP) during initiation of transcription?
i. TFIIB ii. TFIIH
iii. TFIIF iv. TFIIE
(A) i, ii, iii, iv (B) i, iii, iv, ii
(C) v, ii, iv, iii (D) v, ii, iii, iv

241. When RNA polymerase II bind to TATA-binding protein (TBP) during initiation of transcription?
(A) During the binding of TFIIB to TATA-binding protein
(B) During the binding of TFIIE to TATA-binding protein
(C) During the binding of TFIIH to TATA-binding protein
(D) During the binding of TFIIF to TATA-binding protein

242. Find the correct match between transcription factors and their function

Transcription factor	Function
i. TFIIF	(A) Binds to TBP; recruits Pol II–TFIIF complex
ii. TFIIE	(B) Stabilizes binding of TFIIB and TBP to the promoter
iii. TFIIH	(C) Unwinds DNA at promoter (helicase activity); phosphorylates Pol II (within the CTD); recruits nucleotide-excision repair proteins
iv. TFIIA	(D) Binds tightly to Pol II; binds to TFIIB and prevents binding of Pol II to nonspecific DNA sequences
v. TFIIB	E. Recruits TFIIH; has ATPase and helicase activities

(A) i = D, ii = E, iii = C, iv = B, v = A
(B) i = A, ii = B, iii = C, iv = D, v = E
(C) i = E, ii = D, iii = C, iv = B, v = A
(D) i = B, ii = C, iii = E, iv = A, v = E

243. Which of the following transcription factor has DNA helicase activity that promotes the unwinding of DNA near the RNA start site?
(A) TFIIB (B) TFIIH
(C) TFIIF (D) TFIIA

244. Which of the following transcription factor has ATPase and helicase activities?
(A) TFIIB (B) TFIIA
(C) TFIIF (D) TFIIE

245. Which of the following transcription factor has ATPase, kinase activity and helicase activities?
(A) TFIIB (B) TFIIH
(C) TFIIF (D) TFIIE

246. Which of the following protein is part of the complex pTEFb (*positive transcription elongation factor b*) and also phosphorylate the CTD?
(A) Transcription factors
(B) Cyclin-dependent kinase 9
(C) cAMP receptor protein (CRP)
(D) Zinc finger proteins

247. In which phase of transcription is the phosphorylation of the CTD (carboxylterminal domain) and it affects the interactions between the transcription complex and other enzymes involved in processing the transcript?

(A) Elongation (B) Initiation
(C) Termination (D) Transcript releasing

248. Which of the transcription factor first released during synthesis of the initial 60 to 70 nucleotides of RNA and RNA Pol II enters the elongation phase of transcription?

(A) TFIIB (B) TFIIH
(C) TFIIF (D) TFIIE

249. Which of the transcription factor released during synthesis of the initial 60 to 70 nucleotides of RNA and RNA Pol II enters the elongation phase of transcription?

(A) First TFIIE and then TFIIB
(B) First TFIIB and then TFIIH
(C) First TFIIE and then TFIIH
(D) First TFIIA and then TFIIB

250. What is the length of RNA during RNA strand initiation and promoter clearance after releasing TFIIE and TFIIH and entering RNA Pol II in the elongation phase of transcription?

(A) 8 to 10 nucleotides
(B) 60 to 70 nucleotides
(C) 20 nucleotides
(D) 16 to 27 nucleotides

251. Which of the following transcription factor is not released during transcription initiation and remains associated with Pol II throughout elongation phase of transcription?

(A) TFIIB (B) TFIIH
(C) TFIIF (D) TFIIE

252. Which of the following proteins greatly enhance the the activity of the RNA polymerase II during transcription elongation?

(A) Elongation factors
(B) Cyclin-dependent kinase 9
(C) cAMP receptor protein (CRP)
(D) Zinc finger proteins

253. Which of the following protein is not include in elongation factors that suppress pausing during transcription and also coordinate interactions between protein complexes involved in the posttranscriptional processing of mRNAs?

(A) ELL† and p-TEFb (B) CDK 9
(C) SII (TFIIS) (D) Elongin (SIII)

254. What happen when once the RNA transcript is completed and transcription is terminated Pol II is .... recycled and ready to initiate another transcript?

(A) Methylated (B) Phosphorylated
(C) Dephosphorylated (D) Acetylated

255. Which of the following transcriptuion factor that is a complex of about 12 proteins, regulates the transcription at Pol II promoters by TBP and certain *T*BP-*a*ssociated *f*actors (TAFs)?

(A) TFIIB (B) TFIIH
(C) TFIIF (D) TFIID

256. Which of the following some of subunits of transcription factor are essential components of the separate nucleotide-excision repair complex?

(A) TFIIB (B) TFIIH
(C) TFIIF (D) TFIID

257. Which of the following human disease can produce through genetic loss of transcription factor TFIIH subunits?

(A) Xeroderma pigmentosum and Cockayne's syndrome
(B) Adenosine deaminase (ADA) deficiency and Lesch-Nyhan syndrome
(C) Leukemia and Gout disease in joints
(D) Sickel cell anemia and Cancer-susceptibility syndrome

258. What are the symptoms of human disease Cockayne's syndrome that produced through genetic loss of transcription factor TFIIH subunits?

(A) Elevate the concentration of uric acid in the blood and tissues?
(B) The joints become inflamed, painful, and arthritic
(C) The abnormal deposition of sodium urate crystals
(D) Arrest growth, photosensitivity, and neurological disorders

259. Which of the following antibiotic inhibit the elongation of RNA strands by RNA polymerase in both bacteria and eukaryotes?
(A) Actinomycin D (B) Rifampicin
(C) α–amanitin (D) Puromycin

260. How antibiotic actinomycin D inhibit the elongation of RNA strands by RNA polymerase in both bacteria and eukaryotes?
(A) Deform the DNA by insertion into the ds DNA between successive GqC base pairs
(B) Preventing the promoter clearance step of transcription
(C) Disrupts mRNA formation
(D) Block the activity of RNA polemeraese II

261. Which of the following antibiotics prevent the movement of the polymerase along the template by deforming the DNA?
(A) Actinomycin D (B) Rifampicin
(C) α–amanitin (D) Puromycin

262. Which of the following antibiotics is used to identify cell processes that depend on RNA synthesis because it inhibits RNA elongation in intact cells as well as in cell extracts?
(A) α–amanitin
(B) Rifampicin
(C) Actinomycin D
(D) Puromycin

263. Which of the following antibiotic act similar to Actinomycin D with same the process of inhibition of RNA synthesis?
(A) Chloramphenicol
(B) Cycloheximide
(C) Streptomycin
(D) Acridine

264. Which of the following antibiotic inhibit bacterial RNA synthesis by binding to the *β* subunit of bacterial RNA polymerases?
(A) α–amanitin
(B) Rifampicin
(C) Actinomycin D
(D) Puromycin

265. Which of the following antibiotic prevent the promoter clearance step of transcription?
(A) Rifampicin (B) α–amanitin
(C) Actinomycin D (D) Puromycin

266. What is the function of antibiotics Rifampicin, α–amanitin and Actinomycin D?
(A) All inhibit DNA synthesis
(B) All inhibit protein synthesis
(C) All inhibit amino acid synthesis
(D) All inhibit RNA synthesis

267. Which of the following antibiotic does not inhibit RNA synthesis?
(A) α–amanitin
(B) Rifampicin
(C) Actinomycin D
(D) Puromycin

268. Which of the following micro-organism produce antibiotic α-amanitin disrupts mRNA formation in animal cells by blocking RNA Polymerase II and RNA Polymerase III?
(A) Mushroom *Amanita phalloides*
(B) Mold *Streptomyces alboniger*
(C) Fungus *Aspergillus niger*
(D) Bacteria *Clostridium welchi*

269. Which of the following antibiotic is produced by mushroom *Amanita phalloides* has evolved a very effective defense mechanism against predators?
(A) α–amanitin
(B) Rifampicin
(C) Actinomycin D
(D) Puromycin

270. Which of the following the type of RNA polymerase is not sensitive to antibiotic α-amanitin that produced by mushroom *Amanita phalloides*?
(A) RNA polymerase I and Fungal RNA polymerase
(B) RNA polymerase II and RNA polymerase III
(C) RNA polymerase I and bacterial RNA polymerase
(D) RNA polymerase I, II and III

271. Which of the following RNA polymerase of mushroom *Amanita phalloides* is not sensitive to antibiotic α-amanitin?
(A) RNA polymerase I
(B) RNA polymerase II
(C) RNA polymerase III
(D) RNA polymerase α

272. How many proteins with distict functions are derived from cleaving the single large polypeptide (polyprotein) that produced by transcript containg *gag* and *pol* genes in retroviruses?

(A) Six (B) Ten
(C) Two (D) 100

273. Which of the following enzyme is responsible for the production of DNA from viral RNA?

(A) RNA-dependent DNA polymerase
(B) DNA-dependent RNA polymerase
(C) DNA-dependent DNA polymerase
(D) RNA-dependent RNA polymerase

274. Which of the following element contained in reverse transcriptases?

(A) Zn^{2+} (B) Mn^{2+}
(C) Mg^{2+} (D) Ca^{2+}

275. What is main difference between enzymes polynucleotide phosphorylase and DNA polymerase?

(A) DNA polymerase is not template-dependent
(B) Polynucleotide phosphorylase uses 5'-diphosphates of ribonucleosides as substrat
(C) Polynucleotide phosphorylase is template-dependent
(D) DNA polymerase act on deoxyribonucleoside 5'-diphosphates

276. Which of the following is used as substrates to form polynuclotide chains?

(A) 5'-diphosphates of ribonucleosides
(B) Deoxyribonucleoside 5'-diphosphates
(C) 5'-triphosphates of ribonucleosides
(D) Deoxyribonucleoside 5'-triphosphates

277. Which of the following enzyme is not template dependent used for synthesizing polynuclotides?

(A) DNA polymerase III
(B) Polynucleotide phosphorylase
(C) RNA polymerase II
(D) ALl of the above

278. Who presicted the existence of reverse transcriptases in RNA viruses?

(A) Arthur Kornberg (1959)
(B) Marianne Grunberg-Manago and Severo Ochoa (1955)
(C) Howard Temin (1962) and David Baltimore (1970)
(D) Carl Woese, Francis Crick and Leslie Orgel (1960s)

279. What is the probable function of polynucleotide phosphorylase in the cell?

(A) Degradation of mRNAs to nucleoside triphosphates
(B) Degradation of mRNAs to nucleoside diphosphates
(C) To form a polynucleotide chain using template
(D) To form a specific base sequence

280. Why polynucleotide phosphorylase can be used in the laboratory?

(A) To prepare RNA polymers with many different base sequences and frequencies
(B) To prepare DNA polymers with many different base sequences and frequencies
(C) To prepare amino acid polymers with many different base sequences and frequencies
(D) To prepare glucose polymers with many different base sequences and frequencies

281. Who discovered that that genetic information can flow "backward" from RNA to DNA?

(A) Arthur Kornberg (1959)
(B) Marianne Grunberg-Manago and Severo Ochoa (1955)
(C) Howard Temin (1962) and David Baltimore (1970)
(D) Watson and Crick (1953)

282. Which of the following enzyme is responsible for the production of DNA from viral RNA?

(A) Reverse transcriptase
(B) DNA polymerase III
(C) RNA polymerase II
(D) DNA polymerase III and RNA polymerase II

283. Which of the following was the first nucleic acid–synthesizing enzyme discovered?

(A) Polynucleotide phosphorylase
(B) DNA polymerase

(C) Pyridoxal phosphate
(D) Thiamine pyrophosphate

284. Which of the following represent enzyme reverse transcriptase?
(A) RNA-dependent DNA polymerase
(B) RNA-dependent RNA polymerase
(C) DNA-dependent DNA polymerase
(D) DNA-dependent RNA polymerase

285. What is the general size of single stranded RNA viral genome?
(A) ~10,000 nucleotides
(B) ~10,00 nucleotides
(C) ~100 nucleotides
(D) ~1,00,000 nucleotides

286. Which of the following function is related to reverse transcriptase?
(A) Synthesis of a dsDNA strand from ssRNA
(B) Synthesis of a DNA strand from DNA
(C) Synthesis of a RNA strand from RNA
(D) Synthesis of a DNA strand from RNA

287. Which of the following steps show the correct sequence of the process of reverse transcription?
i. Replaces RNA from RNA-DNA hybrid with DNA
ii. Degrades the RNA strand of the viral RNA-DNA hybrid
iii. The reverse transcriptase catalyzes viral RNA
iv. The synthesis of a DNA strand complementary to the viral RNA
(A) ii, iv, iii and ii (B) i, ii, iii and iv
(C) iii, iv, ii and i (D) iv, iii, ii and i

288. What is called the RNA viruses that contain reverse transcriptases?
(A) Transposons (B) Retroviruses
(C) PPLO (D) Crystal viruses

289. Which of the following steps show the correct sequence of the process of formation of new RNA viruses through reverse transcription?
i. Reverse transcriptase catalyzes the synthesis of a DNA strand complementary to the viral RNA
ii. Degrades the RNA strand of the viral RNA-DNA hybrid and replaces it with DNA
iii. Incorporation of duplex DNA into into the genome of the eukaryotic host cell
iv. Integrated viral genes can be activated and transcribed and the gene products viral proteins and the viral RNA genome
v. Packaging of RNA genome into viral proteins as new RNA viruses
(A) i,iii,ii,v and iv (B) v,iv,iii,ii and i
(C) i,ii,iii,iv and v (D) ii,iii,iv and v

290. What is meaning of word '*retro*' in retroviruse that derived from Latin prefix?
(A) Backward (B) Forword
(C) Behind (D) Reverse

291. How can be pushed in the direction of breakdown of the polyribonucleotide that formed by polynucleotide phosphorylase contains the usual 3',5'-phosphodiester linkages *in vivo*?
(A) To change the reaction temprature
(B) To decrease the phosphate concentration
(C) To increase the phosphate concentration
(D) To increase the enzyme concentration

292. Which of the following organisms are the sources of most RNA-dependent DNA polymerases?
(A) Bacteria
(B) ssDNA viruses
(C) RNA viruses
(D) dsDNA viruses

293. Which of the following gene is not found in retroviruses?
(A) *gag* (B) *pol*
(C) *env* (D) *rec*

294. Which enzyme catalyses the following reaction and discovered by Marianne Grunberg-Manago and Severo Ochoa in 1955?

$$(NMP)_n + NDP \leftrightarrow (NMP)_{n+1} + Pi$$

Lengthened Polynucleotide

(A) Polynucleotide phosphorylase
(B) DNA polymerase
(C) Pyridoxal phosphate
(D) Thiamine pyrophosphate

295. Which of following correct match between genes and their function in retroviruses?

Gene	Function
i. *gag*	A. Encodes the protease that cleaves the long polypeptide
ii. *pol*	B. Encodes the proteins of the viral envelope
iii. *env*	C. Make up the interior core of the viral particle

(A) i = A, ii = B, iii = C
(B) i = C, ii = B, iii = A
(C) i = B, ii = A, iii = C
(D) i = C, ii = A, iii = B

296. Which of the following gene of reverse transcriptase enzyme has two subunits α (*M*r 65,000) and β (*M*r 90,000) in retroviruses?
(A) *pol* (B) *env*
(C) *gag* (D) *rec*

297. What is the role of long terminal repeat (LTR) sequences of a few hundred nucleotides that present at each end of the linear RNA genome in retroviruses?
(A) Facilitate integration of the viral chromosome into the host DNA
(B) Contain promoters for viral gene expression.
(C) Encodes the proteins of the viral envelope
(D) Both (A) and (B)

298. Which of the following reactions are catalyzed by reverse transcriptases?
(A) RNA-dependent DNA synthesis
(B) RNA degradation
(C) DNA-dependent DNA synthesis
(D) All of the above

299. Who discovered the enzyme DNA polymerase?
(A) Arthur Kornberg (1959)
(B) Marianne Grunberg-Manago and Severo Ochoa (1955)
(C) Howard Temin and David Baltimore (1962)
(D) Carl Woese, Francis Crick and Leslie Orgel (1960s)

300. Which of the following enzyme does not contain Zn^{2+}?
(A) DNA polymerases
(B) RNA polymerases
(C) Reverse transcriptases
(D) Polynucleotide phosphorylase

Answer-Key

Q	Ans	Q	Ans	Q	Ans	Q	Ans	Q	Ans	Q	Ans	Q	Ans	Q	Ans
1.	(A)	41.	(B)	81.	(A)	121.	(D)	161.	(C)	201.	(C)	241.	(D)	281.	(C)
2.	(C)	42.	(C)	82.	(B)	122.	(B)	162.	(A)	202.	(D)	242.	(A)	282.	(A)
3.	(B)	43.	(A)	83.	(C)	123.	(C)	163.	(C)	203.	(B)	243.	(B)	283.	(A)
4.	(D)	44.	(D)	84.	(B)	124.	(A)	164.	(B)	204.	(A)	244.	(D)	284.	(A)
5.	(B)	45.	(A)	85.	(A)	125.	(A)	165.	(D)	205.	(A)	245.	(B)	285.	(A)
6.	(C)	46.	(C)	86.	(B)	126.	(C)	166.	(B)	206.	(A)	246.	(B)	286.	(D)
7.	(C)	47.	(B)	87.	(A)	127.	(D)	167.	(A)	207.	(B)	247.	(B)	287.	(C)
8.	(C)	48.	(A)	88.	(B)	128.	(C)	168.	(C)	208.	(B)	248.	(D)	288.	(B)
9.	(A)	49.	(D)	89.	(C)	129.	(A)	169.	(A)	209.	(D)	249.	(C)	289.	(C)
10.	(D)	50.	(D)	90.	(B)	130.	(C)	170.	(C)	210.	(A)	250.	(B)	290.	(A)
11.	(D)	51.	(D)	91.	(D)	131.	(B)	171.	(A)	211.	(C)	251.	(C)	291.	(C)
12.	(D)	52.	(D)	92.	(B)	132.	(D)	172.	(D)	212.	(C)	252.	(A)	292.	(C)
13.	(B)	53.	(A)	93.	(A)	133.	(A)	173.	(A)	213.	(B)	253.	(B)	293.	(D)
14.	(D)	54.	(C)	94.	(C)	134.	(D)	174.	(B)	214.	(C)	254.	(C)	294.	(A)
15.	(D)	55.	(A)	95.	(D)	135.	(B)	175.	(C)	215.	(D)	255.	(D)	295.	(D)
16.	(A)	56.	(A)	96.	(A)	136.	(D)	176.	(A)	216.	(D)	256.	(B)	296.	(A)
17.	(C)	57.	(B)	97.	(B)	137.	(D)	177.	(A)	217.	(B)	257.	(A)	297.	(D)
18.	(C)	58.	(D)	98.	(C)	138.	(B)	178.	(C)	218.	(C)	258.	(D)	298.	(D)
19.	(D)	59.	(A)	99.	(C)	139.	(A)	179.	(C)	219.	(A)	259.	(A)	299.	(A)
20.	(B)	60.	(A)	100.	(B)	140.	(C)	180.	(B)	220.	(C)	260.	(A)	300.	(D)
21.	(A)	61.	(A)	101.	(A)	141.	(D)	181.	(C)	221.	(A)	261.	(A)		
22.	(B)	62.	(A)	102.	(C)	142.	(C)	182.	(D)	222.	(A)	262.	(C)		
23.	(C)	63.	(D)	103.	(D)	143.	(B)	183.	(B)	223.	(D)	263.	(D)		
24.	(A)	64.	(C)	104.	(B)	144.	(D)	184.	(A)	224.	(C)	264.	(B)		
25.	(D)	65.	(B)	105.	(D)	145.	(C)	185.	(B)	225.	(B)	265.	(A)		
26.	(B)	66.	(C)	106.	(A)	146.	(B)	186.	(C)	226.	(B)	266.	(D)		
27.	(C)	67.	(A)	107.	(B)	147.	(C)	187.	(C)	227.	(A)	267.	(D)		
28.	(C)	68.	(B)	108.	(D)	148.	(D)	188.	(D)	228.	(C)	268.	(A)		
29.	(B)	69.	(D)	109.	(C)	149.	(A)	189.	(C)	229.	(D)	269.	(A)		
30.	(D)	70.	(B)	110.	(A)	150.	(C)	190.	(A)	230.	(A)	270.	(C)		
31.	(A)	71.	(D)	111.	(B)	151.	(D)	191.	(A)	231.	(C)	271.	(B)		
32.	(D)	72.	(C)	112.	(C)	152.	(B)	192.	(A)	232.	(B)	272.	(A)		
33.	(C)	73.	(A)	113.	(A)	153.	(A)	193.	(D)	233.	(C)	273.	(A)		
34.	(D)	74.	(A)	114.	(C)	154.	(D)	194.	(B)	234.	(A)	274.	(A)		
35.	(A)	75.	(C)	115.	(C)	155.	(D)	195.	(C)	235.	(A)	275.	(D)		
36.	(D)	76.	(B)	116.	(B)	156.	(B)	196.	(D)	236.	(A)	276.	(A)		
37.	(A)	77.	(D)	117.	(A)	157.	(A)	197.	(B)	237.	(B)	277.	(B)		
38.	(A)	78.	(A)	118.	(C)	158.	(D)	198.	(A)	238.	(A)	278.	(C)		
39.	(B)	79.	(B)	119.	(A)	159.	(C)	199.	(B)	239.	(C)	279.	(B)		
40.	(C)	80.	(C)	120.	(D)	160.	(D)	200.	(C)	240.	(B)	280.	(A)		

7

Protein Synthesis and Metabolism

Translation is a biological process to synthesis of proteins in which sequence of mRNA is converted into the sequence of amino aids that are the building blocks of proteins. Initially, DNA is transcribed into mRNA in nucleus. This RNA molecule then moves from the nucleus to the cell cytoplasm, where protein synthesis takes place in ribosome called as the factory of protein synthesis. Proteins are the final fundamental biological and functional molecules which involed in all the biological and metabolic processes.

Genetic code

The relationship between the sequence of amino acid in a polypeptide chain and nucleotide sequence of DNA or m-RNA is called genetic code. George Gamow postulated that sets of three bases must be employed to encode the 20 standard amino acids used by living cells to build proteins. The Crick, Brenner *et al.* experiment first demonstrated that codons consist of three DNA bases. Marshall Nirenberg and Heinrich J. Matthaei were the first to elucidate the nature of a codon in 1961 and used a cell-free system to translate a poly-uracil RNA sequence (i.e., UUUUU...) and discovered that the polypeptide that they had synthesized consisted of only the amino acid phenylalanine.

The main proble of genetic code was to determine the exact number of nucleotide in a codon which codes for one amino acid. There are four types of nitrogen bases in m-RNA (A, U, G, C) for 20 types of amino aicds. If genetic code is singlet i.e. codon is the combination of only one nitrogen base, then only four codons are possible A, C, G, and U ($4^1 = 4 \times 1 = 4$). These are insufficient to code for 20 types of amino acids. If genetic code is doublet (i.e. codon is the combination of two bases) then 16 codons are formd ($4^2 = 4 \times 4 = 16$). 16 codons are also insufficient for 20 amino acids. Therefore, Gamow (1954) pointed out the possibility of three letter code (triplet code). Genetic code is triple i.e. one codon consists of three nitrogen bases ($4^3 = 4 \times 4 \times 4 = 64$). 64 codons are sufficient to code 20 amino acids.

First Letter	Second Letter: T		C		A		G		Third Letter
T	TTT	Phe	TCT	Ser	TAT	Tyr	TGT	Cys	T
	TTC		TCC		TAC		TGC		C
	TTA	Leu	TCA		TAA	Stop	TGA	Stop	A
	TTG		TCG		TAG	Stop	TGG	Trp	G
C	CTT	Leu	CCT	Pro	CAT	His	CGT	Arg	T
	CTC		CCC		CAC		CGC		C
	CTA		CCA		CAA	Gln	CGA		A
	CTG		CCG		CAG		CGG		G
A	ATT	Ile	ACT	Thr	AAT	Asn	AGT	Ser	T
	ATC		ACC		AAC		AGC		C
	ATA		ACA		AAA	Lys	AGA	Arg	A
	ATG	Met	ACG		AAG		AGG		G
G	GTT	Val	GCT	Ala	GAT	Asp	GGT	Gly	T
	GTC		GCC		GAC		GGC		C
	GTA		GCA		GAA	Glu	GGA		A
	GTG		GCG		GAG		GGG		G

- Codon is the code needed

- Codon is the code needed to convert mRNA into protein language.
- Amino acids are coded by groups of 3 nucleotides, called Codons.
- Each codon consist 3 nitrogen bases which code for one amino acid.
- Genetic code is universal.
- There are 4x4x4 = 64 codons which code 20 amino acids.
- AUG is a START codon as well as the codon for methionine.
- Three stop codons are UAG (amber), UGA (opal) and UAA (ochre).
- More than one codon can code for the same amino acid.
- Each amino acid is coded by more than 1 codon.
- Four codons have special functions in the genetic code
- one start AUG codes for methionine

Protein synthesis (Translation)

20 types of amino acids participate in protein synthesis. Amino acid reacts with ATP to form "amino acid AMP complex" which is also as 'activated amino acid'.

$$\text{Amino acid} + \text{ATP} \xrightarrow{\text{Amino acyl tRNA synthatase}} \text{Amino acyl AMP-enzyme complex} + \text{PP}$$

This reaction is catalysed by a specific amino acyl t-RNA synthatase enzyme that a separate amino acyl t-RN synthatse enzyme for each kind of amino acid. Specific activated amino acid is recognized by t-RNA. Now amino acid attach to amino acid attachment site of its specific t-RNA and AMP and enzyme are separate from it. This is called charging of t-RNA. Amino acid t-RNA complex is also called 'charged t-RNA'. Now this moves to the ribosome for protein synthesis.

Amino acid AMP-enzyme comples +t-RNA → Amino acyl t-RNAcomplex + AMP + enzyme

Protein synthesis involves three steps: Initiation, elongation and termination.

Translation is the process in which cellular ribosomes create proteins from mRNA.

In protein synthesis following chemical reactions involve

1 The energy of the charged tRNA is converted into a peptide bond linking the amino acid to another one on the ribosome.

2. New amino acids are linked by peptide bond to the growing polypeptide chain. This process continues until aa_n (the final amino acid) is added.

3. The complete process involves mRNA, ribosomes, several additional protein factors, enzymes and inorganic ions.

- Ribosomes consist of two subunits in prokaryotes (sediment as 50S as large subunit and 30S as small subunit and associate to form a 70S particle. The eukaryotic counterparts are 60S and 40S for the large and small subunits and 80S for the complete ribosome.
- Ribosomes contain specific sites that enable them to bind to the mRNA, the tRNAs, and specific protein factors required for protein synthesis.
- The mRNA binds to the 30S small subunit.
- The tRNAs bind to two sites on the ribosome. These sites overlap the subunits. The **A site** is the entry site for an aminoacyl-tRNA (a tRNA carrying a single amino acid).

- The peptidyl-tRNA carrying the growing polypeptide chain binds at P site.
- Each new amino acid is added by the transfer of the growing chain to the new aminoacyl-tRNA forming a new peptide bond.
- The deacylated tRNA is then released from the P site and the ribosome moves one codon along the message, transferring the new peptidyl-tRNA to the P site and leaving the A site vacant for the next incoming aminoacyl-tRNA.
- Thus, protein synthesis divided into three distinct steps

Initiation

- Initiation requires mRNA, ribosomes, specific tRNA molecules and the participation of several factors (initiation factors IF1, IF2, and IF3).
- In *E. coli* and in most other prokaryotic organisms, the first amino acid in any newly synthesized polypeptide is *N*-formylmethionine.
- It is inserted not by $tRNA^{Met}$, however, but by an initiator tRNA called $tRNA^{fMET}$.
- This initiator tRNA has the normal methionine anticodon but inserts *N*-formylmethionine rather than methionine.
- In *E. coli,* AUG (methionine) and GUG (Valine) and on rare occasions UUG (Leucine) serve as initiation codons.
- When initiation codons in the initiation position recognized by *N*-formylMet-tRNA, and methionine appears as the first amino acid in the chain.
- The first step in initiation is the binding of the mRNA to the 30S subunit. The binding is stimulated by the initiation factor IF3.
- The initiation factor IF2 binds to GTP and to the initiator fMet-tRNA and stimulates the binding of fMet-tRNA to the initiation complex leading the fMet-tRNA into the P site.
- A ribosomal protein splits the GTP bound to IF2 helping to drive the assembly of the two ribosomal subunits. At this stage, the factors IF2 and IF3 are released.

Ribosome-binding sites

- John Shine and Lynn Dalgarno first noticed that true initiation codons were preceded by sequences that paired well with the 3′ end of 16S rRNA. There is a short but variable separation between the Shine-Dalgarno sequence and the initiation codon.
- The 16S rRNA results in ribosome-mRNA complexes leading to protein initiation in the presence of fMet-tRNA.

Elongation

- Elongation factor EF-Tu mediates the entry of amino-acyl-tRNAs into the A site.
- To do so, EF-Tu first binds to GTP. This activated EF-Tu–GTP complex binds to the tRNA.
- Hydrolysis of the GTP of the complex to GDP helps drive the binding of the aminoacyl-tRNA to the A site at which point the EF-Tu is released and leav the new tRNA in the A site.
- Elongation factor EF-Ts mediates the release of EF-Tu–GDP from the ribosome and the regeneration of EF-Tu–GTP.
- In the translocation step, the polypeptide chain on the peptidyl-tRNA is transferred to the aminoacyl-tRNA on the A site in a reaction catalyzed by the enzyme peptidyltransferase.

- The ribosome then translocates by moving one codon along the mRNA which, goes in the 5′ → 3′ direction.
- This step is mediated by the elongation factor EF-G and is driven by splitting a GTP to GDP.
- This action releases the uncharged tRNA from the P site and transfers the newly formed peptidyl-tRNA from the A site to the P site.

Termination

- Translation is terminated by three chain-termination codons are UAG, UAA and UGA. These three triplets are not recognized by a tRNA, but instead by protein factors, termed release factors (RF1, RF2 and RF3).
- RF1 recognizes the triplets UAA and UAG, and RF2 recognizes UAA and UGA.
- RF3 helps to catalyze chain termination.
- When the peptidyl-tRNA is in the P site, the release factors bind to the A site.
- The polypeptide is then released from the P site and the ribosomes dissociate into two subunits in a reaction driven by the hydrolysis of a GTP molecule.

Nonsense suppressor mutations

- Suppressors of the nonsense mutations defined by Brenner and co-workers.
- Many of these nonsense suppressor mutations are known to alter the anticodon loop of specific tRNAs in such a way as to allow recognition of a nonsense codon in mRNA. Thus, an amino acid is inserted in response to the nonsense codon, and translation continues past that triplet.
- The amber mutation replaces a wild-type codon with the chain-terminating nonsense codon UAG which results in prematurely cutting off the protein at the corresponding position.
- The suppressor mutation in this case produces a $tRNA^{Tyr}$ with an anticodon that recognizes the mutant UAG stop codon. The suppressed mutant thus contains tyrosine at that position in the protein.

Protein product and processing

- After translation of mRNA into its protein product, protein processing continued.
- Membrane proteins or proteins that are secreted from the cell are synthesized with a short leader peptide, called a signal sequence, at the amino-terminal (N-terminal) end. This signal sequence is a stretch of 15 to 25 amino acids, most of which are hydrophobic. It allows for recognition by factors and protein receptors that mediate transport through the cell membrane.
- In this process, the signal sequence is cleaved by a peptidase
- Several small peptide hormones, such as corticotropin (ACTH), result from the specific cleavage of a single, large polypeptide precursor.

Protein splicing

- An extraordinary process that splices out an internal segment of certain proteins has been described in both prokaryotes and eukaryotes. This internal segment is termed an intervening protein sequence, or IVPS.
- The essential facet of this process is the formation of a new peptide bond between the two sequences flanking the IVPS.

- This reaction is autocatalytic and can take place in vitro.
- All IVPS segments contain an endonuclease activity, although this activity is unrelated to the protein-splicing reaction.

Inhibitors of protein synthesis

- Tetracyclin: The primary target is the 30S ribosome, preventing binding of aminoacyl tRNA to the acceptor (A) site on mRNA.
- Chloramphenicol: It binds to the 50S ribosome preventing attachment to the acceptor (A) site on mRNA and causes Gray baby syndrome.
- Streptogramins: This is a combination of two drugs quinupristin and dalfopristin. Both are protein synthesis inhibitors working at the 50S ribosomal subunit.
- Oxazolidinones: Linezolid is the sole member of this new class of bacteriostatic agents. It is active at the 50S ribosomal subunit. It inhibits formation of the initiation complex.

Multiple Choice Questions

1. Overall how many different macromolecules cooperate to synthesize polypeptides?
(A) 10 (B) 150
(C) 300 (D) 600

2. Protein synthesis can account for .... of the chemical energy used by a cell for all biosynthetic reactions.
(A) Up to 50% (B) Up to 30%
(C) Up to 70% (D) Up to 90%

3. How many ribosomes, protein synthesis–related protein factors and enzymes, and tRNA molecules in a typical bacterial cell can account for more than 35% of the cell's dry weight?
(A) 100,000 = ribosomes, 15,000 = protein synthesis–related protein factors and enzymes and 200,00 = tRNA molecules
(B) 15,000 = ribosomes, 100,000 = protein synthesis–related protein factors and enzymes and 200,000 = tRNA molecules
(C) 100,000 = ribosomes, 200,000 = protein synthesis–related protein factors and enzymes and 15,000 = tRNA molecules
(D) 15,000 = ribosomes, 200,000 = protein synthesis–related protein factors and enzymes and 100,000 = tRNA molecules

4. At which temperature a polypeptide of 100 residues is synthesized in an *Escherichia coli* cell in about 5 seconds.
(A) 94°C (B) 50°C
(C) 37°C (D) 70°C

5. Who identified that ribosomes has the sites of protein synthesis through his experiment on rats using radioactive amino acids?
(A) Paul Zamecnik and his colleagues
(B) Mahlon Hoagland and Zamecnik
(C) Robert Holley
(D) Francis Crick

6. Who discovered transfer RNA (tRNA)?
(A) Paul Zamecnik and his colleagues
(B) Mahlon Hoagland and Zamecnik
(C) Robert Holley
(D) Francis Crick

7. Who proposed the clover leaf structure of transfer RNA (tRNA)
(A) Paul Zamecnik and his colleagues
(B) Robert Holley
(C) Mahlon Hoagland and Zamecnik
(D) Francis Crick

8. Which of the following is heat-stable soluble RNA?
(A) mRNA (B) mRNA
(C) tRNA (D) siRNA

9. Which of the following enzyme catalyzes the reaction to form aminoacyl-tRNA?
(A) Aminoacyl-tRNA synthetases
(B) Acyl-CoA:cholesterol acyltransferase
(C) Aminotransferases
(D) Peptidyl transferase

10. How many amino acids were encoded by four nucleotides (A,T,G,C) that carry genetic information?
(A) 10 (B) 300 (C) 20 (D) 100

11. What is called the process of the formation of the amino acid sequence of a polypeptide from the nucleotide sequence of an mRNA?
(A) Transcription
(B) Reverse transcription
(C) Replication (D) Translation

12. How many nucleotide residues of DNA are necessary to encode each amino acid?
(A) Two (B) Four
(C) Three (D) One

13. Three nucleotides encode one codon for each amino acid Why cannot encode one codon by two or four nucleotides?
(A) Codons of two nucleotides insufficient to encode 20 amino acids ($4^2 = 16$)
(B) Codons of four nucleotides are more to encode 20 amino acids ($4^4 = 256$)
(C) Three are more sufficient to encode 20 amino acids ($4^3 = 64$)
(D) All of the above

14. What is codon that codes for a specific amino acid?
(A) A duplet of nucleotides
(B) A triplet of nucleotides
(C) A triplet of nucleosides
(D) A duplet of nucleosides

15. What is reading frame?

(A) It established by a specific first codon in the sequence and a new codon begins every three nucleotide residues

(B) It established by a specific first codon in the sequence and a new codon begins every three nucleoside residues

(C) It established by a specific first codon in the sequence and a new codon begins every two nucleotide residues

(D) It established by a specific first codon in the sequence and a new codon begins every three amino acid residues

16. In principle, how many reading frames are possible from any given single-stranded DNA or mRNA sequence?

(A) Two (B) Three

(C) Four (D) Only one

17. Who found the three letter code words for each amino acid?

(A) Marshall Nirenberg and Heinrich Matthaei (1961)

(B) H. Gobind Khorana (1964)

(C) Mahlon Hoagland and Zamecnik (1950s)

(D) Nirenberg and Philip Leder (1964)

18. What happen after incubation of synthetic polyuridylate, poly(U) and GTP, ATP with an *E. coli* extract and a mixture of the 20 amino acids in 20 different tubes? Each tube contains a different radioactively labeled amino acid.

(A) No amino acid encoded by any codon

(B) All amino acids encoded by only one codon

(C) Only one radioactive amino acid phenylalanine encoded by the triplet UUU

(D) All of the above are right

19. What happen? if polynucleotide phosphorylase, which catalyzes the formation of RNA polymers, If it is presented with a mixture of five parts ADP and one part CDP?

(A) It makes only poly(A)

(B) It makes only poly(C)

(C) It makes a polymer of about five-sixths of the cytidylate residues and one-sixth adenylate

(D) It makes a polymer of about five-sixths of the adenylate residues and one-sixth cytidylate

20. Who conduct a experiment on isolated *E. coli* ribosomes for binding a specific aminoacyl-tRNA in the presence of the synthetic polynucleotide messenger?

(A) Marshall Nirenberg and Heinrich Matthaei (1961)

(B) H. Gobind Khorana (1964)

(C) Mahlon Hoagland and Zamecnik (1950s)

(D) Nirenberg and Philip Leder (1964)

21. What happen? if the ribosomes are incubated with poly(C) and any aminoacyltRNA together, except phenylalanyl-tRNAPhe?

(A) It recognizes only the CCC triplets in poly(C)

(B) It recognizes only the UUU triplets in poly(U)

(C) It recognizes only the UUU triplets in poly(A)

(D) It recognizes only the UUU triplets in poly(T)

22. Who first discovered chemical methods to synthesize polyribonucleotides with defined, repeating sequences of two to four bases.

(A) Marshall Nirenberg and Heinrich Matthaei

(B) Har Gobind Khorana

(C) Mahlon Hoagland and Zamecnik

(D) Nirenberg and Philip Leder

23. How many codons code 20 amino acid residues?

(A) 64 (B) 3

(C) 61 (D) 27

24. How many codons do not code any amino acids?

(A) 64 (B) 3

(C) 61 (D) 27

25. What is called the codons that not code any amino acids?

(A) Initiation codons

(B) Elongation codons

(C) Start codons

(D) Termination codons

26. What happen If the initial reading frame is off by one or two bases, or if translation somehow skips a nucleotide in the mRNA?
(A) A "missense" protein with a garbled amino acid sequence
(B) A "sense" protein with a garbled amino acid sequence
(C) Both "missense and sense" protein with a garbled amino acid sequence
(D) A "missense" protein with a progressive amino acid sequence

27. Which of the following mostly act as initiation codon for the beginning of a polypeptide in all cells?
(A) UAA (B) UAG
(C) UGA (D) AUG

28. Which of the following does not act as termination codon for the stoping of a polypeptide in all cells?
(A) UAA (B) AUG
(C) UGA (D) UAG

29. Which of the following amino acid residue is coded by initiation codon AUG?
(A) Metheonin (B) Histidine
(C) Threonin (D) Tryptophan

30. What are called UAA, UAG, and UGA codons used for the stopping of a polypeptide in all cells?
(A) Termination codons
(B) Stop codons
(C) Nonsense codons
(D) All of the above

31. Which of the following act as termination codons that normally signal the end of polypeptide synthesis and do not code for any known amino acids in all cells?
(A) UAA (B) UAG
(C) UGA (D) All of the above

32. What is open reading frame (ORF)?
(A) A reading frame with a termination codon among 50 or more codons
(B) A reading frame without a termination codon among 50 or more codons
(C) A reading frame without a initiation codon among 50 or more codons
(D) A reading frame with a initiation codon among 50 or more codons

33. What is the use of open reading frame (ORF)?
(A) To find genes among the huge background of nongenic DNA
(B) To show the sequence of nucleotide in DNA
(C) To find the process of protein synthesis
(D) To find the cDNA sequence for gene expression

34. How many nucleotides require for making 100 codons?
(A) 100 (B) 61
(C) 300 (D) 1000

35. How many nucleotides required for coding 100 amino acids?
(A) 100 (B) 61
(C) 300 (D) 1000

36. How many amino acids are coded by 1000 nucleotides?
(A) 333 (B) 61
(C) 300 (D) 1000

37. What are degenerated codons?
(A) An amino acid specified by only one codon
(B) A nucleotide used to form more than one codon
(C) Termination codons are also called degenerate codons
(D) An amino acid specified by more than one codon

38. When an amino acid may be specified by more than one codon, so the code is described as
(A) Universal code (B) Degenerate code
(C) Termination code (D) Uniform code

39. Which of the following amino acids code by only one codon?
(A) Methionine and tryptophan
(B) Methionine and tryrosin
(C) Tryptophan and Tyrosin
(D) Phenyl alanine and tryptophan

40. Which of the following amino acids are coded by maximum codons (six codons)?
(A) Methionine, valine and tryptophan
(B) Tyrosin, tryptophan and phenyl alanine
(C) Leucine, Serine and Arginine
(D) Isoleucine, Serine and Arginine

41. Which of the following amino acid has three codons?
(A) Methionine (B) Tryrosin
(C) Leucine (D) Isoleucine

42. Which f the following exception of the genetic code is nearly universal?
(A) A few minor variations in mitochondria
(B) Some bacteria
(C) some single-celled eukaryotes
(D) All of the above

43. What is difference between several different codons that specify one amino acid?
(A) At the second base position in codons at the 3' end
(B) At the third base position in codons at the 3' end
(C) At the first base position in codons at the 3' end
(D) At the third base position in some codons and second base position in other

44. Which of the following symbols are used to different codons that code one amino acid according to wobble hypothesis?
(A) XYAU or XYAC
(B) AGXY or UCXY
(C) XYCG or XYAU
(D) XYAG or XYUC

45. Transfer RNAs base-pair with mRNA codons at a three-base sequence on the tRNA called ....
(A) Codon (B) Anticodon
(C) Common Triplet (D) Antisense codon

46. In which direction read any codons of mRNA?
(A) 3' to 5' direction
(B) 5' to 2' direction
(C) 5' to 3' direction
(D) Both 5' to 3' and 3' to 5' direction

47. How pair between codon and anticodon?
(A) Third base of the codon in mRNA pairs with first base of the anticodon
(B) First base of the codon in mRNA pairs with third base of the anticodon
(C) First base of the codon in mRNA pairs with first base of the anticodon
(D) First base of the codon in mRNA pairs with second base of the anticodon

48. Which of the following nucleotide involved in anticodon of tRNA base-pair with codon in mRNA, If the anticodon triplet of a tRNA recognized only one codon triplet through Watson-Crick base pairing at all three positions?
(A) Inosinate (B) Hypoxanthin
(C) Xanthin (D) All of the above

49. Which of the following base is a part of nucleotide inosinate that is involved in anticodon of tRNA base-pair with codon in mRNA, If the anticodon triplet of a tRNA recognized only one codon triplet through Watson-Crick base pairing at all three positions?
(A) Thymine (B) Hypoxanthin
(C) Xanthin (D) Uracil

50. Which of the following base pairing is much weaker than the hydrogen bonds of Watson-Crick base pairs?
(A) Inosinate with uracil, cytosine and adenine
(B) Uracil with adenine
(C) Cytosine with guanine
(D) Adenine with thymine

51. In yeast, one tRNAArg has the anticodon (5')ICG, which recognizes three arginine codons that are ....
(A) (5')CGA (B) (5')CGU
(C) (5')CGC (D) All of the above

52. Who proposed wobble hypothesis?
(A) Marshall Nirenberg and Heinrich Matthaei
(B) H. Gobind Khorana
(C) Crick
(D) Nirenberg and Philip Leder

53. Which of the following relationship is not proposed according to wobble hypothesis by Crick?
(A) The first two bases of an mRNA codon always form strong Watson-Crick base pairs with the corresponding bases of the tRNA anticodon
(B) The first base of the anticodon determines the number of codons recognized by the tRNA
(C) When an amino acid is specified by several different codons, the codons that

differ in either of the first two bases require different tRNAs

(D) A minimum of 31 tRNAs are required to translate all 61 codons

54. Which of the following cell organells carry out the process of the particular amino acid sequence of a protein is constructed through the translation of information encoded in mRNA?

(A) Ribosomes
(B) Lysosomes
(C) Nucleolus
(D) Endoplasmic reticulum

55. Which of the following molecules act as adapter molecule that recognize codons and insert amino acids into their appropriate sequential positions in the polypeptide?

(A) mRNA (B) rRNA
(C) tRNA (D) dsDNA

56. The codon AUG signals .... of translation. The triplets UAA, UAG, and UGA are signals for ....

(A) Initiation and termination
(B) Elongation and termination
(C) Initiation and elongation
(D) Initiation, elongation and termination

57. What is called the third base position in each codon is much less specific than the first and second/

(A) Bubble (B) Confusion
(C) Wobble (D) Vibrate

58. Which of the first stage of protein synthesis?

(A) Folding and Posttranslational Processing
(B) The nascent polypeptide is lengthened by successive amino acid units
(C) Large ribosomal subunit then binds to form an initiation complex
(D) Activation of amino acids

59. Choose the correct sequence of the stages of protein synthesis?

i. Folding and Posttranslational Processing
ii. The nascent polypeptide is lengthened by successive amino acid units
iii. Activation of amino acids
iv. Large ribosomal subunit then binds to form an initiation complex
v. Completion of polypeptide chain signaled by a termination codon in the mRNA

(A) i, ii, iii, iv, v
(B) iii, iv, ii, v, i
(C) iii, iv, i, v, ii
(D) iii, iv, ii, iv, i

60. When is completed the activation of amino acids to facilitate formation of a peptide bond during translation?

(A) Before termination stage
(B) After termination stage
(C) Before initiation stage
(D) During elongation stage

61. Which of the following first fundamental chemical requirements before initiation of translation must necessary for the synthesis of a polypeptide with a defined sequence?

(A) Carboxyl group of each amino acid must be activated
(B) The nascent polypeptide is lengthened by successive amino acid units
(C) Large ribosomal subunit then binds to form an initiation complex
(D) A link between each new amino acid and the information in the mRNA

62. Which of the following reaction takes place in the cytosol during translation?

(A) Synthesis of ribosomes
(B) Attachment of the right amino acid to the right tRNA
(C) The binding of each incoming aminoacyl-tRNA to growing polypeptide chain
(D) Releasing the new synthesized polypeptide chain

63. Which of the following reaction is not take place in ribosome during translation?

(A) The initiating aminoacyl-tRNA basepairs with the mRNA codon AUG
(B) Attachment of the right amino acid to the right tRNA
(C) The binding of each incoming aminoacyl-tRNA to growing polypeptide chain
(D) Releasing the new synthesized polypeptide chain

64. Which of the following Mg^{2+}-dependent activating enzymes require for activation of carboxyl group of each amino acid during translation?
(A) Aminoacyl-tRNA synthetases
(B) Peptidyl transferase
(C) RNA polymerase.
(D) Protease

65. Which of the following stage is called charging stage beacause this stage requires ATP energy and Mg^{2+}-dependent activating enzymes 'aminoacyltRNA synthetase'?
(A) Releasing the new synthesized polypeptide chain
(B) A link between each new amino acid and the information in the mRNA
(C) The binding of each incoming aminoacyl-tRNA to growing polypeptide chain
(D) Attachment of the right amino acid to the right tRNA

66. Choose the correct sequence of reaction performed during initiation stage of protein synthesis?
i. The mRNA binds to the smaller ribosomal subunit and to initiating aminoacyl-tRNA
ii. The large ribosomal subunit binds to form an initiation complex
iii. The initiating aminoacyl-tRNA basepairs with the mRNA codon AUG
iv. The mRNA codon AUG signals the beginning of the polypeptide

(A) i, ii, iii, iv (B) i, iii, ii, iv
(C) i, ii, iii, iv (D) iv, iii, ii, i

67. What are the requirements for the completion of initiation stage of protein synthesis?
(A) GTP and cytosolic proteins
(B) ATP and ribosomal proteins
(C) Only cytosolic proteins
(D) ATP, ribosomal proteins and cytosolic proteins

68. Which of the following mRNA codon is basepaired with the initiating aminoacyl-tRNA and signals the starting of polypeptide?
(A) UAA (B) UAG
(C) UGA (D) AUG

69. What are called cytosolic proteins that promote the basepairing between the initiating aminoacyl-tRNA and the mRNA codon AUG that signals the beginning of the polypeptide?
(A) Elongation factors
(B) Initiation factors
(C) Transcription factors
(D) Release factors

70. In which stage of translation, the nascent polypeptide is lengthened by covalent attachment of successive amino acid units?
(A) Elongation
(B) Initiation
(C) Activation of amino acids
(D) Termination

71. What are called cytosolic proteins that require for the nascent polypeptide is lengthened by covalent attachment of successive amino acid units?
(A) Elongation factors
(B) Initiation factors
(C) Transcription factors
(D) Release factors

72. How facilitated the binding of each incoming aminoacyl-tRNA and the movement of the ribosome along the mRNA during elongation stage of translation?
(A) By the hydrolysis of ATP
(B) By the activation of amino acids
(C) By the hydrolysis of GTP
(D) By the release factors

73. What are called cytosolic proteins that require for the completion of the polypeptide chain is signaled by a termination codon in the mRNA and releasing of the new polypeptide from the ribosome?
(A) Elongation factors
(B) Initiation factors
(C) Transcription factors
(D) Release factors

74. Which of the following cell organelle is connected with newly synthesized polypeptide chain?
(A) Lysosome (B) Ribosome
(C) Peroxysome (D) Glyoxysome

75. Which of the following processes help to achieve the biologically active form and proper three-dimensional conformation of newly synthesized polypeptide chain?
(A) By the hydrolysis of ATP
(B) By the activation of amino acids
(C) Movement of the ribosome along the mRNA
(D) Folding and post-translational processing

76. Which of the following enzymatic process is not included in post-translation processing before or after folding of new polypeptides?
(A) Removal of one or more amino acids
(B) Addition of acetyl, phosphoryl, methyl or carboxyl group to amino acid residues
(C) Activation of carboxyl group of each amino acid
(D) Proteolytic cleavage and/or attachment of oligosaccharides or prosthetic groups

77. Which of the following major key components help in protein biosynthesis?
(A) Ribosome and tRNAs
(B) Endoplasmic reticulum and rRNA
(C) Cytosol and nucleolus
(D) Glyoxysome and mRNA

78. How many ribosomes are contained in each *E. coli* cell?
(A) 50,000 or more ribosomes
(B) 15,000 or less ribosomes
(C) 50,00,00 or more ribosomes
(D) 15,000 or more ribosomes

79. What is the composion of bacterial ribosomes?
(A) About 95% rRNA and 5% protein
(B) About 65% rRNA and 35% protein
(C) About 60% rRNA, 5% tRNA and 35% protein
(D) About 60% rRNA and 40% protein

80. What is the diameter of bacterial ribosomes?
(A) About 5 nm (B) About 100 nm
(C) About 18 nm (D) About 23 nm

81. Which of the following unequal subunits of bacterial ribosomes combined sedimentation coefficient of 70S?
(A) 30S and 50S (B) 40S and 60S
(C) 23S and 5.8S (D) 70S and 80S

82. Which of the following type of ribosome is found in bacteria based on sedimentation coefficient?
(A) 60S (B) 40S
(C) 80S (D) 70S

83. Who discovered that ribosomes are the complexes responsible for protein synthesis?
(A) Zamecnik
(B) Thomas Cech
(C) Grunberg-Manago
(D) Ochoa

84. Who demonstrated that both ribosomal subunits can be broken down into their RNA and protein components and reconstituted in vitro?
(A) Zamecnik
(B) Palade
(C) Grunberg-Manago
(D) Masayasu Nomura

85. What is molecular weight of bacterial ribosomes (70S) ?
(A) ~2.7 million (B) ~4.2 million
(C) ~1.8 million (D) ~0.9 million

86. Which of the following rRNA units found in 50S subunit of bacterial ribosome?
(A) 5S and 28S rRNAs
(B) 5S 23S and 16S rRNAs
(C) 5S and 23S rRNAs
(D) 5S and 16S rRNAs

87. Which of the following subunits of bacteriasl ribosome form the structural core in 50S subunit?
(A) 5S and 28S rRNAs
(B) 5S and 23S rRNAs
(C) 5S, 16S and 23S rRNAs
(D) 5S and 16S rRNAs

88. Which of the following types of proteins overhang into the rRNA core of the ribosome to stabilize its structure?
(A) Globular domains
(B) Filamentous domain
(C) Z-finger structure
(D) Snakelike protein extensions

89. What is the diameter of the ribosomes of eukaryotic cells other than mitochondrial and chloroplast ribosomes?

(A) About 23 nm (B) About 100 nm
(C) About 18 nm (D) About 10 nm

90. Which of the following type of ribosome is found in the eukaryotic cells based on sedimentation coefficient?
(A) 50S (B) 30S
(C) 80S (D) 70S

91. Which of the following unequal subunits of ribosomes of the eukaryotic cells combined sedimentation coefficient of 80S?
(A) 30S and 50S (B) 40S and 60S
(C) 23S and 5.8S (D) 70S and 80S

92. How many proteins are contained in eukaryotic ribosomes?
(A) More than 50 same proteins
(B) More than 40 different proteins
(C) More than 80 different proteins
(D) More than 60 different proteins

93. Which of the following eukaryotic cell organess have autonomous ribosomes like bacterial ribosomes?
(A) Mitochondria and chloroplasts
(B) Nucleus and Endoplasmic reticulum
(C) Mitochondria, Nucleus and chloroplasts
(D) Lysosomes and golgi complex

94. Which of the following type of ribosome is found in the mitochondria and chloroplasts of the eukaryotic cells based on sedimentation coefficient?
(A) 50S (B) 30S
(C) 80S (D) 70S

95. Where found tRNA in eukaryotic cells?
(A) Nucleus (B) Cytosol
(C) ER (D) Golgy complex

96. How many nucleotide residues found in tRNA of bacteria and in the cytosol of eukaryotes?
(A) Between 90 and 100 nucleotide residues
(B) Between 30 and 50 nucleotide residues
(C) Between 73 and 93 nucleotide residues
(D) Between 38 and 59 nucleotide residues

97. What is the molecular weight of tRNA of bacteria and in the cytosol of eukaryotes?
(A) 24,000-31,000 (B) ~2.7 million
(C) ~1.8 million (D) ~0.9 million

98. Which of the following cell organelles contain distinctive smaller tRNAs?
(A) Nucleus and golgi body
(B) Mitochondria and chloroplasts
(C) Lysosomes and ribosomes
(D) Peroxysome and glyoxysome

99. Which of the following amino acid initiate protein synthesis?
(A) Methionine (B) Histidine
(C) Tryptophan (D) Cystein

100. Which of the following codon specifies an *amino-terminal* methionine residue to begin protein synthesis?
(A) GUG (B) AGG
(B) GAG (D) AUG

101. Which of the following first nucleic acid to be completely sequenced that contains 76 nucleotide residues with 10 modified bases?
(A) Yeast alanine tRNA
(B) Yeast tyrosine tRNA
(C) Yeast arginine tRNA
(D) Yeast cystein tRNA

102. How many nucleotides contained in first nucleic acid yeast alanine tRNA to be completely sequenced with 10 modified bases?
(A) 96 nucleotide residues
(B) 67 nucleotide residues
(C) 76 nucleotide residues
(D) 10 nucleotide residues

103. How many modified bases are present in first nucleic acid yeast alanine tRNA to be completely sequenced that contains 76 nucleotide residues?
(A) 96 modified bases
(B) 76 modified bases
(C) 67 modified bases
(D) 10 modified bases

104. What are the features of the 5' and 3' ends of most tRNAs, respectively?
(A) A guanylate (pG) residue and trinucleotide sequence CCA
(B) A guanylate (pG) residue and poly(A) tail
(C) Poly (A) tail and trinucleotide sequence CCA

(D) Trinucleotide sequence CCA and a guanylate (pG) residue

105. Which of the following RNA shows cloverleaf structure in its two dimension structure?
(A) mRNA (B) rRNA
(C) tRNA
(D) Both rRNA and tRNA

106. The hydrogen-bonding pattern of all tRNAs forms a cloverleaf structure in ....
(A) Three dimension (B) Two dimension
(C) One dimension (D) Multi dimension

107. How many arms contained in clover leaf structure of tRNA in two dimensions?
(A) Six (B) Five
(B) Three (D) Two

108. How shows tRNA in three dimensions?
(A) Clover leaf like
(B) Twiested
(C) Linear double straded
(D) Three dimensional structure is not possible

109. Which of the following part is not included in the clover leaf structure of tRNA?
(A) Amino acid acceptor arm
(B) Anticodon arm
(C) Poly(A) tail
(D) TΨC arm

110. Which of the following two arms of tRNA are critical for its adaptor function?
(A) Amino acid arm and TΨC arm
(B) Amino acid arm and anticodon arm
(C) Poly(A) tail and D arm
(D) TΨC arm and D arm

111. Which of the following arm is present at 3' end of tRNA and can carry a specific amino acid esterified by its carboxyl group to the 2'- or 3'-hydroxyl group of the A residue?
(A) Amino acid arm (B) Anticodon arm
(C) D arm (D) TΨC arm

112. Which of the following arm of tRNA contains the unusual nucleotide dihydrouridine (D)?
(A) Amino acid arm (B) Anticodon arm
(C) D arm (D) TΨC arm

113. Which of the following arm of tRNA contains ribothymidine (T) and pseudouridine (Ψ)?
(A) Amino acid arm (B) Anticodon arm
(C) D arm (D) TΨC arm

114. Which of the following type of RNA contains ribothymidine (T)?
(A) rRNA (B) mRNA
(C) tRNA (D) snRNA

115. Which of the following nucleotide derivative is not present generally in any type of RNA except tRNA?
(A) Thymine (B) Ribothymidine
(C) Uracil (D) Cytosine

116. Which of the following nucleotide derivative is not present generally in any type of RNA except tRNA?
(A) Thymine (T) (B) Guanine (G)
(C) Uracil (U) (D) Pseudouridine (Ψ)

117. Which of the following nucleotide derivative has an unusual carbon–carbon bond between the base and ribose?
(A) Thymine (T)
(B) Guanine (G)
(C) Uracil (U)
(D) Pseudouridine (Ψ)

118. Which of the following arm of tRNA interacts with the large-subunit Rrna?
(A) Amino acid arm (B) Anticodon arm
(C) TΨC arm (D) D arm

119. Which of the following stage of protein synthesis is take place in cytosol?
(A) Attachment of aminoacyl-tRNA synthetases to the correct amino acids
(B) Addition of amino acids to the carboxyl-terminal end of the growing polypeptide,
(C) Peptide Bond Formation
(D) Movement of the ribosome toward the 3' end of the mRNA

120. Which of the following enzymes esterify the 20 amino acids to their corresponding tRNAs?
(A) Aminoacyl-tRNA synthetases
(B) Peptidyl transferase
(C) RNA polymerase.
(D) Protease

121. Which of the following reaction catalyzed by an aminoacyl-tRNA synthetase during protein synthesis?

(A) Amino acid + tRNA + ATP $\xleftrightarrow[\text{Light}]{Mg^{+2}}$ aminoacyl-tRNA + AMP + PPi

(B) $2H_2O + 2NADP^+ \leftrightarrow 2NADPH + 2H^+ + O_2$

(C) 2 Glycine + $NAD^+ + H_2O \leftrightarrow$ serine + $CO_2 + NH_3 + NADH + H^+$

(D) $(dNMP)n + dNTP \rightarrow (dNMP)_{n+1} + PPi$
DNA Lengthened DNA

122. Which of the following organism contain two types of tRNA specific for methionine are designated tRNAMet and tRNAfMet?

(A) Yeast (B) Human
(C) Bacteria (D) Virus

123. Which of the following amino acid is specified by tRNAfMet (tRNA*N*-formylmethionine) in bacteria?

(A) (5')GUG (B) AUG
(B) GAG (D) AGG

124. Which of the following enzyme catalyze the attachment of methionine to tRNAfMet that coded by (5')AUG initiation codon in bacterial?

(A) Aminoacyl-tRNA synthetases
(B) Peptidyl transferase
(C) RNA polymerase.
(D) Met-tRNA synthetase

125. Which of the following reaction catalyzed by an Met-tRNA synthetase during protein synthesis?

(A) Amino acid + tRNA + ATP $\xleftrightarrow[\text{Light}]{Mg^{+2}}$ aminoacyl-tRNA + AMP + PPi

(B) Methionine + $tRNA^{fMet}$ + ATP $\leftrightarrow$ Met-$tRNA^{fMet}$ + AMP + PPi

(C) 2 Glycine + $NAD^+ + H_2O \leftrightarrow$ serine + $CO_2 + NH_3 + NADH + H^+$

(D) $(dNMP)n + dNTP \rightarrow (dNMP)_{n+1} + PPi$
DNA Lengthened DNA

126. What are the requirements for the initiation of polypeptide synthesis in bacteria?

(A) 30S ribosomal subunit
(B) 50S ribosomal subunit and Mg^{2+}
(C) GTP (D) All of the above

127. What are the requirements for the initiation of polypeptide synthesis in bacteria?

(A) mRNA coding for the polypeptide
(B) A set of three proteins (IF-1, IF-2 and IF-3)
(C) Initiating fMet-tRNAfMet
(D) All of the above

128. Which of the following step is not involved in the formation of the initiation complex for the synthesis of polypeptide in bacteria?

(A) 30S ribosomal subunit binds initiation factors (IF-1 and IF-3) and mRNA
(B) Complex is joined by both GTP-bound IF-2 and the initiating fMet-tRNAfMet
(C) The appropriate incoming aminoacyl-tRNA binds to GTP-bound EF-Tu complex
(D) Large complex combines with the 50S ribosomal subunit

129. Which of the following first step is not involved in the formation of the initiation complex for the synthesis of polypeptide in bacteria?

(A) 30S ribosomal subunit binds initiation factors (IF-1 and IF-3) and mRNA
(B) Complex is joined by both GTP-bound IF-2 and the initiating fMet-tRNAfMet
(C) The appropriate incoming aminoacyl-tRNA binds to GTP-bound EF-Tu complex
(D) Large complex combines with the 50S ribosomal subunit

130. Which of the following subunit of ribosome bind to initiation factors (IF-1 and IF-3) to form the initiation complex for the synthesis of polypeptide in bacteria?

(A) 50S (B) 60S
(C) 30S (D) 40S

131. Where add amino acids of the growing polypeptide when protein synthesis proceeded?

(A) Carboxyl-terminal end
(B) Amino-terminal end
(C) Anticodon site
(D) TψC site

132. Who determined that Protein synthesis begins at the amino-terminal end and proceeds by the stepwise addition of amino acids to the carboxyl-terminal end of the growing polypeptide?
(A) Robert W. Holley (1965)
(B) Günter Blobel (1970)
(C) Howard Dintzis (1961)
(D) George Palade (1970)

133. How many tRNA are found for coding methionine (5'AUG codon) in all organisms?
(A) Two
(B) One
(C) Three
(D) Four

134. Methionine-tRNA codes for .... during protein synthesis.
(A) Only for initiation codon (5'AUG)
(B) Initiation codon (5'AUG) and Met residue in an internal position in a polypeptide
(C) Termination codons (5' UAA, UAG, and UGA)
(D) No one codon

135. What is the function of initiation factors (IF-1 and IF-3) during the formation of the initiation complex for the synthesis of polypeptide in bacteria?
(A) To guide the initiating (5')AUG to its correct position by the Shine-Dalgarno sequence in the mRNA
(B) To prevent the 30S and 50S subunits from combining prematurely
(C) To form a peptide bond between the two amino acids
(D) To release the newly synthesized polypeptide chain

136. Who identified the Shine-Dalgarno sequence that guide the initiating (5')AUG to its correct position in the mRNA in bacteria?
(A) John Shine and Lynn Dalgarno
(B) Günter Blobel and George Palade
(C) Masayasu Nomura
(D) Robert W. Holley

137. What is the function of Shine-Dalgarno sequence in protein synthesis process of bacteria?
(A) To guide the initiating (5')AUG to its correct position by the Shine-Dalgarno sequence in the mRNA
(B) To prevent the 30S and 50S subunits from combining prematurely
(C) To form a peptide bond between the two amino acids
(D) To release the newly synthesized polypeptide chain

138. Which of the consensus sequence guide the initiating (5')AUG to its correct position in the mRNA in protein synthesis process of bacteria?
(A) Shine-Dalgarno sequence
(B) Palindromes sequences
(C) mirror repeat sequences
(D) cozak sequences

139. What is called the consensus sequence 5' AGGAGGU3' that guide the initiating (5')AUG to its correct position in the mRNA in protein synthesis of bacteria?
(A) Shine-Dalgarno sequence
(B) Palindromes sequences
(C) Mirror repeat sequences
(D) cozak sequences

140. What is the size of the consensus sequence Shine-Dalgarno sequence that guide the initiating (5')AUG to its correct position in the mRNA in protein synthesis of bacteria?
(A) 1 to 5bp
(B) 15 to 20bp
(C) 3bp
(D) 8 to 10bp

141. Which of the following four to nine nucleotide residues are present in the consensus sequence Shine-Dalgarno sequence that guide the initiating (5')AUG to its correct position in the mRNA in protein synthesis of bacteria?
(A) Purines
(B) Pyrimidines
(C) Pyrimidine derivatives
(D) Purine derivatives

142. Which of the following sequence base-pairs with a complementary pyrimidine-rich sequence near the 3' end of the 16S rRNA of the 30S ribosomal subunit during the initiation of protein synthesis in bacteria?
(A) Mirror repeat sequences
(B) Palindromes sequences
(C) Shine-Dalgarno sequence
(D) Kozak sequences

143. Which of the following rRNA of the 30S ribosomal subunit help in the base-pairing of shine-dalgarno sequence with a complementary pyrimidine-rich sequence?
(A) 23S rRNA (B) 16S rRNA
(C) 18S rRNA (D) 28S rRNA

144. In which stage of protein synthesis formed a mRNA-rRNA interaction?
(A) Elongation (B) Termination
(C) Initiation (D) Releasing

145. Which of the following sites are contained in bacterial riosomes?
(A) Aminoacyl (A) site
(B) Peptidyl (P) site (C) Exit (E) site
(D) All of the above

146. Which of the following site of bacterial ribosome is largely confined to the only 50 subunit?
(A) Aminoacyl (A) site
(B) Peptidyl (P) site (C) Exit (E) site
(D) Both (A) and (B)

147. Which of the following ribosomal subunits contribute to the characteristics of the A and P sites of bacterial ribosomes?
(A) 50S (B) 60S and 40S
(C) 30S (D) Both a and c

148. At which site of bacterial ribosome, the initiating (5')AUG is positioned and fMettRNAfMet can bind?
(A) Aminoacyl (A) site
(B) Peptidyl (P) site (C) Exit (E) site
(D) Both (A) and (B)

149. Initiation codon (5')AUG code fMettRNAfMet that first bind to which site of bacterial ribosome during intiation of protein synthesis?
(A) Aminoacyl (A) site
(B) Peptidyl (P) site (C) Exit (E) site
(D) Both (A) and (B)

150. During the subsequent elongation stage, all incoming aminoacyl-tRNAs including the Met-tRNAMet that binds to interior AUG codons, bind first to .... of bacterial ribosome?
(A) Aminoacyl (A) site
(B) Peptidyl (P) site (C) Exit (E) site
(D) Both (A) and (B)

151. From which site of bactefrial ribosome, the "uncharged" tRNAs leave during elongation of protein synthesis?
(A) Aminoacyl (A) site
(B) Peptidyl (P) site
(C) Exit (E) site
(D) Both (A) and (B)

152. What is the function of initiation factor IF-1 after binding at the A site in bacterial protein synthesis?
(A) To prevent tRNA binding at this site
(B) To prevent the 30S and 50S subunits from combining prematurely
(C) To form a peptide bond between the two amino acids
(D) To release the newly synthesized polypeptide chain

153. In which stage of prokaryotic protein synthesis, the anticodon of initiating fMet-tRNAfMet pairs correctly with the mRNA's initiation codon?
(A) Elongation (B) Termination
(C) Initiation (D) Releasing

154. Which of the following energy molecule is used during initiation of protein synthesis of bacteria?
(A) ATP (B) GTP
(C) NADPH (D) NADP

155. Which of the following structure containing the mRNA and the initiating fMettRNAfMet, is produced after completion initiation of protein synthesis of bacteria?
(A) A functional 50S ribosome initiation complex
(B) A functional 30S ribosome initiation complex
(C) A functional 80S ribosome initiation complex
(D) A functional 70S ribosome initiation complex

156. Which of the following site of bacterial ribosomes paly a important role to form a functional 80S ribosome initiation complex that ready for elongation?
(A) Aminoacyl (A) site
(B) Peptidyl (P) site (C) Exit (E) site
(D) Both (A) and (B)

157. How eukaryotic mRNAs are bound to the ribosome as a complex during initation of protein synthesis?

(A) By the poly(A) binding protein (PAB) at the 3' end,

(B) 5' cap

(C) By ligase enzyme

(D) All of the above

158. How many initiation factrtos for protein synthesis found in prokaryoptic and eukaryotic cells, respectively?

(A) Three and at least four

(B) Both have three

(C) Three and at least nine

(D) At least nine and three

159. Which of the following complex includes the proteins eIF4E, eIF4G, and eIF4A, binds to the 5' cap through eIF4E during initation of protein synthesis in eukaryotes?

(A) elF4F (B) EF-Ts

(C) eIF4B (D) EF-Tu

160. Which of the following protein initiation factor has an RNA helicase activity in eukaryotic cells?

(A) eIF4E (B) eIF4G

(C) eIF4A (D) eIF4B

161. What is the function of the RNA helicase activity of eIF4A contained in the eIF4F complex involved in signaling the beginning of the reading frame during initiation of protein synthesis of eukaryotes?

(A) To facilitat translational regulation of gene expression

(B) To eliminate secondary structure in the 5' untranslated portion of the mRNA

(C) To detect the initiating (5')AUG within the mRNA

(D) To facilitate the binding of initiating Met-tRNAMet to 40S ribosomal subunit

162. Chhose the correct matching between prokaryotic initiation factors of protein synthesis?

Sl.	Initiation factor	Sl.	Function
1.	IF-1	A.	prevents premature association of 50S subunit
2.	IF-2	B.	Prevents premature binding of tRNAs to A site
3.	IF-3	C.	Facilitates binding of fMet-tRNAfMet to 30S ribosomal subunit

(A) 1 – A, 2 – B and 3 – C

(B) 1 – B, 2 – A and 3 – C

(C) 1 – C, 2 – B and 3 – A

(D) 1 – B, 2 – C and 3 – A

163. In which stage peptide bonds are formed during protein synthesis in both prokaruotic and eukaryotic cells?

(A) Elongation and termination

(B) Elongation

(C) Initiation and elongation

(D) Termination

164. What are the requirements to continue elongation stage of protein synthesis in bacteria?

(A) Aminoacyl-tRNAs

(B) A set of three soluble cytosolic proteins

(C) GTP. (D) All of the above

165. Which of the following protein factors are involved in intiation stage of protein synthesis in bacteria?

(A) RF-1, RF-2 and RF-3

(B) IF-1, IF-2 and IF-3

(C) EF-Tu, EF-Ts and EF-G

(D) eIF2, eIF2B, eIF3, eIF4A, eIF4B, eIF4E, eIF4G, eIF5 and eIF6

166. Which of the following protein factors are involved in intiation stage of protein synthesis in eukaryotic cells?

(A) RF-1, RF-2 and RF-3

(B) IF-1, IF-2 and IF-3

(C) EF-Tu, EF-Ts and EF-G

(D) eIF2, eIF2B, eIF3, eIF4A, eIF4B, eIF4E, eIF4G, eIF5 and eIF6

167. Which of the following protein factors are involved in elongation stage of protein synthesis in bacteria?

(A) eEF1α, eEF1βγ, and eEF2

(B) IF-1, IF-2 and IF-3

(C) EF-Tu, EF-Ts and EF-G

(D) eIF2, eIF2B, eIF3, eIF4A, eIF4B, eIF4E, eIF4G, eIF5 and eIF6

168. Which of the following protein factors are involved in elongation stage of protein synthesis in eukaryotic cells?
(A) eEF1α, eEF1βγ, and eEF2
(B) IF-1, IF-2 and IF-3
(C) EF-Tu, EF-Ts and EF-G
(D) eIF2, eIF2B, eIF3, eIF4A, eIF4B, eIF4E, eIF4G, eIF5 and eIF6

169. Which of the following protein factors are involved in elongation stage of protein synthesis in bacteria?
(A) eEF1α, eEF1βγ, and eEF2
(B) IF-1, IF-2 and IF-3
(C) EF-Tu, EF-Ts and EF-G
(D) RF-1, RF-2 and RF-3

170. What is the sequence to continue elongation stage of protein synthesis in bacteria?
i. Binding of an Incoming Aminoacyl-tRNA
ii. Translocation
ii. Peptide Bond Formation
iv. Proofreading on the Ribosome
(A) ii, i, iii, iv (B) i, iii, ii, iv
(C) iii, i, iv, i (D) iv, ii, i, iii

171. What is the sequencial steps to bind an Incoming Aminoacyl-tRNA in elongation stage of protein synthesis in bacteria?
i. The appropriate incoming aminoacyl-tRNA binds to a complex of GTP-bound EF-Tu
ii. The resulting aminoacyltRNA–EF-Tu–GTP complex binds to the A site of the 70S initiation complex
iii. The GTP is hydrolyzed and an EF-Tu–GDP complex is released from the 70S ribosome
v. The EF-Tu–GTP complex is regenerated in a process involving EF-Ts and GTP.
(A) ii, i, iii, iv (B) i, iii, ii, iv
(C) i, ii, iii, iv (D) iv, ii, i, iii

172. Which of the following groups participate to form a peptide bond between the two amino acids bound by their tRNAs to the A and P sites on the ribosome in bacteria?
(A) Initiating *N*-formylmethionyl group and amino group
(B) Initiating *N*-formylmethionyl group and carboxylic group
(C) Uncharged and deacylated tRNAfMet and amino group
(D) Uncharged and deacylated tRNAfMet and carboxylic group

173. Which of the following group act as nucleophile that displaces the tRNA in the P site to form the peptide bond in elongation stage of protein synthesis in bacteria?
(A) α-amino group of the amino acid in the A site
(B) α-amino group of the amino acid in the P site
(C) Uncharged tRNAfMet in the A site
(D) Deacylated tRNAfMet in the P site

174. Which of the following enzyme catalize the peptide bond formation?
(A) Aminoacyl-tRNA synthetases
(B) Peptidyl transferase
(C) Translocase
(D) Met-tRNA synthetase

175. Which of the following enzyme catalyse the movement of the ribosome along the mRNA?
(A) Aminoacyl-tRNA synthetases
(B) Peptidyl transferase
(C) Translocase
(D) Met-tRNA synthetase

176. When ribosome move from codon to codon along the mRNA toward the 3' end during protein synthesis in bacteria?
(A) After changing in the two-dimensional conformation of the entire ribosome
(B) After changing in the three-dimensional conformation of the entire ribosome
(C) After dissociation 30S subunit from ribosome
(D) After dissociation 50S subunit from ribosome e

177. The α-amino group of the amino acid in the A site acts as a nucleophile which displac the tRNA in the P site to form the peptide bond This reaction produces a .... in the A site,
(A) dipeptidyltRNA
(B) Aminoacyl-tRNA
(C) Met-tRNA
(D) All of the above

178. How many GTPs are hydrolyzed to GDP and Pi for each amino acid residue correctly added to the growing polypeptide when the ribosome moves from codon to codon along the mRNA toward the 3' end?

(A) One (B) Two

(C) Three

(D) ATP hydroled instead of GTP

179. Which of the following bond is formed between tRNA and the carboxyl terminus of the growing polypeptide activates the terminal carboxyl group for nucleophilic attack by the incoming amino acid?

(A) Peptide bond (B) Covalent bond

(C) Ester linkage (D) Glycosidic bond

180. Which of the following site is not found in eukaryotic ribosomes?

(A) Aminoacyl (A) site

(B) Peptidyl (P) site

(C) Exit (E) site

(D) Both (A) and (B)

181. What is the difference between eukaryotic and prokaryotic ribosomes?

(A) Eukaryotic ribosomes do not have an P site

(B) Eukaryotic ribosomes do not have an A site

(C) Prokaryotic ribosomes do not have an E site

(D) Eukaryotic ribosomes do not have an E site

182. From which site of eukaryotic ribosomes, uncharged tRNAs are expelled directly?

(A) Aminoacyl (A) site

(B) Peptidyl (P) site

(C) Exit (E) site

(D) Both (A) and (B)

183. Why uncharged tRNAs are expelled directly from the P site of eukaryotic ribosomes?

(A) Because eukaryotic ribosomes do not have an E site

(B) Because eukaryotic ribosomes have only P site, not A and E site

(C) Uncharged tRNAs are not expelled, whereas used in posttranslation mechanism

(D) None of the above

184. Which does make an important contribution to the rate and fidelity of the overall biosynthetic process during the first step of elongation in bacterial cells?

(A) GTPase activity of EFTu

(B) Translocase

(C) RNA polymerase

(D) Aminoacyl-tRNA Synthetases

185. Incorrect aminoacyl-tRNAs normally dissociate from the .... during existing doth the EF-Tu–GTP and EF-Tu–GDP complexes for a few milliseconds before they dissociate?

(A) Exit (E) site

(B) Peptidyl (P) site

(C) Aminoacyl (A) site

(D) tRNA

186. What happen? if the GTP analog guanosine 5'-*O*-(3-thiotriphosphate) (GTPYS) is used in place of GTP that slowly hydrolysed?

(A) Reducee the fidelity but inprove the rate of protein synthesis

(B) Decrease the proofreading intervals

(C) Balance both speed and fidelity of protein synthesis

(D) Improve the fidelity but reduce the rate of protein synthesis

187. What is the requirement for clearly optimization of the process of protein synthesis through evolution?

(A) Reducee the fidelity but inprove the rate of protein synthesis

(B) Decrease the proofreading intervals

(C) Balance both speed and fidelity of protein synthesis

(D) Improve the fidelity but reduce the rate of protein synthesis

188. What is established by the proofreading mechanism on the ribosome?

(A) Proper codon-anticodon pairing

(B) Identity of the amino acid attached to a tRNA

(C) Balance both speed and fidelity of protein synthesis

(D) All of the above

189. What happen, if a tRNA is successfully aminoacylated with the wrong amino acid?

(A) Incorrect amino acid is efficiently incorporated into a protein in response to whatever codon is normally recognized by the tRNA
(B) Protein synthesis blocked
(C) New amino acids will not be added to growing polypeptide chain
(D) Incorrect amino acid is removed

190. Which of the following one of codons signal to terminate protein synthesis?
(A) UAA (B) UAG
(C) UGA (D) All of the above

191. What happen, if mutations occur in a tRNA anticodon?
(A) A termination codon is inserted
(B) It is beneficial for cell
(C) it helps to continue protein synthesis
(D) It is hartmful for cells but protein synthesis is continued

192. Where termination codon is occupied in bacteria?
(A) Exit (E) site
(B) Peptidyl (P) site
(C) Aminoacyl (A) site
(D) tRNA

193. Which of the following function is done by termination or release factors (RF-1, RF-2, and RF-3) in bacteria after occupying once termination codon?
(A) Hydrolysis of the terminal peptidyltRNA bond
(B) Release of the free polypeptide and the last tRNA from the P site
(C) Dissociation of the 70S ribosome into its 30S and 50S subunits
(D) All of the above

194. Which termination codon recognized by termination factor RF-1 in bacteria?
(A) UAA
(B) UAG
(C) UGA
(D) Both (A) and (B)

195. Which termination codon recognized by termination factor RF-2 in bacteria?
(A) UAA (B) UAG
(C) UGA (D) Both (B) and (C)

196. Which of the following termination factor binds at a termination codon and induces peptidyl transferase to transfer the growing polypeptide to a water molecule rather than to another amino acid?
(A) RF-1 (B) RF-2
(C) RF-3 (D) Both a and b

197. Which of the following enzyme is induced by either RF-1 or RF-2 that binds at a termination codon to to transfer the growing polypeptide to a water molecule rather than to another amino acid?
(A) GTPase
(B) Translocase
(C) Peptidyl transferase
(D) Aminoacyl-tRNA Synthetases

198. What is the function of releasinf factor RF-3 during protein synthesis?
(A) To change in the three-dimensional conformation of the entire ribosome
(B) To release the ribosomal subunit
(C) Recognizes all three termination codons
(D) To balance both speed and fidelity of protein synthesis

199. Which of the following termination factor recognize all three termination codons in eukaryotes?
(A) RF-1 (B) RF-2
(C) RF-3 (D) eRF

200. Formation of each aminoacyl-tRNA uses two high-energy phosphate groups that provided by .... during protein synthesis?
(A) ATP (B) GTP
(C) NADP (D) NADPH

201. Which of the energy molecule is consumed each time an incorrectly activated amino acid is hydrolyzed by the deacylation activity of an aminoacyl-tRNA synthetase?
(A) ATP (B) GTP
(C) NADP (D) NADPH

202. How much energy require to generate a peptide bond?
(A) 122 kJ/mol
(B) about -21 kJ/mol
(C) thus -101 kJ/mol
(D) 30.5 kJ/mol

203. How much net free-energy change during peptide bond synthesis?
(A) 122 kJ/mol (B) about -21 kJ/mol
(C) -101 kJ/mol (D) 30.5 kJ/mol

204. How much a standard free energy of hydrolysis contained in a peptide bond?
(A) 122 kJ/mol (B) About -21 kJ/mol
(C) -101 kJ/mol (D) 30.5 kJ/mol

205. What is clled the large clusters of 10 to 100 ribosomes that are very active in protein synthesis and isolated from both eukaryotic and bacterial cells?
(A) Nucleosome (B) Polysome
(C) Ribozyme (D) Peroxisome

206. What are polysomes?
(A) Large clusters of 10 to 100 ribosomes
(B) A cluster of 8 to 10 Nuclosomes
(C) Ribosomes with enzymatic activity
(D) A cluster of RNAs that participate in protein synthesis

207. In which direction, messenger RNAs are synthesized and translated in bacteria?
(A) Both in same 3' to 5' direction
(B) Trancription in 5' to 3' direction and translation in 3' to 5' direction
(C) Both in same 5' to 3' direction
(D) Trancription in 3' to 5' direction and translation in 5' to 3' direction

208. Which of the following bonds help to maintain protein in its native conformation during or after its synthesis?
(A) Hydrogen bonds
(B) Van der Waals
(C) Ionic and hydrophobic interactions.
(D) All of the above

209. What happen, when mRNA is converted into protein through translation process?
(A) Linear or threedimensional genetic message in the mRNA is converted into the one-dimensional structure of the protein
(B) Non-linear or one-dimensional genetic message in the mRNA is converted into the one-dimensional structure of the protein
(C) Linear or one-dimensional genetic message in the mRNA is converted into the threedimensional structure of DNA
(D) Linear or one-dimensional genetic message in the mRNA is converted into the threedimensional structure of the protein

210. What is called the folding and processing of newly synthesized polypeptide chains into its biologically active form in both prokaryotic and eukaryotic cells?
(A) Posttracriptional modifications
(B) Proofreading
(C) Posttranslational modifications
(D) Confirmational modification

211. Which of the following first step of posttranslational modifications included folding and processing of proteins?
(A) Loss of signal sequences
(B) Amino-terminal and carboxyl-terminal modifications
(C) Modification of individual amino acids
(D) Addition of isoprenyl groups

212. Which of the following first residue is inserted in all polypeptides in bacteria?
(A) *N*-formylmethionine
(B) Methionine
(C) *N*-formylcystein
(D) Valine

213. Which of the following first residue is inserted in all polypeptides in eukaryotes?
(A) *N*-formylmethionine
(B) Methionine
(C) *N*-formylcystein
(D) Histidine

214. The amino group of the amino-terminal residue is ……………………after translation in as many as 50% of eukaryotic proteins?
(A) *N*-acetylated (B) N-fomylated
(C) N^9 Isotope labelled (D) Ionized

215. Which of the following enzymes ultimately remove signal sequences of the 15 to 30 residues at the amino-terminal end of some proteins play a role in directing the protein to its ultimate destination in the cell?
(A) Translocase
(B) Specific peptidases
(C) Peptidyl transferase
(D) Aminoacyl-tRNA Synthetases

216. Which of the following essential nutrients are provided by the milk protein casein?
(A) Calcium (B) Phosphate
(C) Amino acids (D) All of the above

217. Which of the following group of certain Ser, Thr and Tyr residues of some proteins are enzymatically phosphorylated by ATP?
(A) Carboxyl group (B) Amino group
(C) Hydroxyl group (D) Nitrogen group

218. Why polypeptide negatively charged during protein folding and processing?
(A) Hydroxyl groups of amino acids are enzymatically phosphorylated by ATP
(B) Proteins bind to Ca^{2+}
(C) Amino acids have carboxylic group
(D) Amino acids have amino group

219. Which of the following protein in which certain Ser, Thr and Tyr residues are enzymatically phosphorylated by ATP during protein folding and processing?
(A) Egg albumin (B) Human serum
(C) Milk protein casein
(D) Blood-clotting protein prothrombin

220. Which of the following protein contain extra carboxyl groups in glutamic acid (Glu) residues in its amino-terminal region that introduced by an enzyme that requires vitamin K?
(A) Egg albumin
(B) Human serum
(C) Milk protein casein
(D) Blood-clotting protein prothrombin

221. Which of the following amino acid of blood-clotting protein prothrombin contain extra carboxyl group in its amino-terminal region that introduced by an enzyme that requires vitamin K?
(A) Serine (B) Threonine
(C) Tyrosine (D) Glutamic acid

222. What is main role of extra carboxylic group in glutamic acid of blood protein prothrombin in its amino-terminal region that introduced by an enzyme that requires vitamin K?
(A) Initiate the clotting mechanism
(B) Remove high cholesterol from body
(C) Decrease platelets count
(D) Increase hemoglobin concentration

223. Which of the following side chains of glycoproteins are attached with covalently during or after synthesis of the polypeptide?
(A) Hydroxyl group
(B) Phospholipid side chains
(C) Carbohydrate side chains
(D) nucleotide side chain

224. Which of the following is required to initiate the blood clotting mechanism?
(A) Zn^{2+} (B) Ca^{2+}
(C) Mg^{2+} (D) Ba^{2+}

225. Which of the following residues occur in some muscle proteins and in cytochrome *c*?
(A) Monoethyl- and dimethyllysine
(B) Trimethyl- and diethyllysine
(C) dimethyl- and trimethyllysine
(D) Monomethyl- and dimethyllysine

226. Which protein of most species contain one trimethyllysine residue at a specific position?
(A) Calmodulin
(B) cytochrome *c*.
(C) Muscle proteins
(D) blood protein prothrombin

227. Why the carboxyl groups of some Glu residues undergo methylation in some proteins?
(A) To remove negative charge
(B) To increase acidity
(C) To release the ribosomal subunit
(D) To protect the native conformation of the protein molecule

228. Which of the following proteins act as lubricating proteoglycans, contain oligosaccharide side chains that is attached enzymatically to Asn residues (*N*-linked oligosaccharides) and Ser or Thr residues (*O*-linked oligosaccharides)?
(A) Calmodulin (B) cytochrome *c*.
(C) Mucous membrane proteins
(D) blood protein prothrombin

229. Which bond is formed between the isoprenyl group and a Cys residue of the protein?
(A) Thioether bond
(B) Glycosidic bond
(C) Phosphodiester bond
(D) Ionic bond

230. How derived isoprenyl groups that linked with a cys residue of protein?
(A) Phospholipid biosynthetic pathway
(B) Blood clotting mechanism
(C) Cholesterol biosynthetic pathway
(D) Nucleotide biosynthetic pathway

231. What is called the pyrophosphorylated intermediates of the cholesterol biosynthetic pathway that derives isoprenyl group linked to a Cys residue of the protein?
(A) Farnesyl pyrophosphate
(B) Dolichol pyrophosphate
(C) guanosine tetraphosphate
(D) Isopentenylpyrophosphate

232. Which of the following protein products of the *ras* oncogenes and proto-oncogenes found in the nuclear matrix an lost its carcinogenic activity after isoprenylation?
(A) Nas (B) Lac
(C) Tus (D) Ras

233. After wich process, Ras lost its carcinogenic activity that produced by *ras* oncogenes and proto-oncogenes found in the nuclear matrix?
(A) Methylation (B) Acetylation
(C) Isoprenylation (D) Phosphorylation

234. Which of the following stimulating process of Ras protein is a part of inhibition of posttranslation modification and used in cancer chemotherapy?
(A) Methylation (B) Acetylation
(C) Isoprenylation (D) Phosphorylation

235. What is use of isoprenyl group that is pyrophosphorylated intermediates of the cholesterol biosynthetic pathway?
(A) Blood clotting mechanism
(B) Help to anchor the protein in a membrane
(C) Remove high cholesterol from body
(D) Increase hemoglobin concentration

236. Which prosthetic groups is found in acetyl-CoA carboxylase for its activity?
(A) Biotin molecule (B) Heme group
(C) Carboxylic group (D) FAD and FMN

237. Which prosthetic groups is found in hemoglobin or cytochrome *c* for its activity?
(A) Biotin molecule (B) Heme group
(C) Carboxylic group (D) FAD and FMN

238. What is called the conversion of initially synthesized large, inactive precursor polypeptides into smaller, active forms?
(A) Formation of disulfide cross-links
(B) Addition of Prosthetic Groups
(C) Proteolytic processing
(D) Modification of Individual Amino Acids

239. Which residues participate to form interchain disulfide bridges after folding into their native conformations in some proteins?
(A) Histidine residues
(B) Tryptophan residues
(C) Cystein residues
(D) Prolein residues

240. What function of disulfide bonds in proteins of eukaryotes?
(A) To remove negative charge
(B) Modification of Individual Amino Acids
(C) Help to anchor the protein in a membrane
(D) Help to protect the native conformation from denaturation

241. Which of the following antibiotic fir inhibiton of protein synthesis made by the mold *Streptomyces alboniger*?
(A) Tetracyclines
(B) Puromycin
(C) Chloramphenicol
(D) cycloheximide

242. Which of the following organism produce a antibiotic 'puromycin' that inhibit protein synthesis?
(A) Mushroom *Amanita phalloides*
(B) Mold *Streptomyces alboniger*
(C) Fungus *Aspergillus niger*
(D) Bacteria *Clostridium welchi*

243. How puromycin inhibit the protein synthesis forming peptide bond and producing peptidyl-puromycin?
(A) By blocking the A site on the ribosome
(B) By blocking peptidyl transfer in bacteria
(C) By blocking the peptidyl transferase of 80S eukaryotic ribosomes
(D) By causing misreading of the genetic code

244. Which of the following antibiotic inhibit protein synthesis in bacteria by blocking the A site on the ribosome with preventing the binding of aminoacyl-tRNAs?

(A) Tetracyclines (B) Chloramphenicol
(C) Rifampicin (D) α–amanitin

245. Which of the following antibiotic inhibit protein synthesis by bacterial (and mitochondrial and chloroplast) ribosomes by blocking peptidyl transfer?

(A) Streptomycin (B) Chloramphenicol
(C) Rifampicin (D) α–amanitin

246. Which of the following antibiotic does not affect cytosolic protein synthesis in eukaryotes, wheras inhibit protein synthesis by bacterial (and mitochondrial and chloroplast) ribosomes by blocking peptidyl transfer?

(A) Streptomycin (B) α–amanitin
(C) Rifampicin (D) Chloramphenicol

247. Which of the following toxin blocks the peptidyl transferase of 80S eukaryotic ribosomes but not that of 70S bacterial (and mitochondrial and chloroplast) ribosomes

(A) Tetracyclines (B) Puromycin
(C) Chloramphenicol (D) cycloheximide

248. Which of the following antibiotic causes misreading of the genetic code in bacteria at relatively low concentrations and inhibits initiation of protein synthesis at higher concentrations?

(A) Streptomycin (B) α–amanitin
(C) Rifampicin (D) Chloramphenicol

249. Which of the following toxin catalyzes the ADP-ribosylation of a diphthamide residue of eukaryotic elongation factor eEF2, thereby inactivating it?

(A) Streptomycin (B) Diphtheria toxin
(C) Rifampicin (D) Chloramphenicol

250. Which of the following toxic protein inactivates the 60S subunit of eukaryotic ribosomes by depurinating a specific adenosine in 23S rRNA?

(A) Streptomycin
(B) Diphtheria toxin
(C) Rifampicin
(D) Ricin

251. Which of the following toxic protein produced by castor bean and inactivate inactivates the 60S subunit of eukaryotic ribosomes by depurinating a specific adenosine in 23S rRNA?

(A) Streptomycin (B) Ricin
(C) Rifampicin (D) Diphtheria toxin

252. Who first postulated the function of a short sequence of amino acids called a signal sequence that directs a protein to its appropriate location in the cell in many targeting pathways?

(A) Gunter Blobel (1970)
(B) Efraim Racker (1960)
(C) John E. Walker (2003)
(D) Melvin Calvin (1950)

253. What are the features of all signal sequences that directs a protein to its appropriate location in the cell in many targeting pathways?

(A) About 10 to 15 hydrophobic amino acid residues
(B) One or more positively charged residues usually near the amino terminus
(C) A short sequence at the carboxyl terminus that is relatively polar
(D) All of the above

254. How many amino acid residues found in length of signal sequences that direct a protein to its appropriate location in the cell in many targeting pathways?

(A) From 10 to 15 amino acid residues
(B) From 3 to 6 amino acid residues
(C) From 13 to 36 amino acid residues
(D) From 31 to 63 amino acid residues

255. Who demonstrated that proteins with these signal sequences are synthesized on ribosomes attached to the ER?

(A) George Palade (B) Efraim Racker
(C) John E. Walker (D) Melvin Calvin

256. The targeting pathway under posttranslation modification begins with initiation of protein synthesis ....

(A) on bounded ribosome
(B) In nucleaus
(C) On free ribosome
(D) In lysosome

257. Which of the energy molecule hydrolyzed during the dissociation SRP from the ribosome accompanied by hydrolysis of GTP in both signal recognition particle (SRP) and the SRP receptor?
(A) ATP (B) GTP
(C) NAD (D) NADP

258. Where the signal sequence is removed by a signal peptidase after completing protein synthesis?
(A) ER lumen (B) Golgybody
(C) Lysosomes (D) Valuole

259. Which process plays a key role in protein targeting?
(A) Methylation (B) Isoprenylation
(C) Glycosylation (D) Phosphorylation

260. Which of the following antibiotic mimics the structure of UDP-*N*-acetylglucosamine and blocks protein glycosylation?
(A) Streptomycin (B) α–amanitin
(C) Rifampicin (D) Tunicamycin

261. Which amino acid residue linked to oligosaccharides in many glycoproteins by glucosilation?
(A) Asparagine (B) Serine
(C) Aspartate (D) Histidine

262. Where occurs *O*-glycosilation of proteins rhat do not enter the endoplasmic reticulum?
(A) In nucleus
(B) In mitochondria or golgi complex
(C) In golgi complex or cytosol
(D) In Vocuoles or cytosol

263. What is called the undetermined feature of the three-dimensional structure of the hydrolase that is recognized by a phosphotransferase which phosphorylates certain mannose residues in the oligosaccharide in golgi complex?
(A) Signal patch (B) Signal sequence
(C) Marker (D) Indicator

264. Which of the following residues in *N*-linked oligosaccharide is the structural signal that targets the protein to lysosomes?
(A) Glucose 6-phosphate
(B) Mannose 6-phosphate
(C) Fructose 6-phosphate
(D) Glucose 1,6-diphosphate

265. Why mitochondria and chloroplast protein targeting pathways begin only after a precursor protein that completely synthesized and released from the ribosome?
(A) Mitochondria and chloroplasts contain DNA
(B) Mitocondria is power-house of cell and chloroplast is photosynthetic complex
(C) Most of their proteins are encoded by nuclear DNA
(D) All of the above

266. Which cellular structure require to molecular communication between the nucleus and the cytosol for the movement of macromolecules?
(A) Nucelolus (B) Cell membrane
(C) Nuclear pores (D) Genetic material

267. Where RNA molecules are synthesized in cell?
(A) Mitochondria (B) Ribosome
(C) Choloroplast (D) Nucleus

268. Where ribosomes are synthesized in cell?
(A) Mitochondria (B) Nucleolus
(C) Choloroplast (D) Nucleus

269. Which of the following is not a nuclear protein that are synthesized in the cytosol and imported into the nucleus?
(A) RNA polymerases
(B) DNA polymerases
(C) Histones
(D) Peptidyl transferase

270. Where are a variety of nuclear proteins *viz.*, RNA and DNA polymerases, histones, topoisomerases, proteins that regulate gene expression?
(A) Nucleoplasm (B) Nucleolus
(C) Cytosol (D) Nucleus

271. What is called the signal sequence that targets a protein to the nucleus and is not removed after the protein arrives at its destination?
(A) Consensus sequences
(B) Nuclear Localization Sequence (NLS)
(C) Autonomously Replicating Sequences (ARS)
(D) Short Interspersed Repeat Sequences (SINEs)

272. What is the size of Nuclear Localization Sequence (NLS) that targets a protein to the nucleus and is not removed after the protein arrives at its destination?
(A) Four to eight amino acid residues
(B) fourteen to eighteen amino acid residues
(C) Ten amino acid residues
(D) Fifteen to fifty amino acid residues

273. Which of the following proteins that cycle between the cytosol and the nucleus during mediatation of nuclear importation?
(A) Importin α (B) Importin β
(C) A small GTPase (D) All of the above

274. What is called a small GTPase protein that cycle between the cytosol and the nucleus during mediatation of nuclear importation?
(A) Trp (B) Ran
(C) Lac (D) Rec

275. Which of the following protein are imported into cells from the surrounding medium in eukaryotes?
(A) Lowdensity lipoprotein (LDL)
(B) Iron-carrying protein transferring
(C) Circulating proteins
(D) All of the above

276. Which of the following protein forms closed polyhedral structures as lattice on the cytosolic side that contains coated pits concentrating endocytic receptors in preference to other cell-surface proteins?
(A) Ubiquitin (B) Clathrin
(C) Creatine (D) Serotonin

277. What is the function of protein degradation?
(A) It prevents the buildup of abnormal or unwanted proteins
(B) It permits the recycling of amino acids
(C) It promotes the healing of duodenal ulcers by inhibiting secretion of gastric acid
(D) Both a and b

278. What is the half-lives of eukaryotic proteins?
(A) 30 seconds
(B) 30 seconds to many days
(C) 30 days
(D) 30 minutes

279. What is the half-lives of erythrocytes?
(A) About 110 hours
(B) About 110 minutes
(C) About 110 seconds
(D) About 110 days

280. Which of the following enzyme called Lon is used to degrade many proteins in *E. coli*?
(A) ATP-dependent protease
(B) Protein Kinase
(C) α_1-Antiproteinase
(D) Ferrireductase

281. How many ATP molecules are hydrolyzed for the cleaving of every peptide bond?
(A) One (B) Three
(C) Two (D) Four

282. How many amino acid residues contained in highly conserved protein 'Ubiquitin' in eukaryotic cells?
(A) 100 (B) 69
(C) 98 (D) 76

283. Which of the following a large complex degrade ubiquitinated proteins?
(A) Nucleosomes (B) 26S proteasome
(C) Phagolysosomes (D) Enhanceosome

284. Which disease associated with the formation of characteristic proteinaceous structures in neurons?
(A) Renal disease
(B) Cystic fibrosis
(C) Parkinson's disease
(D) Liddle's syndrome

285. Which disease caused by a too-rapid degradation of a chloride ion channel due to resultant loss of function?
(A) Renal disease
(B) Cystic fibrosis
(C) Parkinson's disease
(D) Liddle's syndrome

286. In which disease a sodium channel in the kidney is not degraded, leading to excessive Na^+ absorption and early-onset hypertension?
(A) Renal disease
(B) Cystic fibrosis
(C) Parkinson's disease
(D) Liddle's syndrome

287. Which of the following sequence direct many proteins after synthesis to particular locations in the cell through targeting mechanism and generally found at the amino terminus of a newly synthesized protein?
(A) Consensus sequences
(B) Peptide signal sequence
(C) Autonomously Replicating Sequences (ARS)
(D) Short Interspersed Repeat Sequences (SINEs)

288. Which recognize one class of signal sequence in eukaryotic cells and binds the signal sequence as soon as it appears on the ribosome and transfers the entire ribosome and incomplete polypeptide to the ER?
(A) Signal recognition particle (SRP)
(B) Autonomously Replicating Sequences (ARS)
(C) Short Interspersed Repeat Sequences (SINEs)
(D) Ubiquitin

289. Which of the following statement is not true about protein targeting and degradation?
(A) Proteins targeted to mitochondria and chloroplasts in eukaryotic cells and destined for export in bacteria also make use of an amino-terminal signal sequence
(B) Proteins targeted to the nucleus have an internal signal sequence that is not cleaved once the protein is successfully targeted
(C) All cells eventually degrade proteins by using specialized proteolytic systems.
(D) Defective proteins for rapid turnover are not degraded by an ATP-dependent system

290. What is Ubiquitin?
(A) A highly conserved transposon
(B) A highly conserved carbohydrate
(C) A highly conserved protein
(D) A highly conserved amino acid

291. Who proposed the structure of yeast tRNA in 1965?
(A) H. Gobind Khorana
(B) Marshall Nirenberg
(C) Paul Zamecnik
(D) Robert W. Holley

292. Wich of the following RNA show cloverleaf conformation in which intrastrand base pairing is maximal?
(A) mRNA (B) tRNA
(C) rRNA (D) miRNA

293. The hnRNA is processed within the ....
(A) Cytoplasm (B) Nucleus
(C) Ribosome
(D) Endoplasmic reticulum

294. What is the characteristic of the genetic code?
(A) Degeneracy (B) Universal
(C) Nonoverlapping (D) All of the above

295. The set of tRNA molecules in mitochondria from lower and higher eukaryotes, including humans, reads **four** codons differently from the tRNA molecules in the cytoplasm. That is:
(A) AUA = Met
(B) UGA = Trptophan
(C) AGA and AGG = Terminator codons
(D) All of the above

296. Both strands of DNA are produced by .... which is accomplished by looping the DNA of the lagging strand bringing together the two points of polymerization.
(A) A single symmetric DNA polymerase III dimer
(B) A single asymmetric DNA polymerase II dimer
(C) A single asymmetric DNA polymerase III dimer
(D) A single asymmetric DNA polymerase I dimer

297. What is difference between the template strands used in ρ-dependent and ρ-independent termination of transcription, respectively?
(A) A highly conserved string of three A residues and a CA-rich sequence
(B) CA-rich sequence and A highly conserved string of three three CA residues
(C) ρ-dependent termination lack the sequence of repeated A residues and a CA-rich sequence
(D) None of the above

298. Which of the final stage of protein synthesis?
(A) Folding and Posttranslational Processing
(B) The nascent polypeptide is lengthened by successive amino acid units
(C) Large ribosomal subunit then binds to form an initiation complex
(D) Activation of amino acids

299. Which of the following reaction require energy as ATP with Mg^{2+}-dependent activating enzymes during translation?
(A) Releasing the new synthesized polypeptide chain
(B) A link between each new amino acid and the information in the mRNA
(C) The binding of each incoming aminoacyl-tRNA to growing polypeptide chain
(D) Activation of carboxyl group of each amino acid

300. Which of the following protein factor catalyze the movement of the ribosome along the mRNA?
(A) Aminoacyl-tRNA synthetases
(B) Peptidyl transferase
(C) EF-G
(D) Met-tRNA synthetase

Answer-Key

Q	Ans	Q	Ans	Q	Ans	Q	Ans	Q	Ans	Q	Ans	Q	Ans	Q	Ans
1.	(C)	41.	(D)	81.	(A)	121.	(A)	161.	(B)	201.	(A)	241.	(B)	281.	(C)
2.	(D)	42.	(D)	82.	(D)	122.	(C)	162.	(D)	202.	(A)	242.	(B)	282.	(D)
3.	(B)	43.	(B)	83.	(A)	123.	(B)	163.	(B)	203.	(C)	243.	(A)	283.	(B)
4.	(C)	44.	(D)	84.	(D)	124.	(D)	164.	(D)	204.	(B)	244.	(A)	284.	(C)
5.	(A)	45.	(B)	85.	(A)	125.	(B)	165.	(B)	205.	(B)	245.	(B)	285.	(B)
6.	(C)	46.	(C)	86.	(C)	126.	(D)	166.	(D)	206.	(A)	246.	(D)	286.	(D)
7.	(B)	47.	(B)	87.	(B)	127.	(D)	167.	(C)	207.	(C)	247.	(D)	287.	(B)
8.	(C)	48.	(A)	88.	(D)	128.	(C)	168.	(A)	208.	(D)	248.	(A)	288.	(A)
9.	(A)	49.	(B)	89.	(A)	129.	(A)	169.	(D)	209.	(D)	249.	(B)	289.	(D)
10.	(C)	50.	(A)	90.	(C)	130.	(C)	170.	(B)	210.	(C)	250.	(D)	290.	(C)
11.	(D)	51.	(D)	91.	(B)	131.	(A)	171.	(C)	211.	(B)	251.	(B)	291.	(D)
12.	(C)	52.	(C)	92.	(C)	132.	(C)	172.	(A)	212.	(A)	252.	(A)	292.	(B)
13.	(D)	53.	(D)	93.	(A)	133.	(A)	173.	(A)	213.	(B)	253.	(D)	293.	B
14.	(B)	54.	(A)	94.	(D)	134.	(B)	174.	(C)	214.	(A)	254.	(C)	294.	D
15.	(A)	55.	(C)	95.	(B)	135.	(B)	175.	(C)	215.	(B)	255.	(A)	295.	D
16.	(B)	56.	(A)	96.	(C)	136.	(A)	176.	(B)	216.	(D)	256.	(C)	296.	C
17.	(A)	57.	(C)	97.	(A)	137.	(A)	177.	(A)	217.	(C)	257.	(B)	297.	(C)
18.	(C)	58.	(D)	98.	(B)	138.	(A)	178.	(B)	218.	(A)	258.	(A)	298.	(A)
19.	(D)	59.	(B)	99.	(A)	139.	(A)	179.	(C)	219.	(C)	259.	(C)	299.	(D)
20.	(D)	60.	(C)	100.	(D)	140.	(D)	180.	(C)	220.	(D)	260.	(D)	300.	(C)
21.	(A)	61.	(A)	101.	(A)	141.	(A)	181.	(D)	221.	(D)	261.	(A)		
22.	(B)	62.	(B)	102.	(C)	142.	(C)	182.	(B)	222.	(A)	262.	(C)		
23.	(C)	63.	(B)	103.	(D)	143.	(B)	183.	(A)	223.	(C)	263.	(A)		
24.	(B)	64.	(A)	104.	(A)	144.	(C)	184.	(A)	224.	(B)	264.	(B)		
25.	(D)	65.	(D)	105.	(C)	145.	(D)	185.	(C)	225.	(D)	265.	(A)		
26.	(A)	66.	(C)	106.	(B)	146.	(C)	186.	(D)	226.	(A)	266.	(C)		
27.	(D)	67.	(A)	107.	(B)	147.	(C)	187.	(C)	227.	(A)	267.	(D)		
28.	(B)	68.	(D)	108.	(B)	148.	(B)	188.	(D)	228.	(C)	268.	(B)		
29.	(A)	69.	(B)	109.	(C)	149.	(B)	189.	(A)	229.	(A)	269.	(D)		
30.	(D)	70.	(A)	110.	(B)	150.	(A)	190.	(D)	230.	(C)	270.	(C)		
31.	(D)	71.	(A)	111.	(A)	151.	(C)	191.	(A)	231.	(A)	271.	(B)		
32.	(B)	72.	(C)	112.	(C)	152.	(A)	192.	(C)	232.	(D)	272.	(A)		
33.	(A)	73.	(D)	113.	(D)	153.	(C)	193.	(D)	233.	(C)	273.	(D)		
34.	(C)	74.	(B)	114.	(C)	154.	(B)	194.	(D)	234.	(C)	274.	(B)		
35.	(C)	75.	(D)	115.	(B)	155.	(D)	195.	(D)	235.	(B)	275.	(D)		
36.	(A)	76.	(C)	116.	(D)	156.	(B)	196.	(D)	236.	(A)	276.	(B)		
37.	(D)	77.	(A)	117.	(D)	157.	(A)	197.	(C)	237.	(B)	277.	(D)		
38.	(B)	78.	(D)	118.	(C)	158.	(C)	198.	(B)	238.	(A)	278.	(B)		
39.	(A)	79.	(B)	119.	(A)	159.	(A)	199.	(D)	239.	(C)	279.	(D)		
40.	(C)	80.	(C)	120.	(A)	160.	(C)	200.	(B)	240.	(D)	280.	(A)		

8

Regulation of Gene Expression

The mechanism stimulates the expression of certain genes and inhibits other genes called regulation of gene expression. It is a complex mechanism that is used by cells to increase or decrease the production of specific gene products (protein or RNA). It is possible only of the organism of regulating gene activity by allowing some to function and others to restrain their activity through switching on and switching off system. This means, the genes are turned 'on' or 'off'. A set of genes is switched on when enzymes are required to metabolites a new substrate. The enzymes produced by these genes metabolise the substrate. The molecules of metabolite that come to switch on of the genes are termed as inducers and the phenomenon is called induction. Similarly, certain genes which are in their switch on state, continue to synthesize a metabolite till the later is produced in amount more than required or else. It is supplied to the cell from outside. In other words, certain genes continue to express themselves till the end product of inhibits or repress their expression. Inhibition by an end product is known as feedback inhibition.

Operon cocept

In 1961, two French microbiologists Francis Jocob and Jacques Monad at te Pasteur Institute in Paris, proposed a mechanism called 'Operon model' for the regulation of gee action in *E. coli*. An operon is a part of genetic material of DNA which acts as a single regulated unit having one or more structural genes, an operator gene, a promoter gene and a regulator gene.

Operons are of two types (i) inducible operon system (ii) repressible operon system

1. Inducible operon system *(Lac* operon of *E. coli)*

An inducible operon system normally remains in switched off condition and begins to work only when the substance to be metabolized by it is present in the cell. Inducible operon generally occurs in catabolic pathways, eg. *Lac* operon of *E. coli*.

Active repressor + inducer = inactive repressor

An inducible operon system contains four types of gene

Structural genes: An operon system may have one or more structural genes. Each structural gene of an operon is called cistron. The lac operon (lactose operon) of E. coli contains three structural genes. These genes transcribe a polycistronic mRNA molecule (a single stretch of mRNA covering all the three genes) that helps in the synthesis of three enzymes.

a. *LacZ* encodes β-galactosidase, an intracellular enzyme that cleaves the disaccharide lactose into glucose and galactose.

b. *LacY* encodes β-galactoside permease, a transmembrane symporter that pumps β-galactosides including lactose into the cell using a proton gradient in the same direction.

c. *LacA* encodes β-galactoside transacetylase (LacA), an enzyme that transfers an acetyl group from acetyl-CoA to β-galactosides.

Operator gene: It lies adjacent to the structural genes and directly controls the synthesis of mRNA over the structural genes. It is switched of by the present of a repressor. An inducer can take away the repressor and switch on the gene that directs the structural genes to transcribe.

Promoter gene: the gene is the site for initial binding of RNA polymerase. When the operator gene is turned on, the enzyme RNA polymerase moves over it and reaches the structural genes to perform transcription.

Regulator gene: it produces a repressor that binds to operator gene and stops the working of the operator gene.

Repressor: it is a protein, produced by the regulator gene. It binds to the operator gene so that the transcription of structural gene stops. Repressor has two allosteric sites (1) operator gene site and (2) effective molecule site (inducer/corepressor).

Inducer: it is a chemical substrate (hormone or some other metabolite) which after coming in contact with repressor forms an inducer-repressor complex. This complex cannot bind with the operator gene, which is thus switched on. The free operator gene allows the structural gene to transcribe mRNA to synthesize the corresponding enzymes.

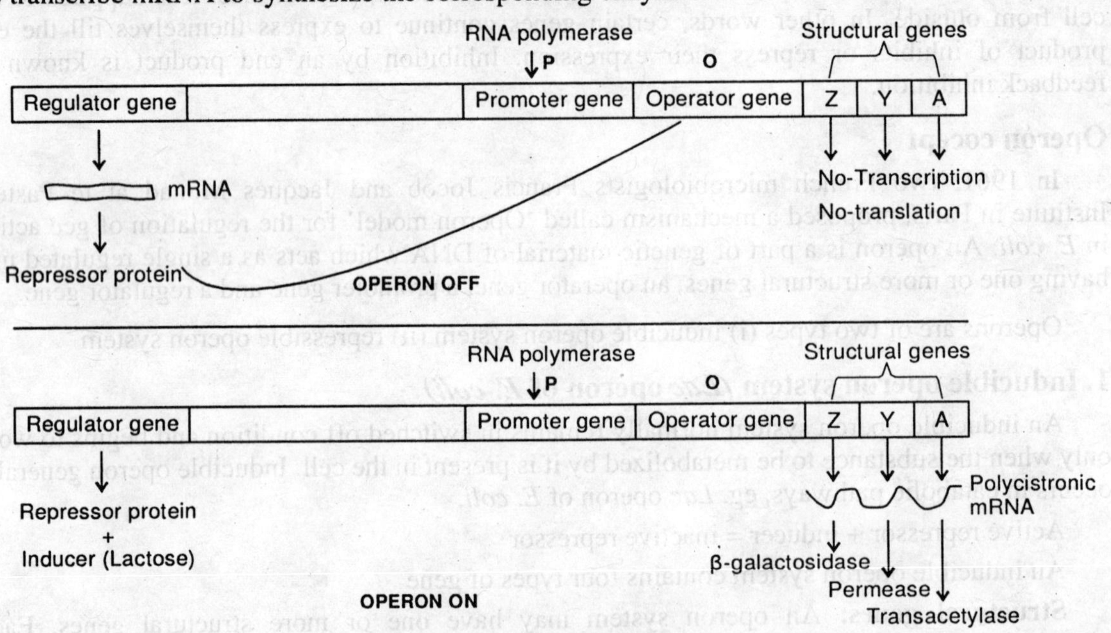

Inducible lactose operon system

The inducer for lac operon of *E. coli* is lactose (in fact allolactose an isomer of lactose). When the sugar lactose is added to the culture of *E. coli*, a few molecules of lactose get into the bacterial cells by the action of the enzyme permease. A small amount of its enzyme is present in the cell even when the operon is not working. These few lactose molecules are then converted into an active form which acts as an inducer and binds to the repressor protein. The inducer-repressor complex fails to join with the operator and expression of genes is turned on. The three structural genes are expressed as three enzymes to metabolize lactose. Allolactose is real inducer of lac operon. Gratuitous inducer → some molecules resemble with natural inducer but are not metabolize by the enzyme. Example: Isopropyl thiogalactoside (IPTG) which resembles lactose and thus has the property of induction. Such inducer which induces enzyme synthesis, without getting metabolize are called gratuitous inducer.

2. Repressible operon (Tryptophan operon of *E. coli*)

A repressible operon system is normally in it's switch on state and contribute to synthesize a metabolite till the latter is produced in amount more than required, or else it becomes available to the cell from outside. Repressible operon system is commonly found in anabolic pathway. eg. Tryptophan operon in *E.coli*.

Inactive repressor + co-repressor = active repressor

The *trp* operon in *E. coli* was discovered by Jacques Monod and colleagues in 1953. It was the first repressible operon that is inhibited by a chemical (tryptophan).

Structural genes: These genes are meant for transcription of mRNA which is turn synthesize enzyme. *Trp* operon contains five structural genes: *trp* E, D, C, B and A which encode tryptophan synthetase.

Operator gene (*trp* O): it lie adjacent to the structural genes and controls the functioning of the structural genes. Normally, it is kept switched on, because the apo-repressor produced by the regulator gene does not bind to it. The operator gene is switched off when a co-repressor is available alongwith apo-repressor.

Promoter gene (*try* P): it marks the site at which the RNA polymerase enzyme binds. When the operator gene is switched on, it moves from promoter gene to structural genes for transcription.

Regulator gene (*trp* R): It produces a regulatory protein called apo-repressor for (inactive repressor) possible blocking the activity of operator gene.

Apo-repressor: It is a regulatory protein synthesized by regulator gene. When a co-repressor substrate is available in the cell, the ap-repressor combines with the co-represor to form a apo-repressor co-repressor complex. This complex binds with the operator gene and switches it off. Presence of apo-repressor alone, the operator gene is kept switched on because by itself the apo-repressor is unable to block the working of operator gene.

Co-repressor: It is an end product of rection catalyzed by enzyme produced by the structural genes.

In the presence of tryptophan, some molecules of tryptophan act as co-repressor, co-repressor binds with inactive repressor. Co-repressor repressor complex binds with operator region and prevent the binding of RNA polymerase to the promoter, the *trp*-operon is off.

The repressor molecule has key role in regulation of *lac*-operon. Repressor molecule is active or inactive. Active repressor may be rendered inactive by addition of an inducer while the inactive repressor can be made active by addition of a co-repressor. Because the product of regulator gene the repressor act by shuttling off the transcription of structural genes the operon model, as originally proposed by Jacob and monad is referred as negative control system.

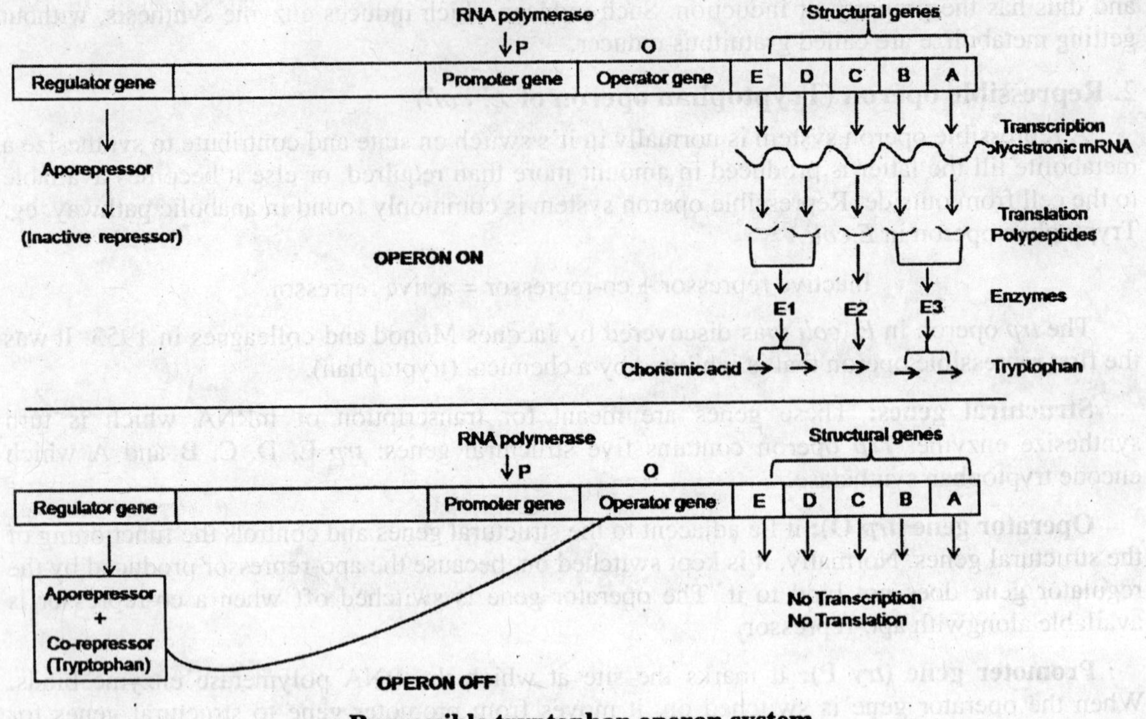

Repressible tryptophan operon system

Gene expression in eukaryotes

In eukaryotes, functionally related genes may not be clustered together constituting an operon. The most popular model is known as 'Britten-Davidson model' or 'Gene-battery model' proposed by Britten and Davidson in 1969. A set of structural genes controlled by one sensor site is termed as battery. Gene-battery model assumes the presence of four classes of sequences.

Producer gene: A producer gene is comparable to structural gene of prokaryotic operon.

Receptor site: a receptor site is comparable to operator gene of bacterial operon and one such receptor site is assumed to be present adjacent to each producer gene.

Integrator gene: it is comparable to regulator gene and is responsible for the synthesis of an activator RNA. It activates the repressor site.

Sensor site: a sensor site regulates the activity of integrator gene. Activator gene can be transcribed only when the sensor site is activated.

The sensor sites are recognized by agents which change the patterns of gene expression like hormones and proteins. When a transcription factor (protein, hormone) bind to the sensor site it causes the transcription of integrator gene.

Key notes

- Barbara McClintock showed interaction between two genetic elements, Activator (*Ac*) and Dissociator (*Ds*), in the color formation of maize seeds related to transposon regulation.
- First discovery of a gene regulation system is widely considered to be the identification in 1961 of the *lac* operon discovered by François Jacob and Jacques Monod.

- In lac operon, lactose metabolism is regulated by some enzymes in *E. coli* in the presence of lactose and absence of glucose.
- Any step of gene expression may be modulated, from the DNA-RNA transcription step to post-translational modification of a protein.
- Gene expression is regulated at different stages:
  - Chromatin domains
  - Transcription
  - Post-transcriptional modification
  - RNA transport
  - Translation
  - mRNA degradation
- In eukaryotes, the accessibility of large regions of DNA can depend on its chromatin structure, which can be altered by histone modifications directed by DNA methylation, ncRNA, or DNA-binding protein.
- These modifications may up or down regulate the expression of a gene.
- Some of these modifications that regulate gene expression are inheritable and are referred to as epigenetic regulation.
- Transcription of DNA is dictated by its structure. The density of its packing is indicative of the frequency of transcription.
- Octameric protein complexes called nucleosomes are responsible for the amount of supercoiling of DNA and these complexes can be temporarily modified by phosphorylation or more permanently modified by methylation.
- Such modifications are considered to be responsible for more or less permanent changes in gene expression levels.
- Methylation of DNA is a common method of gene silencing.
- DNA is typically methylated by methyltransferase enzymes on cytosine nucleotides in a CpG dinucleotide sequence (called "CpG islands").
- Methylation in promoter region of DNA can be achieved through a method called bisulfite mapping.
- Methylated cytosine residues are unchanged by the treatment, whereas unmethylated are changed to uracil.
- The differences are analyzed by DNA sequencing or by methods developed to quantify SNPs, such as Pyrosequencing (Biotage) or MassArray (Sequenom), measuring the relative amounts of C/T at the CG dinucleotide.
- Abnormal methylation patterns are thought to be involved in oncogenesis.
- Histone acetylation is also an important process in transcription.
- Histone acetyltransferase enzymes (HATs) such as CREB-binding protein dissociate the DNA from the histone complex and allow transcription to proceed.
- DNA methylation and histone deacetylation work together in gene silencing.
- Transcription of a gene by RNA polymerase can be regulated by at least five mechanisms:

  - Specificity factors alter the specificity of RNA polymerase for a given promoter or set of promoters making it more or less likely to bind to sigma factors used in prokaryotic transcription.
  - Repressors bind to the Operator (coding sequences on the DNA strand) that are close to or overlapping the promoter region which inhibits RNA polymerase's progress along the strand and block the expression of the gene.
  - General transcription factors position RNA polymerase at the start of a protein-coding sequence and then release the polymerase to transcribe the mRNA.
  - Activators enhance the interaction between RNA polymerase and promoter and encourage the expression of the gene.
  - Enhancers are sites on the DNA helix that are bound by activators in order to loop the DNA bringing a specific promoter to the initiation complex.
  - Silencers are regions of DNA sequences that when bound by particular transcription factors, can silence expression of the gene.
- After the DNA is transcribed into mRNA, there must be some sort of regulation on how much the mRNA is translated into proteins.
- Cells do this by modulating the capping, splicing, addition of a Poly(A) Tail, the sequence-specific nuclear export rates and sequestration of the RNA transcript.
- These processes occur in eukaryotes but not in prokaryotes. This modulation is a result of a protein or transcript that is regulated and may have an affinity for certain sequences.
- Three prime untranslated regions (3'-UTRs) of mRNAs contain regulatory sequences that post-transcriptionally influence gene expression.
- 3'-UTRs contain both binding sites for microRNAs (miRNAs) as well as for regulatory proteins.
- By binding to specific sites within the 3'-UTR, miRNAs can decrease gene expression of various mRNAs by either inhibiting translation or directly causing degradation of the transcript.
- The 3'-UTR may have silencer regions that bind repressor proteins that inhibit the expression of a mRNA.
- The 3'-UTR contains miRNA response elements (MREs) which are sequences to which miRNAs bind. These are prevalent motifs within 3'-UTRs. Among all regulatory motifs within the 3'-UTRs (including silencer regions), MREs make up about half of the motifs.
- As of 2014, the miRBase web site, an archive of miRNA sequences and annotations, listed 28,645 entries in 233 biologic species. Of these, 1,881 miRNAs were in annotated human miRNA loci.
- miRNAs were predicted to have an average of about four hundred target mRNAs (affecting expression of several hundred genes).
- Freidman et al. estimate that >45,000 miRNA target sites within human mRNA 3'-UTRs are conserved above background levels and >60% of human protein-coding genes have been under selective pressure to maintain pairing to miRNAs.
- Direct experiments show that a single miRNA can reduce the stability of hundreds of unique mRNAs.
- The effects of miRNA dysregulation of gene expression are important in cancer,

neuropsychiatric disorders (schizophrenia, bipolar disorder, major depressive disorder, Parkinson's disease, Alzheimer's disease and autism spectrum disorders).

- The translation of mRNA can also be controlled by a number of mechanisms, mostly at the level of initiation.
- Recruitment of the small ribosomal subunit can indeed be modulated by mRNA secondary structure, antisense RNA binding, or protein binding.
- In both prokaryotes and eukaryotes, a large number of RNA binding proteins exist which are directed to their target sequence by the secondary structure of the transcript which may change depending on certain conditions (temperature or presence of a ligand, aptamer).
  - Enzyme induction is a process in which a molecule (a drug) induces (initiates or enhances) the expression of an enzyme.
  - The induction of heat shock proteins in the fruit fly *Drosophila melanogaster*.
  - The Lac operon is an interesting example of how gene expression can be regulated.
  - Viruses possess mechanisms to regulate their gene expression, typically into an early and late phase, using collinear systems regulated by anti-terminators (lambda phage) or splicing modulators (HIV).
  - GAL4 is a transcriptional activator that controls the expression of GAL1, GAL7, and GAL10 (code for the metabolic of galactose in yeast). The GAL4/UAS system has been used in a variety of organisms across various phyla to study gene expression.
  - The colinearity of the Hox gene cluster with their nested antero-posterior patterning.
- Up-regulation is a process that occurs within a cell triggered by a signal which results in increased expression of one or more genes and as a result the protein(s) encoded by those genes.
- Up-regulation occurs when a cell is deficient in some kind of receptor. In this case, more receptor protein is synthesized and transported to the membrane of the cell and the sensitivity of the cell is brought back to normal, reestablishing homeostasis.
- Down-regulation is a process resulting in decreased gene and corresponding protein expression.
- Down-regulation occurs when a cell is overstimulated by a neurotransmitter, hormone or drug for a prolonged period of time, and the expression of the receptor protein is decreased in order to protect the cell.
- Gene Regulation can be summarized by the response of the respective system:
  - Inducible systems - An inducible system is off unless there is the presence of some molecule (called an inducer) that allows for gene expression also called "induce expression". Ex. *Lac* operaon
  - Repressible systems - A repressible system is on except in the presence of some molecule (called a corepressor) that suppresses gene expression also called "repress expression". Ex. *Trp* operon
  - The GAL4/UAS system is an example of both an inducible and repressible system. GAL4 binds an upstream activation sequence (UAS) to activate the transcription of the GAL1/GAL7/GAL10 cassette. On the other hand, a MIG1 response to the presence of glucose can inhibit GAL4 and therefore stop the expression of the GAL1/GAL7/GAL10 cassette.

Techniques to study gene expression at different stages

- Up and down regulation of gene can be analyzed by quantitative PCR and DNA microarray.
- The local chromatin environment of the region can be determined by ChIP-chip analysis by pulling down RNA Polymerase II, Histone 3 modifications, Trithorax-group protein, Polycomb-group protein.
- Epistatic interactions can be investigated by synthetic genetic array analysis.
- Due to post-transcriptional regulation, transcription rates and total RNA levels differ significantly. To measure the transcription rates nuclear run-on assays can be done and newer high-throughput methods are being developed, using thiol labelling instead of radioactivity.
- Alternative splicing can be analyzed with a splicing array or with a tiling array called *DNA microarray.*
- All *in vivo* RNA is found in a complex of RNPs. The quantity of transcripts bound to specific protein can be also analyzed by RIP-Chip.
- Protein levels can be analyzed by Mass spectrometry, which can be compared only to quantitative PCR data, as microarray data is relative and not absolute.
- RNA and protein degradation rates are measured by means of transcription inhibitors (actinomycin D or α-amanitin) or translation inhibitors (Cycloheximide), respectively.

Multiple Choice Questions

1. How many genes are found in human genome?
   (A) About 4000 genes
   (B) About 35000 genes
   (C) Only 23 genes
   (D) About 46 genes
2. Which processes are included in the determination of cellular concentration of a protein?
   (A) Synthesis of the primary RNA transcript (transcription)
   (B) Posttranscriptional modification of mRNA
   (C) Messenger RNA degradation and Protein synthesis (translation)
   (D) All of the above
3. Which of the following process is not included in the determination of cellular concentration of a protein?
   (A) Posttranslational modification of proteins
   (B) DNA replication
   (C) Protein targeting and transport
   (D) Protein degradation
4. What are called the genes that required at all times and expressed at a more or less constant level in virtually every cell of a species or organism?
   (A) Housekeeping genes
   (B) Chimeric genes
   (C) Pseudogenes
   (D) Duplicate genes
5. What is called unvarying expression of a gene?
   (A) Regulated gene expression
   (B) Lipogenic gene expression
   (C) Constitutive gene expression
   (D) Tissue-specific expression
6. What is called the expression of gene products that rise cellular levels and fall in response to molecular signals?
   (A) Regulated gene expression
   (B) Lipogenic gene expression
   (C) Constitutive gene expression
   (D) Tissue-specific expression
7. What is called gene products that increase in concentration under particular molecular circumstances?
   (A) Inducible (B) Repressible
   (C) Suppressible (D) Reducible
8. What is called the process of increasing the expression of gene products that increase in concentration under particular molecular circumstances?
   (A) Reduction (B) Repression
   (C) Domination (D) Induction
9. In the lac operon, under which of the following conditions will the lac genes be transcribed at high levels?
   (A) Low glucose, low lactose
   (B) High glucose, low lactose
   (C) Low glucose, high lactose
   (D) High glucose, high lactose
10. What is called gene products that decrease in concentration in response to a molecular signal?
    (A) Inducible (B) Repressible
    (C) Enhancer (D) Activator
11. What is called the process in which gene products that decrease in concentration in response to a molecular signal?
    (A) Activation (B) Repression
    (C) Stimulation (D) Induction
12. Which of the following lead to repression of the genes for the enzymes that catalyze tryptophan biosynthesis in bacteria?
    (A) Glucose (B) Lactose
    (C) Tryptophan (D) Tyrosine
13. How is also determined the basal rate of transcription initiation at the promoters of nonhousekeeping genes?
    (A) By the promoter sequence
    (B) By regulatory proteins
    (C) By terminator proteins
    (D) by RNA polymerase
14. Which of the following proteins regulate transcription initiation by RNA polymerase?
    (A) Specificity factors (B) Repressors
    (C) Activators (D) All of the above

15. Which of the following proteins alter the specificity of RNA polymerase for a given promoter or set of promoters to regulate transcription initiation by RNA polymerase?
(A) Specificity factors (B) Repressors
(C) Activators (D) Enhancer

16. Which of the following proteins slow down access of RNA polymerase to the promoter to regulate transcription initiation by RNA polymerase?
(A) Specificity factors (B) Repressors
(C) Activators (D) Enhancer

17. Which of the following proteins enhance the RNA polymerase–promoter interaction to regulate transcription initiation by RNA polymerase?
(A) Specificity factors (B) Repressors
(C) Activators (D) Stimulators

18. Which of the following statements regarding regulation of transferrin-receptor protein synthesis is correct?
(A) The iron-responsive element is an iron-binding sequence in the mRNA that encodes transferrin-receptor protein
(B) The iron-responsive element is in the 5' untranslated region of the mRNA that encodes transferrin-receptor protein
(C) When iron is abundant the IRE-binding protein binds to and stabilizes the mRNA that encodes transferrin-receptor protein
(D) When iron is scarce the IRE-binding protein binds to and stabilizes the mRNA that encodes transferrin-receptor protein

19. Which of the following statements, concerning regulation of *trp* operon expression by attenuation, is correct?
(A) Rapid translation of the leader peptide prevents completion of the mRNA transcript
(B) The leader peptide sequence contains no tryptophan residues
(C) Rapid translation of the leader peptide allows completion of the mRNA transcript
(D) The leader peptide sequence encodes enzymes required for tryptophan synthesis

20. Which of the following proteins act as helicase and primase, respectively?
(A) DnaB and DnaG (B) DnaC and DnaA
(C) DnaG and DnaC (D) DnaE and DnaG

21. The role of a metabolite that controls a repressible operon is to ....
(A) Bind to the promoter region and decrease the affinity of RNA polymerase for the promoter
(B) Bind to the operator region and block the attachment of RNA polymerase to the promoter
(C) Bind to the repressor protein and activate it
(D) Increase the production of inactive repressor proteins

22. Which of the following subunit recognize most *E. coli* promoters that control the expression of a set of genes that encode the heat-shock response proteins, when the bacteria are subjected to heat stress?
(A) σ^{70} subunit (B) σ^{32} subunit
(C) ρ subunit (D) ε subunit

23. What is called the binding sites on DNA are generally near a promoter in prokaryotic cells and bound to repressors?
(A) Operators (B) Repressors
(C) Activators (D) Enhancer

24. When blocked the binding of RNA polymerase and its movement along the DNA after binding?
(A) Repressor is present
(B) Activator is present
(C) Repressor is absent
(D) Effector is absent

25. What is called the regulation in which a repressor protein blocks transcription?
(A) Allosteric regulation
(B) Negative regulation
(C) Positive regulation
(D) Metabolic regulation

26. What is called the molecular signal protein that binds to the repressor and causes a conformational change to block transcription process?
(A) Activator (B) Mutagen
(C) Enhancer (D) Effector

27. What is not happen when the molecular signal protein binds to the repressor and causes a conformational change in prokaryotes?
(A) Dissociation of a DNA-bound repressor from the operator
(B) Transcription initiation can proceed unhindered
(C) Inhibiting the activity of a transcription complex at the promoter
(D) Transcription initiation can be proceed on-time

28. What is the function of activators during positive regulation of gene?
(A) Enhance the activity of RNA polymerase at a promoter
(B) Dissociation of a DNA-bound repressor from the operator
(C) Transcription initiation can proceed unhindered
(D) Inhibiting the activity of a transcription complex at the promoter

29. What is called the regulation in which activators bind to DNA and enhance the activity of RNA polymerase at a promoter?
(A) Allosteric regulation
(B) Negative regulation
(C) Positive regulation
(D) Metabolic regulation

30. What is called the eukaryotic activators bind to DNA sites and are quite distant from the promoter affecting the rate of transcription at a promoter?
(A) Metabolites (B) Mutagen
(C) Enhancer (D) Effector

31. What is polycistronic mRNA?
(A) Only one gene on a single transcript
(B) Multiple genes on a single transcript
(C) No genes on a single transcript
(D) Multiple recombinant genes on a single transcript

32. What is operon transcribed as a unit?
(A) A gene cluster and promoter
(B) Promoter is also called operon
(C) A short repetitive DNA sequence
(D) A type of Transposon

33. How were first described many of the principles of prokaryotic gene expression in *E. coli*?
(A) Glucose metabolism
(B) Tryptophan metabolism
(C) Lactose metabolism
(D) Galactose metabolism

34. What is use of lactose in lactose metabolism in *E. coli* during prokaryotic gene expression?
(A) As carbon source
(B) As hydrogen source
(C) As oxygen source
(D) As nitrogen source

35. Who discovered lac operon model in prokaryotic cell (*E. coli*)?
(A) Howard Temin and David Baltimore (1962)
(B) Barbara McClintock (1940)
(C) Marianne Grunberg and Manago Severo Ochoa (1955)
(D) François Jacob and Jacques Monod (1960)

36. In which magazine or journal François Jacob and Jacques Monod published their research a paper on lactose metabolism?
(A) Proceedings of National Academy of Sciences, USA
(B) Proceedings of the French Academy of Sciences
(C) Natural Science Society in Brunn, Austria
(D) Proceedings of Royal Society of London

37. Which of the following two adjacent genes are involved in lactose metabolism and coordinately regulate by a genetic element located at one end of the gene cluster?
(A) β-galactosidase and galactoside permease
(B) β-Glucuronidases and Hyaluronidase
(C) Hydroperoxidase and carboxypeptidase A
(D) γ-Glutamyl transpeptidase and glutathione peroxidase

38. What is the function of enzyme β-galactosidase during *Lac* operon in *E. coli*?
(A) It act as carbon source
(B) It transports lactose into the cell

(C) It cleaves lactose to galactose and glucose
(D) It modifies toxic galactosides to facilitate their removal from the cell

39. What is the function of enzyme galactoside permease during *Lac* operon in *E. coli*?
(A) It act as carbon source
(B) It transports lactose into the cell
(C) It cleaves lactose to galactose and glucose
(D) It modifies toxic galactosides to facilitate their removal from the cell

40. Which of the following is example of negative regulation?
(A) ara operon (B) trp operon
(C) lac operon (D) All of the above

41. For which of the following enzymes, genes work in the lactose (*lac*) operon in *E. coli*?
(A) β-galactosidase (*Z*)
(B) Galactoside permease (*Y*)
(C) Thiogalactoside transacetylase (*A*)
(D) All of the above

42. What is the function of enzyme thiogalactoside transacetylase during *Lac* operon in *E. coli*?
(A) It act as carbon source
(B) It transports lactose into the cell
(C) It cleaves lactose to galactose and glucose
(D) It modifies toxic galactosides to facilitate their removal from the cell

43. Which of the following enzyme appears to modify toxic galactosides to facilitate their removal from the cell during *lac* operon?
(A) β-galactosidase (*Z*)
(B) Galactoside permease (*Y*)
(C) Thiogalactoside transacetylase (*A*)
(D) β-Glucuronidases

44. Which of the following enzyme cleaves lactose to galactose and glucose during *lac* operon?
(A) β-galactosidase (*Z*)
(B) Galactoside permease (*Y*)
(C) Thiogalactoside transacetylase (*A*)
(D) β-Glucuronidases

45. When the *lac* operon genes are repressed?
(A) In the absence of lactose
(B) In the minimum quantity of lactose
(C) In the absence of glucose
(D) In the absence of galactose

46. By which process in the operator or other gene causes gene repression that can be restored by introducing a functional *I* gene encodes a diffusible molecule into the cell on another DNA molecule?
(A) Recombinations (B) Transpositions
(C) Mutations (D) Crossing over

47. What is called the diffusible protein molecule that encoded by *I* gene produced by mutation in operator and in other gene encodes a diffusible molecule that causes gene repression during the study of *lac* operon?
(A) Lac repressor (B) Activator
(C) Lac inducer (D) Lac operator

48. What is Lac repressor?
(A) A antibiotic that blocks protein synthesis
(B) A tetramer of identical monomers that causes gene repression
(C) An enzyme that "breaks open" the lactone ring in the mycotoxin zearalenone
(D) A protein compound that is naturally produced in human breast milk

49. Which of the following secondary binding sites for Lac repressor contained in *lac* operon mutants?
(A) O_1 and O_3 (B) α and β
(C) σ and ρ (D) GAL1 and GAL2

50. Which of the following statement is false about *lac* operon mutants?
(A) The mutator *I* gene is transcribed from its own promoter (PI) independent of the *lac* operon genes
(B) *Lac* operon has two secondary binding sites for the Lac repressor
(C) Secondary binding site (O_2) for Lac repressor is centered near position +410 within the gene encoding β-galactosidase
(D) The Secondary binding site (O_3) for Lac repressor is near position -90 within the gene encoding galactoside permease

51. What happen when the Lac repressor binds to both the main operator and one of the two secondary sites?
(A) Operon is repressed
(B) Transcription initiation is blocked
(C) Intervening DNA looped out
(D) All of the above

52. If the O_2 and O_3 sites are not eliminated by deletion or mutation, Binding of the Lac repressor reduces the rate of transcription initiation by a factor of ……………..
(A) 10^2 (B) 10^5 (C) 10^3 (D) 10

53. When the *lac* operon is induced and an inducer (signal) molecule binds to a specific site on the Lac repressor?
(A) Cells are provided with lactose
(B) Lactose absent in medium
(C) Lac repressor binds to both the main operator
(D) Lactose is not converted in monosaccharides

54. Which act as inducer in lac operon system?
(A) Glucose (B) Lactose
(C) Allolactose (D) Galactose

55. What is allolactose?
(A) A tetramer of identical monomers that causes gene repression
(B) A monosaccharide sugar
(C) An enzyme that "breaks open" the lactone ring in the mycotoxin zearalenone
(D) An isomer of lactose

56. Which of the following statement is not true about lac open system in *E. coli*?
(A) Allolactose is inducer in *lac* operon but are not substrates for β-galactosidase
(B) Lactose is converted to allolactose by β-galactosidase molecules
(C) Isopropylthiogalactoside (IPTG) is one particularly effective and nonmetabolizable inducer of the *lac* operon
(D) Allolactose act as a carbon source for growth in the *lac* operon

57. Which of the following statement is not true?
(A) In the higher eukaryotic cells, almost all protein-encoding genes are transcribed separately
(B) Repressors and activators are DNA-binding proteins used in *Lac* operon
(C) Lac repressor is an enzyme that "breaks open" the lactone ring in the mycotoxin zearalenone
(D) Several β-galactosides structurally related to allolactose are inducers of the *lac* operon

58. Which of the following statement is not true?
(A) Regulatory proteins generally bind to specific DNA sequences
(B) Most regulatory proteins have discrete DNA-binding domains distinguished from cytosine by its protruding methyl group.
(C) Protein-DNA contacts are not possible in the minor groove of the DNA
(D) Amino acid side chains most often hydrogen-bonding to bases in the DNA

59. What is the affinity of regulatory proteins that is higher for targeted specific DNA sequences than for any other DNA sequences?
(A) 10^4 to 10^6 times
(B) 10^{-4} to 10^{-6} times
(C) 10 to 100 times
(D) 10^6 to 10^7 times

60. Which nucleotide can form two hydrogen bonds at N^6 and N^7 positions with glutamine or asparagine than with any other base?
(A) Guanine (B) Cytosine
(C) Thymine (D) Adenine

61. Which nucleotide can form two hydrogen bonds at N^7 and O^6 positions with an arginine residue than with any other base?
(A) Guanine (B) Cytosine
(C) Thymine (D) Adenine

62. What is called the inverted repeats of a short DNA sequence?
(A) Palindromic sequence
(B) Shine-dalgarno sequence
(C) Repetitive sequence
(D) AFLP sequence

63. How many tetramers of Lac repressors are normally contained by an *E. coli* cell
(A) 10 (B) 20
(C) 50 (D) 100

64. To which sequence is separately bound to each of the tethered dimmers of tetramers of the Lac repressor of *E. coli* cell?
(A) Shine-dalgarno sequence
(B) Repetitive sequence
(C) Palindromic operator sequence
(D) AFLP sequence

65. What is the size of the palindromic operator sequence region in the *lac* operon that is separately bound to each of the tethered dimmers of tetramers of the Lac repressor of *E. coli* cell?
(A) 17bp (B) 22bp
(C) 30bp (D) 36bp

66. Which of the following binding motifs play prominent roles in the binding of DNA by regulatory proteins?
(A) Helix-turn- helix and zinc finger
(B) β-α-β loop and α-α corner
(C) Leucine zipper and basic helix-loop-helix.
(D) All of the above

67. How many amino acids comprised in *helix-turn-helix* DNA-binding motif that is crucial to the interaction of many prokaryotic regulatory proteins with DNA?
(A) About 200 amino acids
(B) About 30 amino acids
(C) About 20 amino acids
(D) About 50 amino acids

68. What is called one of the two α-helical segments that interact with the DNA in a sequence-specific way and contained by the helix-turn-helix motif that plays prominent roles in the binding of DNA by regulatory proteins?
(A) Zinc finger (B) β-α-β loop
(C) Leucine zipper
(D) Recognition helix

69. Where the recognition helix of the two α-helical segments is positioned to interact with the DNA in a sequence-specific way and contained by the helix-turn-helix motif?
(A) Minor groove
(B) Major groove
(C) Nitrogenous base
(D) Five carbon sugar

70. How many amino acid residues in a zinc finger of a single polypeptide form an elongated loop held together at the base by a single Zn^{2+} ion that is coordinated to four of the residues (four Cys, or two Cys and two His)?
(A) About 40 amino acids
(B) About 30 amino acids
(C) About 20 amino acids
(D) About 60 amino acids

71. Which is the mouse regulatory protein contains the interaction between DNA and three zinc fingers of a single polypeptide?
(A) Zif268 (B) GCN4
(C) LexA (D) F-Box Proteins

72. Which of the following structure contained by many eukaryotic DNA-binding proteins?
(A) Zinc finger (B) Ion Channels
(C) Lactone ring (D) Crystalline structure

73. How many zinc finger structure found in one DNA-binding protein of the frog *Xenopus*?
(A) 37 (B) 30
(C) 20 (D) 50

74. Zinc fingers can function as RNA-binding motifs in certain proteins that bind eukaryotic mRNAs and act as ....
(A) Translational activator
(B) Trancriptional activators
(C) Translational repressors
(D) Transcriptional reprassors

75. Which of the following a type of DNA-binding domain that identified in a number of proteins that function as transcriptional regulators especially during eukaryotic development?
(A) Heavy-Chain Variable Domain
(B) Homeodomain
(C) Minimized Domains
(D) Miniprotein Domains

76. How many amino acid residues contained in homeodomain that is a type of DNA-binding protein domain because it was discovered in highly conserved homeotic genes
(A) About 40 amino acids
(B) About 30 amino acids
(C) About 20 amino acids
(D) About 60 amino acids

77. What are the function homeotic genes that contain homeodomain that is a type of DNA-binding protein domain of 60 amino acids?
(A) Used for the purification of interferons
(B) Cause a cell to die by self digestion without rupturing and releasing intracellular contents
(C) Regulate the development of body patterns
(D) Rapid manufacture and secretion by the body of the soluble blood serum components in response to an infection

78. What is called the DNA sequence that encode homeodomain that is a type of DNA-binding protein is related to the helix-turn-helix motif?
(A) Pribnow box
(B) TATA box
(C) Shine-Dalgarno bo
(D) Homeobox

79. Which of the statement is not true?
(A) Many eukaryotic transcription factors bind as dimers to the DNA using DNA-binding domains
(B) Leucine zippers have been found in many eukaryotic and a few prokaryotic proteins
(C) The helix-loop-helix motifs of two nucleotide chains interact to form dimmers
(D) DNA binding is mediated by an adjacent short amino acid sequence in regulatory proteins

80. Which of the following structural motifs is the basis for classifying some regulatory proteins into structural families?
(A) Helix-turn- helix and zinc finger
(B) β-α-β loop and α-α corner
(C) Leucine zipper and basic helix-loop-helix
(D) All of the above

81. Which of the following structural motif is an amphipathic α helix with a series of hydrophobic amino acid residues concentrated on one side with the hydrophobic surface forming the area of contact between the two polypeptides of a dimer?
(A) Zinc finger
(B) Leucine zipper
(C) Helix-turn- helix
(D) basic helix-loop-helix

82. Which of the amino acid residues occur at every seventh position an amphipathic α helices of leucine zipper motif?
(A) Leucine (B) Arginine
(C) Histidine (D) Cystein

83. What is the function of regulatory proteins with leucine zippers have a separate DNA-binding domain with a high concentration of basic Lys or Arg residues?
(A) Used for the purification of interferons
(B) Cause a cell to die by self digestion without rupturing and releasing intracellular contents
(C) Interact with the negatively charged phosphates of the DNA backbone
(D) Regulate the development of body patterns

84. What is the function of basic Helix-Loop-Helix structural motif occurs in some eukaryotic regulatory proteins?
(A) Control of gene expression during the development of multicellular organisms
(B) Cause a cell to die by self digestion without rupturing and releasing intracellular contents
(C) Interact with the negatively charged phosphates of the DNA backbone
(D) Regulate the development of body patterns

85. Eukaryotic regulatory proteins share a conserved region of about 50 amino acid residues important in both DNA binding and protein ....
(A) Ionization
(B) Polymerization
(C) Homogenization
(D) Dimerization

86. Which of the following different types of domains for protein-protein interaction have been characterized primarily in eukaryote?
(A) Glutamine-rich
(B) Proline-rich
(C) Acidic domains
(D) All of the above

87. Which of the following enzyme transports lactose into the cell during *lac* operon?
(A) β-galactosidase (*Z*)
(B) Galactoside permease (*Y*)
(C) Thiogalactoside transacetylase (*A*)
(D) β-Glucuronidases

88. Which of the following a specific small molecule called an inducer mediate the dissociation of the repressor from the operator in the lactose (*lac*) operon?
(A) Allolactose (B) Lactose
(C) Glucose (D) Tryptophan

89. Which of the following statement is not true?
(A) Regulatory proteins are DNA-binding proteins that recognize specific DNA sequences
(B) Regulatory proteins also contain domains for protein-protein interactions
(C) Helix-turn-helix, zinc finger and homeodomain are involved in dimerization
(D) Leucine zipper and helix-loop-helix are the example protein-protein interactions

90. CRP and cAMP are involved in the coordinated regulation of many operons that encode enzymes for the metabolism of which secondary sugars?
(A) Lactose and arabinose
(B) Glucose and Galactose
(C) Sucrose and fructose
(D) Sucrose and arabinose

91. What is called a network of operons with a common regulator?
(A) Cistron (B) Intron
(C) Clone (D) Regulon

92. Which of the following bacterial regulon responds to changes in temperature?
(A) Yeast Two-Hybrid Systems
(B) Symbiotic system
(C) Heat-shock gene system
(D) Restriction-modification system

93. Which of the following statement is not true?
(A) Many genes for amino acid biosynthetic enzymes are regulated by transcription attenuation
(B) Trp repressor is a heterodimer containing 107 amino acid residues in each subunit
(C) Different cellular concentrations of tryptophan can vary the rate of synthesis of the biosynthetic enzymes over a 700-fold range
(D) Attenuator made up of sequences 3 and 4 that base-pair to form a A-T-rich stem-and-loop structure

94. How many genes for the enzymes are required to convert chorismate to tryptophan in the *E. coli* tryptophan (*trp*) operon?
(A) One (B) Five
(C) Ten (D) Three

95. The mRNA from the *trp* operon has a half-life of ....
(A) Only about 10 min
(B) Only about 13 min
(C) Only about 3 min
(D) Only about 20 min

96. What will not happen? When tryptophan is abundant binds to the Trp repressor by causing a conformational change.
(A) The repressor to bind to the *trp* operator and inhibit expression of the *trp* operon
(B) The repressor to bind to the *trp* operator and start expression of the *trp* operon
(C) The *trp* operator site overlaps the promoter
(D) The binding of the repressor blocks binding of RNA polymerase

97. In which process transcription is initiated normally but is abruptly halted before the operon genes are transcribed?
(A) Dosage compensation
(B) Polytenization
(C) Transcription attenuation
(D) Translational amplification

98. What is transcription attenuation?
(A) The reduction in virulence of a substance to be used as an immunogen
(B) The loss in energy of an electromagnetic radiation as it passes through matter
(C) The loss in virulence of a pathogenic organism as it is repeatedly subcultured or is let multiply on unnatural hosts
(D) A mechanism for regulating the expression of bacterial operons that encode enzymes involved in amino acid biosynthesis

99. Which process uses uses signals encoded in four sequences within a 162 nucleotide leader region at the 5' end of the mRNA?
(A) Dosage compensation
(B) DNA supercoiling
(C) *Trp* operon attenuation mechanism
(D) Translational amplification

100. How many nucleotides present in leader region at the 5' end of the mRNA that precede the initiation codon of the first gene for *trp* operon attenuation mechanism?
(A) 162 nucleotides (B) 100 nucleotides
(C) 200 nucleotides (D) 93 nucleotides

101. Which of the following factor metabolized directly by glycolysis that affects the expression of the *lac* genes under a bacterium's environment?
(A) Availability of glucose
(B) Availability of lactose
(C) Availability of maltose
(D) Availability of sucrose

102. In the presence of which sugar, expressing the genes for proteins that metabolize sugars such as lactose or arabinose is wasteful?
(A) Fructose (B) Glucose
(C) Mannose (D) Galactose

103. What happens to the expression of the *lac* operon when both glucose and lactose are present?
(A) Strong induction of the *lac* operon requires only lactose to trigger an increase in (cAMP) and increased binding of cAMP to CRP, not glucose
(B) A regulatory mechanism does not restrict expression of the genes required for catabolism of lactose in the presence of glucose
(C) Catabolite repression restricts expression of the genes required for catabolism of lactose in the presence of glucose
(D) The synthesis of cAMP is continued and efflux of cAMP from the cell in the presence of glucose

104. What is called the regulatory mechanism that restricts expression of the genes required for catabolism of lactose, arabinose, and other sugars in the presence of glucose?
(A) Anabolite repression
(B) Anabolite induction
(C) Catabolite induction
(D) Catabolite repression

105. What is the role of cAMP that mediates the effect of glucose in Lac operon?
(A) Coactivator (B) Activator
(C) Repressor (D) Inhibitor

106. Which of the following protein play act as activator that is a homodimer with binding sites for DNA and cAMP?
(A) CD4 Protein
(B) Chloroplast Transit Peptide (CTP)
(C) cAMP receptor protein (CRP)
(D) *cis*-Acting Protein

107. How mediated the binding for DNA and cAMP in the presence of glucose during lac operon gene regulation?
(A) By a helix-turnhelix motif within the protein's DNA-binding domain
(B) By inducing genes in *E. coli* as part of the SOS response to DNA damage
(C) By stimulating RNA transcription 50-fold
(D) By binding of CRP-cAMP to a site near the *lac* promoter

108. What happen if glucose is absent in medium during lac operon?
(A) CRP-cAMP binds to a site near the *lac* promoter
(B) stimulates RNA transcription 50-fold
(C) CRP-cAMP is a positive regulatory element responsive to glucose levels
(D) The binding for DNA and cAMP is mediated

109. When glucose is absent, CRP-cAMP binds to a site near the *lac* promoter and stimulates RNA transcription 50-fol(D) CRP-cAMP is a .... responsive to glucose levels, whereas the Lac repressor is a .... responsive to lactose.
(A) Positive regulatory element and negative regulatory element
(B) Negative regulatory element and Positive regulatory element
(C) Both positive regulatory element
(D) Both negative regulatory element

110. Which of the following subunit of RNA polymerase directly interact with cAMP receptor protein (CRP)?
(A) β subunit (B) α subunit
(C) β' subunit (D) ω subunit

111. What is called a region within a 162 nucleotide leader at the 5' end of the mRNA made up of sequences 3 and 4 and acts as a transcription terminator?
(A) Attenuator (B) Stringent factor
(C) Adaptor (D) Indicator

112. Attenuator acts as a .... in the *trp* operon attenuation mechanism?
(A) Transcription terminator
(B) Transcription Initiator
(C) Transcription elongator
(D) Translation initiator

113. What will be not happened if tryptophan concentrations are high in the *trp* operon attenuation mechanism?
(A) Proceed translation rapidly
(B) Concentrations of charged tryptophan tRNA (Trp-tRNATrp) are high
(C) The ribosome stalls at the two Trp codons
(D) Attenuator structure forms and transcription halts

114. What will be happened if tryptophan concentrations are low in the *trp* operon attenuation mechanism?
(A) The ribosome stalls at the two Trp codons
(B) Concentrations of charged tryptophan tRNA (Trp-tRNATrp) are high
(C) Proceed translation rapidly
(D) Attenuator structure forms and transcription halts

115. Which of the following statement is not true?
(A) When tryptophan concentrations are low, charged tRNATrp is less available
(B) Concentrations of charged tryptophan tRNA (Trp-tRNATrp) are high
(C) The proportion of transcripts that are attenuated declines as tryptophan concentration declines
(D) The *leu* operon leader peptide has seven contiguous Leu residues

116. If the two secondary binding sites O_2 and O_3 for the Lac repressor in *lac* operon are eliminated by deletion or mutation, the binding of repressor to O_1 alone reduces the rate of transcription initiation by a factor of ...
....
(A) 10^2 (B) 105
(C) 10^3 (D) 10

117. What is called the extensive DNA damage in the bacterial chromosome triggers the induction of many distantly located genes?
(A) Transcription attenuation
(B) SOS response
(C) Dosage compensation
(D) Stringent response

118. Which of the following regulatory proteins that coded by induced genes involved in DNA repair?
(A) UvrC and HolB
(B) UmuC and UmuD
(C) RecA protein and LexA repressor
(D) MutH and XthA

119. Which of the repressor protein inhibits transcription of all the SOS genes and induction of the SOS response?
(A) HolB (B) RecA
(C) LexA (D) XthA

120. Which of the peptide bond inactivate LexA repressor after autocleavage reaction to produce two roughly equal protein fragments?
(A) Glu–Gly peptide bond
(B) Trp–Tyr peptide bond
(C) Cys–Ser peptide bond
(D) Ala–Gly peptide bond

121. Which of the following is not a protese but but interaction with LexA facilitates the repressor's self-cleavage reaction?
(A) RecA (B) Rnase
(C) Pepsin (D) Typsin

122. Which of the following has coprotease activity shown by binding to single-stranded DNA can facilitate cleavage of the LexA repressor?
(A) RecA (B) Rnase
(C) Dnase (D) Typsin

123. What is called the regulation coordinated with amino acid concentrations to rRNA synthesis from the seven rRNA operons in the availability of amino acids?
(A) Stringent response
(B) SOS response
(C) Phase variation
(D) Dosage compensation

124. What is not happened? When amino acid concentrations are low during stringent response
(A) rRNA synthesis is halted
(B) Amino acid starvation leads to the binding of uncharged tRNAs to ribosomal A site
(C) To trigger a sequence of events that begins with the binding of an enzyme called stringent factor to the ribosome
(D) Stringent factor catalyzes formation of the unusual nucleotide cytosine tetraphosphate

125. Which of the following unusual nucleotide is catalyzed by stringent factor (RelA protein) at low amino acid concentrations during stringent response?
(A) Cytosine tetraphosphate
(B) Adenosine tetraphosphate
(C) Guanosine tetraphosphate
(D) Thymine tetraphosphate

126. Which of the following organism inhabits the mammalian intestine and moves by rotating the flagella on its cell surface?
(A) *Salmonella typhimurium*
(B) *Treponema pallidum*
(C) *Bacillus subtilis*
(D) *Plasmodium falciparum*

127. Which of the following protein make up the flagella that help to move by rotating the flagella on its cell surface in *Salmonella typhimurium*?
(A) Flagellin (B) Keratin
(C) Collagen (D) Fibroin

128. What are the fundamental differences that distinguish the regulation of gene expression in eukaryotes from that in bacteria?
(A) Activation of transcription is associated with many changes in chromatin structure in eukaryotes
(B) Eukaryotic cells have larger, more complex multimeric regulatory proteins than do bacteria
(C) Transcription in the eukaryotic nucleus is separated from translation in the cytoplasm in both space and time
(D) All of the above

129. Which of the following statement is not true?
(A) The effects of chromosome structure on gene regulation in eukaryotes have no clear parallel in prokaryotes
(B) More condensed chromatin called as heterochromatin is transcriptionally inactive
(C) Less condensed chromatin is called euchromatin
(D) Transcription of a eukaryotic gene is strongly repressed within euchromatin

130. Which of the following enzyme cleave the DNA of carefully isolated chromatin into fragments of multiples of about 200 bp of the nucleosome?
(A) RNase A
(B) RNase H
(C) DNase I
(D) Protease

131. Which of the following type of structural domains is not used in activation by DNA-binding transactivators?
(A) Gal4p
(B) Sp1
(C) CTF1
(D) eIF2B

132. What is the fundamental difference that distinguish the regulation of gene expression in eukaryotes from in bacteria
(A) In eukaryotes, activation of transcription is associated with many changes in chromatin structure in the transcribed region
(B) Eukaryotic cells have larger, more complex multimeric regulatory proteins than do bacteria
(C) Transcription in the eukaryotic nucleus is separated from translation in the cytoplasm in both space and time
(D) All of the above

133. Which of the following statement is not true?
(A) Transcriptionally active chromatin is structurally distinct from inactive chromatin
(B) About 10% of the chromatin in eukaryotic cell is in a more condensed form
(C) More condensed heterochromatin in eukaryotic cell is transcriptionally inactive
(D) Transcription of a eukaryotic gene is strongly repressed when its DNA is condensed within euchromatin

134. What is called about 10% of the chromatin in a typical eukaryotic cell is in a more condensed form than the rest of the chromatin and is transcriptionally inactive?
(A) Heterochromatin
(B) Euchromatin
(C) Degraded chromatin
(D) Extend chromatin

135. What is called less condensed chromatin in a typical eukaryotic cell is transcriptionally active?
(A) Heterochromatin
(B) Euchromatin
(C) Degraded chromatin
(D) Extend chromatin

136. Which of the following structure contains condensed chromatin part called heterochromatin
(A) Chromomeres (B) Matrix
(C) Centriole (D) Centrosome

137. Which of the following mechanisms of translational regulation found in eukaryotes?
(A) Phosphorylation causes the depression of translation in the cell
(B) Binding of protein as translational repressors to mRNA in the 3' untranslated region
(C) Binding proteins disrupt the interaction between eIF4E and eIF4G
(D) All of the above

138. Which of the following enzyme catalyse the phosphorylation of eIF2 that causes the depression of translation in the cell?
(A) Protein kinase
(B) Peptidyltransferase
(C) β-Galactosidase
(D) Thiogalactoside transacetylase

139. Posttranscriptional gene silencing is mediated by
(A) Histon methylation
(B) Protein kinase
(C) Chromatin remodeling
(D) RNA interference

140. Which of the following type of RNA function by interacting with mRNAs at the 3' UTR in posttranscriptional gene silencing?
(A) hnRNA (B) snRNA
(C) miRNA (D) Both b and c

141. Which process involve in posttranscriptional gene silencing?
(A) Translation inhibition
(B) mRNA degradation
(C) Attaching antisense RNA with target mRNA
(D) All of thr above

142. What is the critical role of small RNA molecules?
(A) Protein synthesis
(B) Formation of heterochromatin
(C) mRNA formation
(D) Chromatin remodeling

143. Which of the following structures is formed by micro-RNAs (miRNAs) and small temporal RNAs (stRNAs) are transcribed as precursor RNAs about 70 nucleotides long?
(A) Cruciform like structures
(B) Cone like structures
(C) Hairpin like structures
(D) complex structures

144. Which of the following endonuclase cleave dsRNA to form small temporal RNAs (stRNAs) in RNAi?
(A) Dicer (B) Gal transferase
(C) EDRF(D) (D) RISC

145. What is the size of short duplexes in length that derived after cleaving by dicer in RNAi?
(A) About 100 to 200 nucleotides
(B) About 20 to 25 nucleotides
(C) About 50 nucleotides
(D) About 11 to 15 nucleotides

146. Which of the strand of the miRNA is transferred to the target mRNA to inhibit the translation or degradation of the RNA in posttranscriptional gene silencing?

(A) Palindrome strand
(B) Sense strand
(C) Antisense strand
(D) All of the above

147. What is the sequence of RNA interference technique that is used for gene silencing?

i. Cleavage the duplex by Dicer endonuclease into short segments
ii. Binding of siRNAs to the mRNA
iii. Silencing of the gene for blocking translation
iv. Introduction a duplex RNA molecule into an organism

(A) i, ii, iv, iii (B) ii, iii, i, iv
(C) iv, i, ii, iii (D) iii, i, iv, ii

148. Which technique is used to silence of gene in an organism?

(A) RNA interference (B) Microarray
(C) Transcription (D) Dimerization

149. What is called short segments that formed by cleaving the duplex RNA by dicer in RNAi?

(A) hnRNA (B) tRNA
(C) snRNA (D) Pallindrome RNA

150. Which type of genes specify which organs and appendages will develop in particular body segments?

(A) Segmentation genes
(B) Gap genes
(C) Pair-rule genes
(D) Homeotic genes

151. Which induce *Lac* operon in E. coli?

(A) β-galactosidase
(B) I gene
(C) Lactose
(D) Promoter gene

152. The lac operon is example of

(A) Repressible operon
(B) Overlapping genes
(C) Arabinose operon
(D) Inducible operon

153. In negative operon

(A) Corepressor binds with repressor
(B) Corepressor does not bind with repressor
(C) Corepressor binds with inducer
(D) cAMP have negative effect on lac operon

154. In *E. coli*, during lactose metabolism repressor binds to

(A) Regulator gene (B) Operator gene
(C) Structural gene (D) Promoter gene

155. In lac operon system, a, I, y and z structural genes code respectively for

(A) Repressor protein, permease, β-galactosidase, transacetylase
(B) transacetylase, permease, β-galactosidase, repressor protein
(C) Permease, transacetylase, repressor protein, β-galactosidase
(D) Transacetylase, repressor protein, permease, β-galactosidase

156. Find the coorect match between Lac operon of *E. coli* and their functions

S. No.	Lac operon	S. No.	Funcgion
i.	Structural gene	w	Binding site for repressor protein
ii.	Operator gene	x	Binding site for RNA polymerase
iii.	Promoter gene	y	Induces lactose transport from the medium
iv	Regulator gene	z	Codes for enzyme proteins

(A) i – y, ii – w, iii – x, iv – z
(B) i – y, ii – z, iii – w, iv - x
(C) i – z, ii – w, iii – x, iv - y
(D) i – x, ii – w, iii – z, iv - y

157. Which reaction show that the ADP formed is phosphorylated to ATP by the glycolytic enzymes or through oxidative phosphorylation?

(A) ATP + AMP ↔ ADP
(B) ATP + ADP ↔ 2AMP
(C) ATP + AMP ↔ 2ADP
(D) ATP - AMP ↔ 2ADP

158. What is the number of telomere repeats (n) for most single-celled eukaryotes and generally more than 1,500 in mammals?
(A) In the range of 200 to 1000
(B) in the range of 2 to 10
(C) in the range of 50 to 200
(D) in the range of 20 to 100

159. What is called the study of the properties of an object that do not change under continuous deformations?
(A) Supercoiling
(B) Relaxing of DNA
(C) Topology
(D) Melting study

160. What is a topological property for circular DNA molecules?
(A) Unaffected by deformations of the DNA strands
(B) Unaffected by deformations of the DNA strands
(C) DNA supercoiling
(D) Continuous breakage in DNA strands

161. Which of the following pathways of the synthesis of nucleotides recycle the free bases and nucleosides released from nucleic acid breakdown in cells?
(A) de novo pathways
(B) S_N2 pathway
(C) A simulated folding pathway
(D) Salvage pathways

162. In which form pyrimidine ring is synthesized that attached to ribose phosphate and converted to common pyrimidine nucleotides in de novo nucleic aci , ynthesis?
(A) PRPP
(B) Orotate
(C) 5-phosphoribosylamine
(D) Inosinate

163. Which of the following pathways in which the free bases are intermediates during nucleotide synthesis?
(A) de novo pathways
(B) S_N2 pathway
(C) A simulated folding pathway
(D) Salvage pathways

164. Which of the following metabolic precursor is most important for the synthesis of pyrimidines and purines in de novo pathway in cells?
(A) Phosphoribosyl pyrophosphate (PRPP)
(B) Orotate
(C) 5-phosphoribosylamine
(D) inosinate

165. Which of the following amino acid work as precursor for the synthesis of pyrimidines and purines, respectively, in de novo pathway in cells?
(A) Tryptophan for purines and histidine for pyrimidines
(B) Glycine for purines and aspartate for pyrimidines
(C) Proline for purines and glycine for pyrimidines
(D) Arginine for purines and Tyrosine for pyrimidines

166. Who determined the purine biosynthesis with the origin of the carbon and nitrogen atoms of the purine ring system using isotopic tracer experiments?
(A) Buchanan and G. Robert Greenberg
(B) Rosalind Franklin and Maurice Wilkins
(C) Measolson and Stahal
(D) Gertrude Elion and George Hitchings

167. In which base imidazole ring is found in DNA?
(A) Cytosine (B) Thimine
(C) Uracil (D) Adenine

168. Which is first chemical synthesized, highly unstable with a half-life of 30 seconds at pH 7.5, through donating amino acid glutamate by glutamine that is attached at C-1 of PRPP in de novo purine biosynthesis?
(A) 5-Phospho-β- D-ribosylamine
(B) Formylglycinamide ribonucleotide
(C) 5-Aminoimidazole ribonucleotide
(D) Inosinate

169. How many ATP molecules are consumed in de novo adenylate (AMP) biosynthesis?
(A) 2 (B) 4
(C) 6 (D) 5

170. How many GTP molecules are consumed in de novo adenylate (AMP) biosynthesis?

(A) 2 (B) 1
(C) 6 (D) 5

171. Which is first intermediate with a complete purine ring in each de novo adenylate (AMP) and guanylate (GMP) biosynthesis?

(A) 5-Phospho-β- D-ribosylamine
(B) Formylglycinamide ribonucleotide
(C) 5-Aminoimidazole ribonucleotide
(D) Inosinate

172. In which step NAD^+ is converted into NADH in de novo guanylate (GMP) biosynthesis?

(A) Xanthylate to guanylate
(B) Inosinate to xanthylate
(C) Inonylate to adenylosuccinate
(D) Adenylosuccinate to adenylate

173. The overall rate of de novo purine and pyrimidine nucleotide biosynthesis is regulated by

(A) Forward inhibition
(B) Feedback inhibition
(C) Backward inhibition
(D) Opinion inhibition

174. In first feedback inhibition mechanism of de novo purine biosynthesis, the transfer of an amino group to PRPP to form 5-phosphoribosylamine is catalyzed by the allosteric enzyme .... which is inhibited by the end products IMP, AMP, and GMP?

(A) FGAR amidotransferase
(B) GAR synthetase
(C) IMP synthase
(D) Glutamine-PRPP amidotransferase

175. How many enzymes are separately used in each de novo adenylate (AMP) and guanylate (GMP) biosynthesis?

(A) 12 (B) 14
(C) 11 (D) 13

176. Which enyme is reuired for the conversion of the carbamoyl phosphate into urea synthesis is made in mitochondria of animals?

(A) Carbamoyl phosphate synthetase I
(B) Carbamoyl phosphate synthetase II
(C) Aspartate transcarbamoylase
(D) Dihydroorotase

177. How many ATP molecules are consumed in de novo guanylate (GMP) biosynthesis?

(A) 2 (B) 4
(C) 6 (D) 5

178. Which enyme is reuired for the conversion of carbamoyl phosphate into pyrimidine is made in the cytosol?

(A) Carbamoyl phosphate synthetase I
(B) Carbamoyl phosphate synthetase II
(C) Aspartate transcarbamoylase
(D) Dihydroorotase

179. In which reaction ATP is not consumed for the synthesis of purine and pyrimidine biosynthesis?

(A) 5-Phospho-β- D-ribosylamine to Glycinamide ribonucleotide
(B) Formylglycinamide ribonucleotide to Formylglycinamidine ribonucleotide
(C) Inosinate to Adenylosuccinate
(D) Carboxyaminoimidazole ribonucleotide to SAICAR

180. Regulation of the rate of pyrimidine nucleotide synthesis in bacteria occurs in large part through .... which catalyzes the first reaction in the sequence and is inhibited by CTP, the end product of the sequence?

(A) Aspartate transcarbamoylase (ATCase)
(B) Carbamoyl phosphate synthetase I
(C) carbamoyl phosphate synthetase II
(D) Dihydroorotase

181. How many subunits consisted in the bacterial ATCase molecule?

(A) Three catalytic subunits and six regulatory subunits
(B) Total six catalytic subunits and regulatory subunits
(C) Six catalytic subunits and three regulatory subunits
(D) Six catalytic subunits and six regulatory subunits

182. Which enzyme catalyze the Phosphorylation of AMP to ADP?

(A) Adenylate kinase
(B) Ribonucleotide reductase
(C) Thymidylate synthase
(D) ATCase

183. Which isotopes precursors are used for the study of purine biosynthesis pathway by Buchanan and G. Robert Greenberg in the 1950s?
(A) ^{12}C- or ^{15}N-labeled precursors
(B) ^{14}C- or ^{14}N-labeled precursors
(C) ^{13}C- or ^{15}N-labeled precursors
(D) ^{14}C- or ^{15}N-labeled precursors

184. Which are the bulding blocks of DNA?
(A) Ribonucleotides
(B) Deoxyribonucleotides
(C) Amino acids
(D) Glucose and Fructose

185. How derived deoxyribonucleotides from D-ribose?
(A) By direct oxidation at the 2'-carbon atom of the D-ribose
(B) By direct reduction at the 2'-carbon atom of the D-ribose
(C) By direct reduction at the β-carbon atom of the D-ribose
(D) By direct reduction at the 5'-carbon atom of the D-ribose

186. Adenosine diphosphate (ADP) is reduced to
(A) 2'-deoxyadenosine diphosphate (dADP)
(B) β-deoxyadenosine diphosphate (dADP)
(C) 3'-deoxyadenosine diphosphate (dADP)
(D) 5'-deoxyadenosine diphosphate (dADP)

187. Which of the following enzyme catalyze the reduction of D-ribose to deoxyribonucleotides?
(A) Dihydrofolate reductase
(B) Thioredoxin reductase
(C) Nucleotide reductase
(D) Ribonucleotide reductase

188. Which does donate a pair of hydrogen in the reduction of the D-ribose portion of a ribonucleoside diphosphate to 2'-deoxy-D-ribose?
(A) FADH (B) FADPH
(C) NADPH (D) NADH

189. Which hydrogen-carrying protein helps to donate by NADPH a pair of hydrogen in the reduction of the D-ribose portion of a ribonucleoside diphosphate to 2'-deoxy-D-ribose?
(A) Thioredoxin (B) Allantoin
(C) Xanthine oxidase (D) Allopurinol

190. What is the function of thioredoxin that has a pairs of –SH groups for the conversion of D-ribose portion of a ribonucleoside diphosphate to 2'-deoxy-D-ribose?
(A) Carry oxygen atoms by NADPH
(B) Carry sulphate group by NADPH
(C) Carry hydrogen atoms by NADPH
(D) Carry phosphorus atoms by NADPH

191. Which of the following enzyme is found as dimer form that has two subunits R1 (regulatory sites) and R2 (binuclear iron) and catalyze the reduction of D-ribose to deoxyribonucleotides?
(A) Dihydrofolate reductase
(B) Thioredoxin reductase
(C) Nucleotide reductase
(D) Ribonucleotide reductase

192. Enzyme ribonucleotide reductase contains two active sites at which R1 contributes .... required for activity and R2 contributes ....
(A) Two hydrogen atoms and a stable tyrosyl radical, respectively
(B) Two sulfhydryl groups and a stable tyrosyl radical, respectively
(C) Two a stable tyrosyl radical and sulfhydryl groups, respectively
(D) a sulfhydryl groups and two stable tyrosyl radical, respectively

193. What are differences in four classes (I, II, III and IV) of ribonucleotide reductase that convert D-ribose portion of a ribonucleoside diphosphate to 2'-deoxy-D-ribose?
(A) In the identity of the group supplying the active-site radical
(B) In the cofactors
(C) Both (A) and (B)
(D) Only in the cofactors

194. Which of the following is correct match?

S. No.	Ribonucleotide reductase	Function
i.	Class I	Requires oxygen to regenerate the tyrosyl Radical and functions only in an aerobic environment

ii.	Class II	Have 5'-deoxyadenosylcobalamin rather than a binuclear iron center
iii.	Class III	Contains an iron-sulfur cluster and requires NADPH and S-adenosylmethionine for activity and function in an anaerobic environment
iv.	Class IV	Containing a binuclear manganese center

(A) i, ii & iii (B) i, ii & iv
(C) ii, iii, iv (D) i, ii, iii, iv

195. Which of the following class of ribonucleotide reductase is found in *E. coli*.
i. Class I ii. Class II
iii. Class III iv. Class IV
(A) i, ii & iii (B) i & iii
(C) ii & iv (D) i, ii, iii, iv

196. Which type of molecules is used in the regulation of the activity but substrate specificity of *E. coli* ribonucleotide reductase by inducing enzyme conformations with altered specificities?
(A) Effector molecules
(B) Suppressor molecules
(C) Enhancer molecules
(D) Inducer molecules

197. Which of the following is not the function of regulatory site of R1 subunit of ribonucleotide reductase which convert D-ribose portion of a ribonucleoside diphosphate to 2'-deoxy-D-ribose?
(A) Affects overall enzyme activity
(B) Binds ATP which activates the enzyme and binds dATP that inactivates the enzyme
(C) Reduces the activity of effector molecule
(D) Alters substrate specificity in response to the effector molecule

198. ATP is a general activator for biosynthesis and ribonucleotide reduction. What happen when dATP present in small amounts?
(A) Increases the reduction of pyrimidine nucleotides
(B) Shift the specificity to ADP reduction
(C) Shut down the enzyme
(D) Shifts the specificity to favor reduction of GDP

199. ATP is a general activator for biosynthesis and ribonucleotide reduction. What happen when dATP present at high level?
(A) Shifts the specificity to favor reduction of GDP
(B) Shift the specificity to ADP reduction
(C) Shut down the enzyme
(D) increases the reduction of pyrimidine nucleotides

200. Thymidylate is derived from
(A) dTDP and dUMP
(B) dCDP and dUMP
(C) dTDP and dAMP
(D) dCDP and dTMP

201. Which of the following nitrogen base is not found in DNA?
(A) Thymine (B) Adenine
(C) Cytosine (D) Uracil

202. Which enzyme catalyze the conversion of dUMP to dTMP is catalyzed by
(A) Thymidylate synthase
(B) Dihydrofolate reductase
(C) Xanthine oxidase
(D) Urate oxidase

203. Which of the following produced after degradation of purines and pyrimidines respectively?
(A) Ammonia and Allopurinol
(B) Urea and uric acid
(C) Allopurinol and Uric acid
(D) Uric acid and Urea

204. Which enzyme degrades purine nucleotides by a pathway in which they lose their phosphate in the conversion of AMP or GMP to Adenosine or Guanosine, respectively?
(A) Xanthine oxidase
(B) Adenosine deaminase
(C) 5'- nucleotidase
(D) Urate oxidase

205. Which of following products are produced in the pathway of the degradation of adenine nucleotides?
(A) Adenosine – Inosine – Hypoxanthine – Xanthine – Uric Acid
(B) Inosine – Adenylate – Hypoxanthine – Xanthine – Uric acid
(C) Adenylate – Inosine – Xanthine – Hypoxanthine – Uric Acid
(D) Uric acid – Inosine – Xanthine – Hypoxanthine – Adenylate

206. Different enzymes are involved in the pathway of the degradation of adenine nucleotides.What are the name of enzymes W, X, Y and Z, repectively?

$$\text{Adenosine} \xrightarrow{W} \text{Inosine} \xrightarrow{X} \text{Hypoxanthine} \xrightarrow{Y} \text{Xanthine} \xrightarrow{Z} \text{UricAcid}$$

(A) Adenosine deaminase, Nucleosidase, Xanthine oxidase, Xanthine oxidase
(B) Xanthine oxidase, Nucleosidase, Adenosine deaminase, Xanthine oxidase
(C) Adenosine deaminase, Xanthine Oxidase, Nucleosidase, Hypoxanthine oxidase
(D) Adenosine deaminase, Nucleosidase, Hypoxanthine oxidase, Xanthine oxidase

207. Which of the enzyme is a flavoenzyme with an atom of molybdenum and four iron-sulfur centers in its prosthetic group?
(A) Adenosine Deaminase
(B) Nucleosidase
(C) Hypoxanthine Oxidase
(D) Xanthine Oxidase

208. Which is the electron acceptor in purine nucleotide degradation?
(A) Molecular Hydrogen
(B) Sulphur molecule
(C) Molecular oxygen
(D) Molecular nitrogen

209. Which of following pathway is used in the degradation of guanine nucleotides?
(A) GMP – Guanosine – Guanine – Xanthine – Uric acid
(B) Guanosine – GMP – Guanine – Xanthine – Uric acid
(C) Uric acid – Xanthine – Guanine – Guanosine – GMP
(D) Xanthine – Gunine – Guanosine – GMP – Uric acid

210. What are the names of enzymes W, X, Y and Z, respectively, involved in the pathway of the degradation of Guanine nucleotides.

$$\text{GMP} \xrightarrow{W} \text{Guanosine} \xrightarrow{X} \text{Guanine} \xrightarrow{Y} \text{Xanthine} \xrightarrow{Z} \text{UricAcid}$$

(A) Guanine Deaminase, Xanthine Oxidase, 5'-nucleotidase, Nucleosidase,
(B) 5'-nucleotidase, Nucleosidase, Guanine Deaminase, Xanthine Oxidase
(C) 5'-nucleotidase, Xanthine Oxidase, Guanine Deaminase, Nucleosidase
(D) 5'-nucleotidase, Guanine Deaminase, Nucleosidase, Xanthine Oxidase

211. Which is the excreted end product of purine catabolism in primates, birds, and some other animals?
(A) Urea (B) IAA
(C) Uric acid (D) Ammonia

212. How much excreted uric acid by a healthy adult human?
(A) About 10.6 g/24 h (B) About 5.6 g/24 h
(C) About 0.6 g/24 h (D) About 6 g/24 h

213. Which product is obtained from the degradation of uric acid by the action of urate oxidase in most mammals and many other vertebrates?
(A) Allantoin (B) Allopurinol
(C) Urea (D) Aminopterin

214. Which is the end product of thymine degradation?
(A) Urea (B) Succinyl-CoA
(C) Uric acid (D) Ammonia

215. Which genetic aberration is caused in human purine metabolism that leads to severe immunodeficiency disease in which T lymphocytes and B lymphocytes do not develop properly.
(A) Leukemia
(B) HIV
(C) Adenosine deaminase (ADA) deficiency
(D) Lesch-Nyhan syndrome

216. Deficiency of which enzyme leads to a 100-fold increase in the cellular concentration of dATP that a strong inhibitor of ribonucleotide reductase?
(A) Adenosine Deaminase (ADA)
(B) Nucleosidase
(C) Hypoxanthine Oxidase
(D) Xanthine Oxidase

217. In which cells, high levels of dATP produce a general deficiency of other dNTPs
(A) T lymphocytes (B) B-lymphocyte
(C) Platelets (D) RBCs

218. By which pathways purine and pyrimidine bases are recycled?
(A) De novo pathway
(B) Signaling pathway
(C) CNS pathway
(D) Salvage pathways

219. In the recycling of purine bases of salvage pathways, which enzyme catalyze a single reaction in which free adenine reacts with PRPP to yield the corresponding adenine nucleotide? dAdenine + PRPPO → AMP + PPi
(A) Adenosine phosphoribosyltransferase
(B) Hypoxanthine-guanine phosphoribosyltransferase
(C) Adenosine deaminase
(D) 5'-nucleotidase

220. In the recycling of purine bases of salvage pathways, which enzyme catalyze a single reaction in which free guanine and hypoxanthine reacts with PRPP?
(A) Adenosine phosphoribosyltransferase
(B) Hypoxanthine-guanine phosphoribosyltransferase
(C) Adenosine deaminase
(D) 5'-nucleotidase

221. Which of the following genetic disorder caused by a genetic lack of hypoxanthine-guanine phosphoribosyltransferase activity in male children (in 2 year age)?
(A) Leukemia
(B) Gout disease
(C) Adenosine deaminase (ADA) deficiency
(D) Lesch-Nyhan syndrome

222. What is called the segments of DNA whose migration is suppressed by methylation?
(A) Transposons (B) Intron
(C) Cistron (D) Recon

223. In which eukaryotes Buchanan and Greenberg determined the purine biosynthesis with the origin of the carbon and nitrogen atoms of the purine ring system using isotopic tracer experiments?
(A) Reptiles (B) Birds
(C) Nematodes (D) Human

224. A ribozyme made up of .... is a segment of mRNA in the ciliated protozoan *Tetrahymena thermophila*?
(A) Only intron (B) Only exon
(C) Both (A) and (B) (D) None of above

225. Which is not the symptom Lesch-Nyhan syndrome in male children (in 2 year age)?
(A) Extremely hostile
(B) Hangout tongue from mouth
(C) Mutilate by biting off fingers, toes and lips
(D) Compulsive self-destructive tendencies

226. Which pathway illustrated by the devastating effects of Lesch-Nyhan syndrome
(A) Genetic recombination pathway
(B) Signaling pathway
(C) De novo pathway
(D) Salvage pathways

227. When PRPP levels rise and purines are overproduced by the de novo pathway and results in high levels of uric acid production and gout-like damage to tissue?
(A) In the absence of hypoxanthine-guanine phosphoribosyltransferase
(B) in the increase of urea in body
(C) due to damage T-lymphocytes
(D) due to abnormal deposition of sodium urate crystals

228. Which disease is caused by an elevated concentration of uric acid in the blood and tissues?
(A) Leukemia
(B) Gout disease
(C) Adenosine deaminase (ADA) deficiency
(D) Lesch-Nyhan syndrome

229. What is not the symptom gout disease?
(A) The joints become inflamed, painful, and arthritic
(B) The abnormal deposition of sodium urate crystals
(C) Excess uric acid deposition in the kidney tubules
(D) In the increase of urea in urinal tubes

230. Which of the following drug is used for the treatment of Gout disease in man that inhibits xanthine oxidase enzyme that catalyzes the conversion of purines to uric acid?
(A) Allopurinol
(B) Azaserine
(C) Acivicin
(D) Aminopterin

231. Allopurinol is a substrate of .... which converts allopurinol to oxypurinol (alloxanthine) which inactivates the reduced form of the enzyme by remaining tightly bound in its active site?
(A) Adenosine Deaminase
(B) Nucleosidase
(C) Hypoxanthine Oxidase
(D) Xanthine Oxidase

232. Who was developed allopurinol that allopurinol is a substrate of xanthine oxidase which converts allopurinol to oxypurinol (alloxanthine) which inactivates the reduced form of the enzyme by remaining tightly bound in its active site?
(A) Gertrude Elion and George Hitchings
(B) Buchanan and G. Robert Greenberg
(C) Rosalind Franklin and Maurice Wilkins
(D) Charles Best and J.B. Collip

233. Who developed drug acyclovir which used in treating people with AIDS and other purine analogs used in cancer chemotherapy?
(A) Charles Best and J.B. Collip
(B) John Buchanan
(C) Rosalind Franklin and Maurice Wilkins
(D) Gertrude Elion and George Hitchings

234. Which is a nitrogen donor in at least half a dozen separate reactions in nucleotide biosynthesis?
(A) Aspartin
(B) Glutamine
(C) Tryptophan
(D) Phenyl alanine

235. Who characterized glutamine analog 'azaserine' which is a mechanism-based enzyme inactivator?
(A) Charles Best and J.B. Collip
(B) John Buchanan
(C) Rosalind Franklin and Maurice Wilkins
(D) Gertrude Elion and George Hitchings

236. Which shows as a cancer chemotherapeutic agent which is glutamine analogs?
(A) Allopurinol
(B) Fluorouracil
(C) Acivicin
(D) Aminopterin

237. Choose the correct matching

	Inhibitor	Activity
i.	Azaserine	Glutamine analogs and a mechanism-based enzyme inactivator characterized by John Buchanan
ii.	Acivicin	As a cancer chemotherapeutic agent
iii.	Fluorouracil	Chemotherapeutic agent whom convert to the deoxynucleoside monophosphate FdUMP which binds to and inactivates the enzyme.
iv.	Methotrexate	A folate analog acts as a competitive inhibitor of dihydrofolate reductase
v.	Aminopterin	Inhibitor of nucleotide biosynthesis which is not limited to cancer treatment
vi.	Trimethoprim	An antibiotic binds to bacterial dihydrofolate reductase and used to treat certain urinary and middle ear bacterial infections

(A) i, ii & iii (B) i, ii, iii, iv & v (C) ii, iii, iv & v (D) All of the above

238. Which of the following an antibiotic developed by Hitchings and Elion that binds to bacterial dihydrofolate reductase and used to treat certain urinary and middle ear bacterial infections.
(A) Acivicin (B) Aminopterin
(C) Fluorouracil (D) Trimethoprim

239. Which of the following drug is used to treat African sleeping sickness (African trypanosomiasis) that caused by parasitic protists (trypanosomes) in which lack pathways for de novo nucleotide biosynthesis?
(A) Allopurinol (B) Fluorouracil
(C) Trimethoprim (D) Aminopterin

240. Who determined the origin of the carbon and nitrogen atoms of the purine ring system using isotopic tracer experiments in birds?
(A) Charles Best and J.B. Collip
(B) John Buchanan
(C) Rosalind Franklin and Maurice Wilkins
(D) Gertrude Elion and George Hitchings

241. Which of the following statement is not coorect?
(A) Retroviruses synthesize completory DNA using RNA-dependent DNA polymerase
(B) DNA and RNA synthesis and RNA degradation activities use separate active sites on the protein
(C) Reverse transcriptase requires a primer
(D) Reverse transcriptases have 5' to 3' proofreading exonucleases

242. Which of the following enzyme do not have 5' to 3' proofreading exonucleases?
(A) Reverse transcriptase
(B) DNA polymerase
(C) RNA polymerase
(D) None of the above

243. Which of the following enzyme dose not require a primer?
(A) Reverse transcriptase
(B) DNA polymerase
(C) RNA polymerase
(D) None of the above

244. Which of the following enzyme synthesize new biological molecules (DNA or RNA) in 3' to 5' direction?
(A) Reverse transcriptase
(B) DNA polymerase
(C) RNA polymerase
(D) None of the above

245. What is the error rate in adding nucleotides of enzyme reverse transctase during the syntheses of cDNA from RNA?
(A) About 2 per 20 nucleotides
(B) About 1 per 20,000 nucleotides
(C) About 1 per 1,00,000 nucleotides
(D) About 10 per 20,000 nucleotides

246. In which biological process, error rate of adding nucleotides in forming new polynuclotides chain?
(A) Transcription
(B) Reverse Transcription
(C) DNA replication
(D) Translation

247. What is called the type of DNA synthesized from RNA genome through reverse transcription?
(A) Complementory DNA (cDNA)
(B) Moderatory repeatative DNA
(C) Highly Repeatative DNA
(D) Transposable DNA

248. What is the use of complementary DNA that synthesized from RNA genome through reverse transcription?
(A) To synthesize RNA
(B) To construct genomic library
(C) To clone cellular genes
(D) All of the above

249. Which of the following is the example of retroviruses?
(A) Chicken pox virus
(B) Varicella zoster virus (VZV)
(C) Herpesviruses
(D) RNA tumor viruses

250. Which of the following virus contain an oncogene that can cause the cell to grow abnormally?
(A) Human Immuno vrirus
(B) Adeno viruse
(C) Chicken pox virus
(D) RNA tumor viruses

251. Which of the following retrovirus was discovered that contain an oncogene that can cause the cell to grow abnormally?
(A) Rous sarcoma virus
(B) Human Immuno vrirus
(C) Adeno viruse
(D) Chicken pox virus

252. Who first discovered the Rous sarcoma virus that contain an oncogene that can cause the cell to grow abnormally?
(A) Arthur Kornberg
(B) Grunberg-Manago and Severo Ochoa
(C) F. Peyton Rous
(D) Howard Temin and David Baltimore

253. First on which disease studied by F. Peyton Rous for divering rous sarcoma virus that contain an oncogene that can cause the cell to grow abnormally?
(A) Chicken tumors (B) AIDS
(C) Hepatitis B (D) Chicken pox

254. Which of the following is related to retroviruses?
(A) Tobacco mosaic virus
(B) Human Immunodeficiency vrirus (HIV)
(C) Adeno viruse
(D) Chicken pox virus

255. Which of the following disease is caused by human immunodeficiency vrirus?
(A) Chicken tumors (B) AIDS
(C) Hepatitis B (D) Chicken pox

256. What is AIDS?
(A) Attained invulnerable deficit system
(B) Acquired impervious dearth symptoms
(C) Acquired immune deficiency syndrome
(D) Acceptable internet deficiency scheme

257. Which of the following scientists are related to study on oncogenes that produce tuumor disease through infection of retroviruses?
(A) Arthur Kornberg
(B) Grunberg-Manago and Severo Ochoa
(C) Harold Varmus and Michael Bishop
(D) T.H. Morgan and H.G. Muller

258. Which of the following type of genome is foind in HIV that identified with standard retroviral genes in 1983?
(A) ssRNA (B) ssDNA
(C) dsDNA (D) dsRNA

259. Which of the main type of cells is infected by HIV?
(A) Platelets (B) B lymphocytes
(C) T lymphocytes (D) Nerve cells

260. What are main symptoms of HIV infection in the host organism?
(A) Suppression of the immune system
(B) Damage nerve cells
(C) Blood clotting
(D) Brain hamerage

261. What happen if the reverse transcriptase of HIV is even more (ten times more) error prone than other known reverse transcriptases?
(A) Damage ribonucleotides
(B) High mutation rates
(C) Block protein synthesis
(D) Negetive effects on host genome

262. Which of the following a class of drugs are used to treat HIV-infected individuals?
(A) Protease inhibitors
(B) RNA inhibitors
(C) Antibiotics for Replication inhibitoin
(D) Vaccines for coat-protein formation

263. What are called the eukaryotic DNA transposons that have a structure very similar to retroviruses?
(A) Repeatitive transposons
(B) Polymorphic transposons
(C) Nonmobile transposons
(D) Retrotransposons

264. What are called the nucleotide sequences that encode an enzyme homologous to the retroviral reverse transcriptase and move from one position to another in the cellular genome?
(A) Prokaryotic Transposons
(B) Eukaryotic transposons
(C) Retrotransposons
(D) Psudogene

265. Which of the following gene lack in retroytransposons so, cannot form viral particles?
(A) *gag* (B) *pol*
(C) *env* (D) All of the above

266. What is called the process in which DNA is introduced into a cell by parasites or by other means and promoted insertion of the intron into an identical site in another DNA copy of a homologous gene that does not contain the intron?

(A) Homing (B) Cloning
(C) Sibling
(D) Insertional inactivasion

267. What is called the insertion of the RNA intron into DNA and reverse transcription of the intron, the movement of these introns?

(A) Homing (B) Cloning
(C) Sibling (D) Retrohoming

268. Who are the called molecular parasites?

(A) Pseudogenes (B) Transposons
(C) Chaparons
(D) House keeping genes

269. What is called the structures at the ends of linear eukaryotic chromosomes?

(A) Centromeres (B) Chromomeres
(C) Telomeres (D) Satellite

270. Which of the following structure of eukaryotic chromosomes generally consist of many tandem copies of a short oligonucleotide sequence?

(A) Centrole (B) Telomeres
(C) Chromomeres (D) Chromonemata

271. Which of the following feature is related to many tandem copies of a short oligonucleotide sequence of telomeres in eukaryotic chromosomes?

(A) The sequence usually has the form TxGy in one strand and CyAx in the complementary strand (where x and y are in the range of 1 to 4)
(B) The sequence helps to kill T lymphocytes cells in infected host organmism
(C) They transpose from one position to another in the cellular genome
(D) The sequence has not any introns

272. Which of the following enzyme solve the problem of chromosome shortening in each cell generation at ends?

(A) Polymerase (C) Endonuclease
(C) Transposase (D) Telomerease

273. Which of the following statement is true about DNA replication of telomerase?

(A) DNA replication requires a primer
(B) Telomerase adds telomeres to chromosome ends to solve the DNA replication problem
(C) DNA replication requires a template
(D) DNA template is available for the pairing of an RNA primer at the end of chromosomes

274. What are the main components of enzyme telomerase?

(A) DNA
(B) RNA
(C) Protein
(D) both RNA and protein

275. What is the length size of RNA component of enzyme telomerase?

(A) About 150 nucleotides
(B) About 250 nucleotides
(C) About 350 nucleotides
(D) About 500 nucleotides

276. What is the main feature of RNA component of enzyme telomerase?

(A) It contains about 1.5 copies of the appropriate CyAx telomere repeat
(B) It is about 150 nucleotides long
(C) The region of the RNA acts as a template for synthesis of the TxGy strand of the telomere
(D) All of the above

277. Telomerase acts as .... that provides the active site for RNA-dependent DNA synthesis.

(A) DNA polymerase
(B) RNa polymerase
(C) Restriction endonuclease
(D) Reverse transcriptase

278. Telomere synthesis requires the .... of a chromosome as primer and proceeds in the usual 5' to 3' direction.

(A) 3' end
(B) 5' end
(C) Extra ssDNA segement
(D) Extra ssRNA segement

279. Which of the following enzyme synthesize the complementary CyA*x* strand with using an RNA primer after the extension of the T*x*G*y* strand?
(A) Telomerase
(B) Cellular DNA polymerase
(C) Cellular RNA polymerase
(D) Transposase

280. Which of the following enzymes acts during the formation of the ends of chromosomes?
(A) Telomerase and DNA polymerase
(B) Telomerase and RNA polymerase
(C) DNA polymerase and RNA polymerase
(D) Telomerase and Transposase

281. How protect the sigle-stranded DNA region at 3' end of the TG strand of telomeric DNA in lower eukaryotic chromosomes?
(A) By specific binding proteins
(B) By T-loop formation
(C) By hairpin structure formation
(D) By enzymic activities

282. How protect the sigle-stranded DNA region at 3' end of the TG strand of telomeric DNA in higher eukaryotic chromosomes?
(A) By specific binding proteins
(B) By T-loop formation
(C) By hairpin structure formation
(D) By enzymic activities

283. How form T loop in telomeres of higher eukaryotic chromosomes to protect the sigle-stranded DNA region at the 3' end?
(A) By invasion of the 3' end of the telomere's single strand into the duplex DNA
(B) By base-pairing the telomere's single strand with completery RNA primer
(C) By self base-pairing of the 3' end of the telomere's single strand
(D) By invasion of the 3' end of the telomere's single strand into the ssDNA

284. Which of the following proteins bind with the T-looped DNA in telomeres of mammalian chromosomes?
(A) TRF1 and TRF2
(B) Nickel-binding proteins
(C) Apolipoprotein B and Titin
(D) TFIIa and TFIIH

285. What is main role of T- loop in telomeres of mammalian chromosomes?
(A) T loops protect the 3' ends of chromosomes
(B) Block the DNA replication
(C) Make the transition to the modern DNA-based system
(D) Help to move transposons from one position to another in the cellular genome

286. Which of the following prtozoans loss the telomerase activity in a gradual shortening of telomeres with each cell division?
(A) *Plasmodium* (B) *Parameceum*
(C) *Tetrahymena* (D) *Trypnosoma*

287. Which of the following part of chromosome in human is gradually shortened with age?
(A) Cetromere (B) Telomere
(C) Chromomere (D) Chromonemata

288. What happen if the telomerase reverse transcriptase is introduced into human somatic cells in vitro?
(A) Shortening of telomeres with age
(B) Telomerase activity is restored and the cellular life span decreases
(C) Telomerase activity is lost and the cellular life span increases
(D) Telomerase activity is restored and the cellular life span increases

289. Which of the following eukaryotic virus has RNA genome?
(A) MS2 bacteriophage
(B) R17 bacteriophage
(C) Influenza virus
(D) Qβ bacteriophage

290. Which of the following is *E. coli* bacteriaophage that have RNA genome?
(A) MS2 bacteriophage
(B) R17 bacteriophage
(C) Qβ bacteriophage
(D) All of the above

291. During transcription, histone H3 is methylated by ....
(A) Specific histone methylases
(B) RNA polymerase
(C) Histone deacetylases
(D) Assembly of transcription factors

292. Which is the active form of vitamin B_{12} that participates in intramolecular group transfers between adjacent carbons?
(A) NAD^+
(B) FAD
(C) Adenosine is 5'-deoxyadenosylcobalamin
(D) Ribofavin

293. Which of the following the pathways of the synthesis of nucleotides are used in cells?
i. de novo pathways
ii. S_N2 pathway
iii. A simulated folding pathway
iv. Salvage pathways
(A) i & ii (B) ii & iii
(C) i & iii (D) i & iv

294. Which of the following pathways of the synthesis of nucleotides begin with its metabolic precursor in cells?
(A) de novo pathways
(B) S_N2 pathway
(C) A simulated folding pathway
(D) Salvage pathways

295. Which of the following metabolic precursor does not participate in the synthesis of nucleotides in de novo pathway in cells?
(A) Amino acids
(B) Ribose 5'-phosphate
(C) Tryptophan
(D) CO2 and NH_3

296. Alternative splicing is a process that enables the number of proteins produced by an organism to be vastly greater than its number of genes. How is this possible?
(A) Codons can code for more than one amino acid.
(B) Recombinant technology is able to translate different proteins from the same gene.
(C) The anticodon of tRNA has a wobble effect that allows a variety of translations per gene.
(D) Depending on what sections are treated as introns and exons, different proteins can be made from the same gene

297. A section of mRNA reads as follows: 5' – AUGAAUUGG – 3'. How would this be translated?
(A) Trp-Asn-Met (B) Ile-Asn-Stop
(C) Met-Asn-Trp (D) Start-Asn-Stop

298. During gene expression, an error creates a mutation in the codon GUU. Which of the following is not an example of what could have replaced GUU in order to create a silent mutation.
(A) GUC (B) UGA
(C) GUA (D) GUG

299. The result of transcription is a pre-mRNA molecule that goes on to be processed into an mRNA molecule. Which of the following is not something that occurs during mRNA processing?
(A) Poly-A tail is added to 3' end
(B) Introns are joined together
(C) 5' cap is added to 5' end
(D) Noncoding regions are spliced out

300. The genetic code is universal, which means that gene expression is also universal. What is a difference between prokaryotic and eukaryotic gene expression?
(A) Prokaryotes use the base uracil while eukaryotes use thymine.
(B) In prokaryotes, each step of gene expression occurs in the cytoplasm
(C) In eukaryotes, each step of gene expression occurs in the nucleus.
(D) Prokaryotes do not follow the central dogma

Answer-Key

Q	Ans	Q	Ans	Q	Ans	Q	Ans	Q	Ans	Q	Ans	Q	Ans	Q	Ans
1.	(B)	41.	(D)	81.	(B)	121.	(A)	161.	(D)	201.	(D)	241.	(D)	281.	(A)
2.	(D)	42.	(D)	82.	(A)	122.	(A)	162.	(B)	202.	(A)	242.	(A)	282.	(B)
3.	(B)	43.	(C)	83.	(C)	123.	(A)	163.	(D)	203.	(D)	243.	(C)	283.	(A)
4.	(A)	44.	(A)	84.	(A)	124.	(D)	164.	(A)	204.	(C)	244.	(D)	284.	(A)
5.	(C)	45.	(A)	85.	(D)	125.	(C)	165.	(B)	205.	(A)	245.	(B)	285.	(A)
6.	(A)	46.	(C)	86.	(D)	126.	(A)	166.	(A)	206.	(A)	246.	(C)	286.	(C)
7.	(A)	47.	(A)	87.	(B)	127.	(A)	167.	(D)	207.	(D)	247.	(A)	287.	(B)
8.	(D)	48.	(B)	88.	(A)	128.	(D)	168.	(A)	208.	(C)	248.	(C)	288.	(D)
9.	(A)	49.	(A)	89.	(C)	129.	(D)	169.	(D)	209.	(A)	249.	(D)	289.	(C)
10.	(B)	50.	(D)	90.	(A)	130.	(C)	170.	(B)	210.	(B)	250.	(D)	290.	(D)
11.	(B)	51.	(D)	91.	(D)	131.	(D)	171.	(D)	211.	(C)	251.	(A)	291.	(A)
12.	(C)	52.	(C)	92.	(C)	132.	(D)	172.	(B)	212.	(C)	252.	(C)	292.	(B)
13.	(A)	53.	(A)	93.	(B)	133.	(D)	173.	(B)	213.	(A)	253.	(A)	293.	(D)
14.	(D)	54.	(C)	94.	(B)	134.	(A)	174.	(D)	214.	(B)	254.	(B)	294.	(A)
15.	(A)	55.	(D)	95.	(C)	135.	(B)	175.	(D)	215.	(C)	255.	(B)	295.	(C)
16.	(B)	56.	(D)	96.	(B)	136.	(C)	176.	(A)	216.	(A)	256.	(C)	296.	(D)
17.	(C)	57.	(C)	97.	(C)	137.	(D)	177.	(C)	217.	(A)	257.	(C)	297.	(C)
18.	(D)	58.	(C)	98.	(D)	138.	(A)	178.	(B)	218.	(D)	258.	(A)	298.	(B)
19.	(A)	59.	(A)	99.	(C)	139.	(D)	179.	(C)	219.	(A)	259.	(C)	299.	(C)
20.	(A)	60.	(D)	100.	(A)	140.	(D)	180.	(A)	220.	(B)	260.	(A)	300.	(B)
21.	(C)	61.	(A)	101.	(A)	141.	(D)	181.	(D)	221.	(D)	261.	(B)		
22.	(B)	62.	(A)	102.	(B)	142.	(B)	182.	(A)	222.	(A)	262.	(A)		
23.	(A)	63.	(B)	103.	(C)	143.	(C)	183.	(D)	223.	(B)	263.	(D)		
24.	(A)	64.	(C)	104.	(D)	144.	(A)	184.	(B)	224.	(A)	264.	(C)		
25.	(B)	65.	(A)	105.	(A)	145.	(B)	185.	(B)	225.	(B)	265.	(D)		
26.	(D)	66.	(A)	106.	(C)	146.	(C)	186.	(A)	226.	(D)	266.	(A)		
27.	(D)	67.	(C)	107.	(A)	147.	(A)	187.	(D)	227.	(A)	267.	(D)		
28.	(A)	68.	(D)	108.	(D)	148.	(C)	188.	(C)	228.	(B)	268.	(B)		
29.	(C)	69.	(B)	109.	(A)	149.	(C)	189.	(A)	229.	(D)	269.	(C)		
30.	(C)	70.	(B)	110.	(B)	150.	(D)	190.	(C)	230.	(A)	270.	(B)		
31.	(B)	71.	(A)	111.	(A)	151.	(C)	191.	(D)	231.	(D)	271.	(A)		
32.	(A)	72.	(A)	112.	(A)	152.	(D)	192.	(B)	232.	(A)	272.	(D)		
33.	(C)	73.	(A)	113.	(C)	153.	(A)	193.	(C)	233.	(D)	273.	(B)		
34.	(A)	74.	(C)	114.	(A)	154.	(B)	194.	(D)	234.	(B)	274.	(D)		
35.	(D)	75.	(B)	115.	(D)	155.	(D)	195.	(B)	235.	(B)	275.	(A)		
36.	(B)	76.	(D)	116.	(A)	156.	(C)	196.	(A)	236.	(C)	276.	(D)		
37.	(A)	77.	(C)	117.	(B)	157.	(C)	197.	(C)	237.	(D)	277.	(D)		
38.	(C)	78.	(D)	118.	(C)	158.	(D)	198.	(A)	238.	(D)	278.	(A)		
39.	(B)	79.	(C)	119.	(C)	159.	(C)	199.	(C)	239.	(A)	279.	(B)		
40.	(C)	80.	(C)	120.	(D)	160.	(B)	200.	(B)	240.	(B)	280.	(A)		

9

Techniques and Tools in Molecular Biology

Since few years, plant biotechnology has been used to improve plant traits applying microorganisms and produced benifical products for human and animal welfare. Various techniques like plant tissue culture, genetic transformation and molecular biology techniques included in biotechnology for crop improment.

Plant tissue culture broadly used *in vitro* technique in which various plant parts have been used (tissues, organs, embryos, single cells, protoplasts) to grow complete plant using nutrient media under closely controlled and aseptic conditions. Plant tissue culture involves several culture methods such as callus culture, suspension culture, micropropagation, somatic embryogenesis, somaclonal variation, meristem culture, anther culture, embryo culture, protoplast culture, cryopreservation, and production of secondary metabolites.

Tissue culture techniques and their applications

Plant tissue culture technique	Application
Suspension culture	Production of secondary metabolites
Micropropagation	Multiplication of identicle plants
Somatic embryogenesis	Artificial seed production
Somaclonal variation	Creat variations in plants
Meristem culture	Develop disease free plants
Anther culture	Haploid plant production
Embryo culture and embryo rescue	Distance hybrization
Protoplast culture	Cybrid development
Cryopreservation	Store germplasm collection

Gene transfer methods in plants

Gene transfer method	Important points
Indirect gene transfer	
Agrobacterium–mediated gene transfer	Mostly used method for gene transfer in dicot and monocot plants using Ti plasmid of *Agrobacterium tumefaciens* and Ri plasmid of *Agrobacterium rhizogenes* (natural genetic engineer)
Direct gene transfer	
Chemical mediated gene transfer	Chemicals polyethylene glycol (PEG) and dextran sulphate induce DNA uptake into plant protoplasts.
Microinjection	Used for introducing gene into oocytes, eggs, and the cells of early embryo.
Electroporation	Electrical shocks (high voltage) used to transfer gene into protoplasts/cells/tissues.
Particle gun/Particle bombardment	micro gold or tungsten particles (1-3 micrometers) containing DNA transfer in target tissues/cells and chloroplast using a particle gun. Biolistics gene gun invented by Sanford (1988).

Gene transfer method	Important points
Liposome or Lipofection	Liposomes are circular lipid molecules containing nucleic acids. Lipofection is a very efficient technique used to transfer genes in bacterial, animal and plant cells.

Gel electrophoresis

Tiselius firstly developed the "Tiselius apparatus" gel electrophoresis which is a basic technique used to separate DNA, RNA or proteins on the basis charge and mass of molecules using agarose gel. The molecules move at different speeds depending on their size or mass. Smaller molecules move faster than larger molecules.

Relation between size of linear DNA and agarose gel percentage

Recommended % Agarose	Optimum Resolution for Linear DNA
0.5	1,000–30,000bp
0.7	800–12,000bp
1.0	500–10,000bp
1.2	400–7,000bp
1.5	200–3,000bp
2.0	50–2,000bp

Blotting techniques

Blotting technique includes the transfer biomolecules (DNA, RNA or protein) to on nitrocellulose membrane from agarose gel. The probe is radioactively labeled used for hybridization to DNA to locate location using autoradiography.

1. Southern blot for DNA
2. Nothern blot for RNA
3. Western blot for protein

Enzyme-Linked Immunosorbent Assay (ELISA)

ELISA is a technique used to detect the presence of specific antibodies or antigens in a sample. It is also used to detect presence of virus in samples.

Polymerase Chain Reaction (PCR)

It is discovered by Karry Mullis (1983) used to make thousands of copies of a DNA strand using *Taq* DNA polymerase. The template is provided by the DNA sample to be amplified and the free 3'-OH are provided by site-specific oligonucleotide primers. Three steps are involed in PCR reaction. These steps are repeated from 28-35 times.

1. **Denaturation -** DNA is heated usually at 95oC for 4-5 minuts to separate into single-stranded
2. **Annealing -** Primers bind the appropriate complementary strand at appropriate temperature
3. **Primer Extension -** DNA polymerase extends the primer by its polymerase activity. *Taq* DNA polymerase isolated from the thermophilic ("heat-loving) bacteria *Thermus aquaticus* used for the extension at 72°C

Key notes

- Biotechnology term coined in 1919 by Hungarian engineer Kárl Ereky.
- Gene silencing describes the regulation of gene expression and which has the ability of a cell to prevent the expression of a gene.
- Gene silencing can occur during either transcription or translation.
- Gene silencing is considered gene knockout.
- When genes are silenced, their expression is reduced.
- When genes are knocked out, they are completely erased from the organism's genome and thus, have no expression.
- Types of gene silencing

At transcriptional level

- Genomic Imprinting
- Paramutation
- Transposon silencing (or Histone Modifications)
- Transgene silencing
- Position effect
- RNA-directed DNA methylation

At post-transcriptional level

- RNA interference
- RNA silencing
- Nonsense mediated decay

At Meiotic level

- Transvection
- Meiotic silencing of unpaired DNA

Antisense oligonucleotides

- Antisense oligonucleotides were discovered in 1978 by Paul Zamecnik and Mary Stephenson.
- Oligonucleotides are short nucleic acid fragments that bind to complementary target mRNA molecules when added to the cell.
- These molecules can be composed of single-stranded DNA or RNA and are generally 13-25 nucleotides long.
- The antisense oligonucleotides affect gene expression in two ways: by using an RNase H-dependent mechanism or by using a steric blocking mechanism.
- RNase H-dependent oligonucleotides cause the target mRNA molecules to be degraded while steric-blocker oligonucleotides prevent translation of the mRNA molecule.
- The majority of antisense drugs function through the RNase H-dependent mechanism in which RNase H hydrolyzes the RNA strand of the DNA/RNA heteroduplex.
- This mechanism is thought to be more efficient, resulting in an approximately 80% to 95% decrease in the protein and mRNA expression.

Ribozymes

- Ribozymes are catalytic RNA molecules used to inhibit gene expression.
- These molecules work by cleaving mRNA molecules which silence the genes.
- Sidney Altman and Thomas Cech first discovered catalytic RNA molecules, RNase P and group II intron ribozymes, in 1989 and won the Nobel Prize for their discovery.\
- Several types of ribozyme motifs exist (hammerhead,hairpin, hepatitis delta virus, group I, group II, and RNase P ribozymes).
- Hammerhead, hairpin, and hepatitis delta virus (HDV) ribozyme motifs are generally found in viruses or viroid RNAs. These motifs are able to self-cleave a specific phosphodiester bond on an mRNA molecule.
- Lower eukaryotes and a few bacteria contain group I and group II ribozymes. These motifs can self-splice by cleaving and joining together phosphodiester bonds.
- RNase P ribozyme is found in *Escherichia coli* and is known for its ability to cleave the phosphodiester bonds of several tRNA precursors when joined to a protein cofactor.
- Ribozymes (RNA molecules wih enzymatic activity) bind to a specific site and attack the neighboring phosphate in the RNA backbone with their 2' oxygen, which acts as a nucleophile, resulting in the formation of cleaved products with a 2'3'-cyclic phosphate and a 5' hydroxyl terminal end.

RNA interference

- RNA interference (RNAi) is a natural process used by cells to regulate gene expression.
- It was discovered in 1998 by Andrew Fire and Craig Mello, who won the Nobel Prize for their discovery in 2006.
- The process to silence genes first begins with the entrance of a double-stranded RNA (dsRNA) molecule into the cell, which triggers the RNAi pathway.
- The double-stranded molecule is then cut into small double-stranded fragments by an enzyme called Dicer.
- These small fragments which include small interfering RNAs (siRNA) and microRNA (miRNA), are approximately 21-23 nucleotides in length.
- The fragments integrate into a multi-subunit protein called the RNA-induced silencing complex (RISC) which contains Argonaute proteins that are essential components of the RNAi pathway.
- One strand of the molecule called the "guide" strand that binds to RISC while the other strand, known as the "passenger" strand is degraded.
- The guide or antisense strand of the fragment that remains bound to RISC directs the sequence-specific silencing of the target mRNA molecule.
- The genes can be silenced by siRNA molecules that cause the endonucleatic cleavage of the target mRNA molecules or by miRNA molecules that suppress translation of the mRNA molecule.
- With the cleavage or translational repression of the mRNA molecules, the genes that form them are essentially inactive.

Three prime untranslated regions and microRNAs

- Three prime untranslated regions (3'-UTRs) of mRNAs contain regulatory sequences that post-transcriptionally influence gene expression.
- 3'-UTRs contain both binding sites for microRNAs (miRNAs) as well as for regulatory proteins.
- By binding to specific sites within the 3'-UTR, miRNAs can decrease gene expression of various mRNAs by either inhibiting translation or directly causing degradation of the transcript.
- The 3'-UTR may have silencer regions that bind repressor proteins that inhibit the expression of a mRNA.
- The 3'-UTR contains miRNA response elements (MREs) which are sequences to which miRNAs bind. These are prevalent motifs within 3'-UTRs. Among all regulatory motifs within the 3'-UTRs (including silencer regions), MREs make up about half of the motifs.
- As of 2014, the miRBase web site, an archive of miRNA sequences and annotations, listed 28,645 entries in 233 biologic species. Of these, 1,881 miRNAs were in annotated human miRNA loci.
- miRNAs were predicted to have an average of about four hundred target mRNAs (affecting expression of several hundred genes).
- Freidman et al. estimate that >45,000 miRNA target sites within human mRNA 3'-UTRs are conserved above background levels and >60% of human protein-coding genes have been under selective pressure to maintain pairing to miRNAs.
- Direct experiments show that a single miRNA can reduce the stability of hundreds of unique mRNAs.
- The effects of miRNA dysregulation of gene expression are important in cancer, neuropsychiatric disorders (schizophrenia, bipolar disorder, major depressive disorder, Parkinson's disease, Alzheimer's disease and autism spectrum disorders).

Use of RNAi technology in various plant species against biotic stresses

Crop	Insect/Pathogen	Objective	Targeted genes
Arabidopsis thaliana	*Meloidogyne species*	Utilization of RNAi to silence the parasitism gene	*16D10*
Oryza sativa L	*Magnaporthe grisea* and *Xanthomonas oryzae pv. oryzae*	Functional analysis of a rice homolog SSI2 (OsSSI2) for disease resistance	OsSSI2
Prunus domestica L.	*Plum pox virus* (PPV)	To exploit the role of PTGS (RNAi) for virus resistance in a woody plants	PPV coat protein gene
Gossypium hirsutum	*Helicoverpa armigera*	Silencing a cotton bollworm P450 monooxygenase gene by plant-mediated RNAi	Cytochrome P450 gene (CYP6AE14)
Nicotiana rustica	*Bemisia tabaci*	Enhanced whitefly resistance	v-ATPaseA
Zea mays	*Diabrotica virgifera virgifera* LeConte	Control of coleopteran insect pests	Genes encoding proteins
Medicago sativa	*Acyrthosiphon pisum*	RNAi knockdown of a salivary transcript	C002

Crop	Insect/Pathogen	Objective	Targeted genes
Nicotiana	*Myzus persicae*	To develop the plant-mediated RNAi technology for	*M. persicae* Rack1,
benthamiana and *Arabidopsis thaliana*		aphid resistance	*M. persicae* C002 (MpC002)
Nicotiana rustica	*Helicoverpa armigera*	Improvement of pest resistance in transgenic tobacco plants expressing dsRNA	EcR
Citrus aurantifolia	*Citrus tristeza virus* (CTV)	Transformation to generate transgenic plants carrying the coat protein gene of CTV	CTV-CP
Juglans regia L.	*Agrobacterium tumefaciens*	Application of oncogene silencing technology in the generation of crown gallresistant crops.	Tryptophan mono-oxygenase (iaaM) and isopentenyl transferase (ipt)
Genus *Malus*	*Agrobacterium tumefaciens*	To provide effective method to produce crown gall resistant apple plants	iaaM, iaaH, and ipt
Oryza sativa L.	*Nilaparvata lugens*	Knockdown of midgut genes by dsRNA-transgenic plant-mediated RNA interference	NlHT1, Nlcar, Nltry
Genus *Malus*	*Venturia inaequalis*	To use to hairpin vector approach for resistance against *V. inaequalis*	GFP transgene and tri-hydroxy-naphthalene eeductase gene (THN)

Application of RNAi interference in plant improvements

Application	Trait improvement
Increasing the level of lysine	Reduction of lysine catabolism and improving seed germination generating a dominant high-lysine maize variant by knocking out the expression of the 22-kD maize zein storage protein Resistance of barley to BYDV and producing a rice
Barley and Rice	variety called LGC-1 (low glutenin content 1) by RNAi technology
Banana	Production of banana varieties resistant to the Banana Bract Mosaic Virus (BBrMV) by RNAi Transgenic cotton plants expressing a RNAi construct
Cotton	of the d-cadinene synthase gene of gossypol synthesis fused to a seed-specific promoter caused seed-specific reduction of Gossypol
Jute	Generating jute varieties with low lignin content
Lathyrus sativus	RNAi construct designed to silence the genes encoding the two starch-branching isozymes of amylopectin synthesis RNAi technology can be used to silence the gene(s) responsible for production of BOAA
Tomato	RNAi-mediated suppression of DET1 expression under fruit-specific promoters has recently shown to improve carotenoid and flavonoid levels in tomato fruits with minimal effects on plant growth
Coffee	RNAi technology has enabled the creation of varieties of *Coffee* that produces natural coffee with low or very low caffeine content
Healthier oil	Using RNAi to silence the gene in cotton which codes for the enzyme that converts oleic acid into a different fatty acid

Pest control	Combining Bt technology with RNAi would both enhance product performance and further guard against the development of resistance to Bt proteins
Wood and fruit quality	-Down regulation of lignin biosynthesis pathways. -Producing transgenic hypoallergenic apples and a possible solution for the undesirable separation of juice into clear serum and particulate phase by using RNAi.
Flower color modification	
Gentian	Producing white-flowered transgenic gentians by suppressing the chalcone synthase (*CHS*) gene using antisense technology
Blue Rose	Producing blue transgenic rose by knock–downing the cyanidin genes in rose and carnation by RNAi technology and introduce delphinidin genes

- A DNA microarray (DNA chip or biochip) is a collection of microscopic DNA spots attached to a solid surface.
- DNA microarrays used to measure the expression levels of large numbers of genes simultaneously or to genotype multiple regions of a genome.
- Each DNA spot contains a specific DNA sequence known as probes used to hybridize a cDNA or cRNA (anti-sense RNA) sample (target).
- Probe-target hybridization is usually detected and quantified by detection of fluorophore-, silver-, or chemiluminescence-labeled targets to determine relative abundance of nucleic acid sequences in the target.
- A Southern blot was discovered by Edwin Southern and used for detection of a specific DNA sequence in DNA samples. Southern blotting combines transfer of electrophoresis-separated DNA fragments to a filter membrane and subsequent fragment detection by probe hybridization.
- Western blot (protein immunoblot) is analytical technique to detect specific proteins in a sample of tissue homogenate or extract. It uses gel electrophoresis to separate native proteins by 3-D structure or denatured proteins by the length of the polypeptide. The proteins are then transferred to a membrane (typically nitrocellulose or PVDF), where they are stained with antibodies specific to the target protein. The gel electrophoresis step is included in western blot analysis to resolve the issue of the cross-reactivity of antibodies.
- Northern blot is a technique used in molecular biology research to study gene expression by detection of RNA (or isolated mRNA) in a sample.
- The eastern blot is a biochemical technique used to analyze protein post translational modifications (PTM) such as lipids, phosphomoieties and glycoconjugates. It is most often used to detect carbohydrate epitopes. Thus, eastern blotting can be considered an extension of the biochemical technique of western blotting.
- Lectin blotting used to detect carbohydrate epitopes on proteins or lipids.
- Southwestern blotting was first described by B. Bowen, J. Steinberg and colleagues in 1980. It is a lab technique which involves identifying and characterizing DNA-binding proteins (proteins that bind to DNA) by their ability to bind to specific oligonucleotide probes. The proteins are separated by gel electrophoresis and are subsequently transferred to nitrocellulose membranes similar to other types of blotting.
- During isolation of DNA, CTAB (cetyl trimethylammonium bromide, a cationic detergent) method is most commonoly used for DNA extraction.

- PVP (polyvinylpyrrolidone) is helps in removing phenolics by forming hydrogen bonds.
- Beta -mercaptoethanol is a strong reducing agent which can remove tannins and other polyphenols often present in the crude plant extract.
- EDTA which chelates the Mg^{2+} ions needed for enzymes that degrade DNA termed DNase.
- RNA can be removed by treatment with ribonuclease (RNase) which has been heat treated to inactivate any DNase contaminants.
- Protein is removed by shaking the solution gently with tris-saturated phenol or with a phenol–chloroform mixture which denatures proteins but not nucleic acids.
- DNA is precipitated by isopropanol.
- Contaminants are identified by scanning UV spectrophotometry from 200 to 300 nm. A ratio of 260nm : 280nm of approximately 1.8 indicates that the sample is free of protein contamination, which absorbs strongly at 280 nm.
- Agarose electrophoresis in agarose or polyacrylamide gels is the most usual way to separate DNA and RNA molecules according to size.
- The proteins may be separated by 2-D polyacrylamide gel according to their charge and/or size (isoelectric focusing).
- The first electrophoresis method used to study proteins was the free solution or moving boundary method devised by Tiselius in 1937. The Tiselius apparatus was a rectangular sectional U-tube equipped with sensitive Schlieren optics. It allowed changes in refractive index to be observed as segments of solution with different protein content migrated past. This technique was used in research to measure electrophoretic mobility and to study protein-protein interaction.
- This highly useful electrophoresis equipment was developed by Oliver Smithies and Mario Capecchi for the creation of knockout mice (a particular type of experimental mouse in which a chosen gene has been inactivated).
- Restriction mapping involves the size analysis of restriction fragments produced by several restriction enzymes individually and in combination.
- DNA labelling is of two types, first is 5' end labelling involves a phosphate transfer or exchange reaction where the 5' phosphate of the DNA to be used as the probe is removed and in its place a labelled phosphate, usually ^{32}P, is added. This is usually carried out by using two enzymes; the first, alkaline phosphatase, is used to remove the existing phosphate group from the DNA. a second enzyme, polynucleotide kinase, is added, which catalyses the transfer of a phosphate group (^{32}P-labelled) to the 5' end of the DNA.
- Klenow fragment of DNA polymerase I is used for random primer labelling because it lacks a 5' –3' exonuclease activity.
- The polymerase chain reaction (PCR) developed in 1983 by Kary Mullisis and used to amplify a single copy or a few copies of a piece of DNA across several orders of magnitude, generating thousands to millions of copies of a particular DNA sequence.
- PCR requires several components and reagents like DNA template, primer, Taq polymerase, deoxynucleoside triphosphates (dNTPs), 10x Taq Buffer solution, Bivalent cations magnesium or manganese ions.
- DNA sequencing is the process of determining the precise order of nucleotides within a DNA molecule.

- Frederick Sanger (1955) sequenced first proteins insulin secreted by the pancreas.
- Frederick Sanger developed the chain-termination method in 1977 for DNA sequencing.
- RNA sequencing was one of the earliest forms of nucleotide sequencing. The major landmark of RNA sequencing is the sequence of the first complete gene and the complete genome of Bacteriophage MS2, identified and published by Walter Fiers and his coworkers in 1972 and 1976.
- On October 26, 1990, Roger Tsien, Pepi Ross, Margaret Fahnestock and Allan J Johnston filed a patent describing stepwise ("base-by-base") sequencing with removable 3' blockers on DNA arrays (blots and single DNA molecules).
- In 1996, Pål Nyrén and his student Mostafa Ronaghi at the Royal Institute of Technology in Stockholm published their method of pyrosequencing.
- Allan Maxam and Walter Gilbert published a DNA sequencing method in 1977 based on chemical modification of DNA and subsequent cleavage at specific bases. This method's use of radioactive labeling and its technical complexity discouraged extensive use after refinements in the Sanger methods had been made.
- A parallelized version of pyrosequencing was developed by 454 Life Sciences which has since been acquired by Roche Diagnostics. The method amplifies DNA inside water droplets in an oil solution (emulsion PCR) with each droplet containing a single DNA template attached to a single primer-coated bead that then forms a clonal colony.
- Pyrosequencing uses luciferase to generate light for detection of the individual nucleotides added to the nascent DNA and the combined data are used to generate sequence read-outs. This technology provides intermediate read length and price per base compared to Sanger sequencing on one end and Solexa and SOLiD on the other.
- Solexa Illumina dye sequencing, was founded by Shankar Balasubramanian and David Klenerman in 1998 and developed a sequencing method based on reversible dye-terminators technology and engineered polymerases.

Multiple Choice Questions

1. Which technique is the knockdown of gene expression by small RNA fragments?
(A) Microarray (B) Blotting
(C) RNAi (D) SAGE

2. Who awarded Nobel Prize for the discovery of RNA interference in physiology and medicine?
(A) Andrew Fire and Craig Mello (2000)
(B) Roger Kornberg (2006)
(C) Phil Sharp (1999)
(D) Thomas Tuschl and Greg Hannon (2000)

3. Andrew Fire and Craig Mello (2000) is awarded Nobel Prize for the discovery of
(A) Microarray (B) Blotting
(C) RNAi
(D) Yeast-two hybrid system

4. Which process was elucidated for gene silencing by Fire and Mello (1998)?
(A) Microarray
(B) DNA sequencing
(C) Yeast-two hybrid system
(D) RNAi

5. Gene expression is comprised of .... with distinct roles for DNA and RNA.
(A) Sequencing and silencing
(B) Transcription and Translation
(C) Transformation and Transduction
(D) Blotting and hybridization

6. Who discovered RNA polymerase enzyme (RNAp) and awarded Nobel Prize (2006) in Chemistry?
(A) Andrew Fire and Craig Mello (2000)
(B) Phil Sharp (1999)
(C) Roger Kornberg (2006)
(D) Thomas Tuschl and Greg Hannon (2000)

7. Translation is the synthesis of .... using the mRNA template?
(A) Proteins (B) tRNA
(C) cDNA (D) Sugars

8. Which of the following step of translation is involved in binding of amino acids to transfer RNA (tRNA)?
(A) Elongation (B) Activation
(C) Initiation (D) Termination

9. Which codon does initiate translation process?
(A) UAA (B) UAG
(C) AUG (D) UGA

10. .... is an ideal system for studying gene expression.
(A) *Caenorhabditis elegans*
(B) *Aspergillus niger*
(C) *Tobacco notatum*
(D) *Arabidopsis thaliana*

11. Which is soil living organism and an ideal system for studying gene expression?
(A) Arabidopsis (B) *E. coli*
(C) Rabies virus (D) Roundworm

12. Which technique allows for the direct introduction of macromolecules into the animal unc-22 repression twitching par-1 repression symmetrical cell division?
(A) Microinjection
(B) *Agrobacterium* gene transfer
(C) Liposome
(D) Electroporation

13. Mello and Fire established .... as trigger for RNAi.
(A) ssRNA (B) dsRNA
(C) dsDNA (D) Circular DNA

14. Hannon defines .... as cause of mRNA degradation.
(A) RISC complex (B) Dicer
(C) Antisense RNA (D) RNaseIII

15. Tuschl and Sharp find mRNA is cleaved into .... Fragments of siRNA.
(A) 2-5 nucleotides
(B) 29 nucleotides
(C) 18-20 nucleotides
(D) 21-25 nucleotides

16. Who defined roles of Dicer and Slicer and identified Argonaute in RISC?
(A) Hannon
(B) Tuschl determines
(C) Phil Sharp
(D) Thomas Tuschl

17. Who described the structures of Dicer, dsRNA products, siRNA?
(A) Hannon
(B) Tuschl determines
(C) Phil Sharp (D) Ambros

18. Who discovered the first endogenous short RNA (miRNA) via forward genetics in 1993?
(A) Hannon
(B) Tuschl determines
(C) Phil Sharp (D) Ambros

19. Endogenous miRNAs have inspired the design of ....
(A) Short mRNA
(B) Short small nucleolar RNA
(C) Short hairpin RNAs (shRNAs)
(D) Short Heterogenous RNA

20. A variety of RNA triggers function in the RNAi mechansim result in gene suppression that can be both temporary and permanent.
(A) dsRNA and siRNA
(B) miRNA and shRNA
(C) mRNA, tRNA and rRNA
(D) dsRNA and siRNA, miRNA and shRNA

21. Which is an RNAi-related phenomenon?
(A) Transcriptional gene silencing
(B) Posttranslational gene silencing
(C) Translational gene silencing
(B) Posttranscriptional gene silencing

22. Which of the following plant is first used in RNA silencing through RNAi?
(A) Petunia (B) Arabidopsis
(C) Tomoto (D) Tobacco

23. Which laboratory involved to upregulate the activity of a gene for chalcone synthase (*chsA*) in the production of anthocyanin pigments?
(A) Watson' laboratory
(B) Pasture's laboratory
(C) R. Jorgensen's laboratory
(D) Khorana's laboratory

24. Which of the following gene is first silenced for the production of anthocyanin pigments in transgenic petunia?
(A) Chalcone synthase (*chsA*) gene
(B) wild-type $al1^+$ gene
(C) Gametogenetic defect & sterility
(D) β-glucuronidase transgene

25. What is the role of Chalcone synthase (*chsA*) gene that is silenced by RNai in petunia?
(A) Cause white rust disease in mustard
(B) Involve in the production of anthocyanin pigments
(C) Late ripening in tomato
(D) to boost the production of an orange pigment in *N. crassa*

26. Transgenic petunia plants harboring the *chsA* coding region under the control of a 35S promoter lost ....
(A) Both endogene and transgene chalcone synthase activity
(B) With programmed cell death
(C) Specificity the identities of the anterior AB blastomere
(D) the boosting power of an orange pigment production in *N. crassa*

27. Who coined the term cosuppression to describe the loss of mRNAs of both the endo- and the transgene in petunia?
(A) Hannon
(B) Tuschl determines
(C) Jorgensen (D) Ambros

28. What is the cosuppression?
(A) Inactivation of promoter activity
(B) Degradation of endogene and transgene RNAs
(C) Degradation of short siRNA used in RNAi
(D) Inactivation of RISC complex

29. Posttranscriptional gene silencing (PTGS) could be initiated by ....
(A) Sense transgenes
(B) Antisense transgenes
(C) Both (A) and (B)
(D) It is depend only on biochemical imbalance

30. What are *iaaM* and *ipt*
(A) Oncogenes of agrobacteria
(B) Chalcone synthase (*chsA*) gene
(C) Gametogenetic defect and sterility genes
(D) β-glucuronidase transgenes

31. What is called the phenomena PTGS in found independently in fungal systems?
(A) Splicing (B) Quelling
(C) Cloning (D) Gene inducing

32. Which phenomena is belong to the production of an orange pigment made by the gene *al1* of the fungus *Neurospora crassa*?
(A) Splicing (B) Gene inducing
(C) Methylation (D) Quelling

33. Which organism was used to demonstrate the phenomenon of RNAi first discovered by Fire and Mello?
(A) *Trypanosoma brucei*
(B) *Neurospora crassa*
(C) *Caenorhabditis elegans*
(D) *Chlamydomonas reinhardtii*

34. Which of the following type of RNA was used to demonstrate the biochemical nature of inducers in gene silencing into the body of *Caenorhabditis elegans* by Fire and Mello?
(A) ssRNA (B) dsRNA
(C) Circular dsRNA
(D) Circular dsDNA

35. What is GUS?
(A) β-glucuronidase
(B) Chalcone synthase
(C) Galactosidases
(D) Guanine nucleotide exchange factors

36. What are the critical common components of the RNAi?
(A) The inducer is the dsRNA
(B) The target RNA is degraded in a homologydependent fashion
(C) The degradative machinery requires a set of proteins
(D) All of the above

37. What is the size of siRNA used to silence of gene through RNAi?
(A) 50-100 nucleotides
(B) 10-20 nucleotides
(C) 21-25 nucleotides
(D) 2-10 nucleotides

38. What is siRNA?
(A) Complex of RNA and ribosome
(B) Small RNA fragments of 50-100 nucleotides
(C) Degraded mRNA fragments of 10-11 nucleotides
(D) Breakdown products of an *E. coli* RNase III-like digestion

39. Which of the following statement is not true about siRNA?
(A) Each strand of siRNA has 5'-phosphate and 3'-hydroxyl termini
(B) Each strand of siRNA has 2- to 3-nucleotide 3' overhangs
(C) siRNA is the product of long dsRNA cleaved by RISC
(D) siRNA used in post-transcriptional gene silencing

40. What is dicer?
(A) RNaseIII-like enzyme
(B) RNA with enzymatic activity
(C) Complex of RNA and ribosome
(D) Degraded mRNA fragments

41. Which of the following domains is not contained in dicer?
(A) An aminoterminal helicase domain
(B) Dual RNase III motifs
(C) A mRNA binding domain
(D) A PAZ domain

42. What is PAZ domain of dicer?
(A) A110-amino-acid domain
(B) A dicer Proteins like Piwi, Argo, and Zwille/Pinhead
(C) PAZ shares with the RDE1/QDE2/Argonaute family of proteins
(D) All of the above

43. Which domain of dicer does help in cleavage of dsRNA during RNAi?
(A) An aminoterminal helicase domain
(B) RNase III motifs
(C) A mRNA binding domain
(D) A PAZ domain

44. Ca^{2+}-dependent nuclease that can degrade both DNA and RNA.
(A) S1 Nuclease
(B) Restriction nuclease
(C) Micrococcal nuclease
(D) Mungbean nuclease

45. Which are the classes of quelling-defective mutants in *Neurospora crassa*?
(A) *drh1, drh2, rde4 and rde1*
(B) *unc22, ago2, qde3*
(C) *sgs2, sgs3* and *hen1*
(D) *qde1, qde2,* and *qde3*

46. Which of the following enzyme is used to seal the nicks in DNA?
(A) DNA polymerase
(B) Restriction endonuclease
(C) S1 nuclease
(D) DNA ligase

47. Which enzyme use ATP as energy source for its reaction?
(A) DNA polymerase
(B) DNA ligase
(C) S1 nuclease
(D) Restriction endonuclease

48. How many phosphodiester bonds are formed by DNA ligase in vivo and in vitro, respectively?
(A) One and two (B) Two and one
(C) Two in both (D) One in both

49. Which type of bond is formed by DNA ligase?
(A) Glycosidic (B) Covalent
(C) Phosphodiester (D) Peptide

50. DNA ligase is prepared from *E. coli* cells infected by ....
(A) T2 phage (B) M13 virus
(C) CaMV (D) T4 phase

51. Which type of bond is broken by restriction enzymes?
(A) Glycosidic (B) Covalent
(C) Phosphodiester (D) Peptide

52. Which of the following are the techniques of gene silencing?
(A) RNA splicing and RNAi
(B) SAGE and microarray
(C) RNAi and Antisense-RNA technology
(D) Reverse transcription and RNA capping

53. RNAi' stands for which of the following?
(A) RNA inducer (B) RNA insertion
(C) RNA intron (D) RNA interference

54. Suppose a certain gene contains the double-stranded sequence:

5'- ATGTTTAGCGCC -3'

3'- TACAAATCGCGG -5'

If the top strand is the sense strand and codes for an mRNA whose sequence begins 'ATG', which of the following would be the sequence of the corresponding segment of antisense RNA?
(A) 5'-AUGUUUAGCGCC-3'
(B) 5'-CCGCGAUUUGUA-3'
(C) 5'-GGCGCUAAACAU-3'
(D) 5'-UACAAAUCGCGG-3'

55. The effectors of gene silencing are short double-stranded RNA molecules produced by the action of the enzyme dicer. Approximately what size are these molecules?
(A) 11 bp (B) 22 bp
(C) 75 bp (D) 100 bp

56. Which of the following epigenetic change in gene expression is an inherited change that does not involve any change in the nucleotide sequence of the gene?
(A) RNAi (B) Mutation
(C) Replication (D) Gene splicing

57. Which of the following statements is true of RNA interference?
(A) RNA interference is a normal way for organisms to regulate gene expression
(B) RNA interference can not used to control virus infection in plants
(C) RNA interference occurs only in vertebrates
(D) RNA interference is already used therapeutically for many disorders

58. Which statement best describes the main distinction between the origin of the two classes of small regulatory RNAs: siRNA and miRNA?
(A) siRNAs originate within the cell cytoplasm; miRNAs originate from the cell genome
(B) siRNAs originate from predominantly exogenous dsRNA; miRNAs originate from the cell genome
(C) miRNAs are expressed whenever siRNAs are unable to appropriately degrade RNA sequences
(D) miRNAs are processed from dsRNA viruses, siRNAs are processed from ssRNA viruses

59. Which statement best describes the distinction between the purposes of the two RNAi pathways within the eukaryotic cell?

(A) miRNA-mediated gene silencing represents a back-up pathway should siRNA-mediated gene silencing be unsuccessful at silencing the target dsRNA

(B) siRNA and miRNA-mediated gene silencing pathways must both be active to successfully silence the target dsRNA

(C) miRNA and siRNA-mediated gene silencing represent evolutionary-independent pathways which confer identical silencing mechanisms on the target dsRNA

(D) siRNA-mediated gene silencing represents a cell defence mechanism against exogenous dsRNA; miRNA-mediated gene silencing is an integral gene expression regulation process

60. Which of the following experimental requirements makes an siRNA approach unsuitable for determining the consequence of silencing a gene of interest in an established cell line?

(A) Expression of the gene of interest is integral to cell viability

(B) End point RT-PCR and western blots are the only available validation assays

(C) Lipofectamine is the only available delivery method

(D) The cell line proliferation rate is relatively slow

61. During siRNA-mediated gene silencing within the homo sapiens cell, the RNA-induced Silencing Complex (RISC) is composed of which of the following protein arrangements?

(A) DICER1, TRBP and AGO2

(B) DICER1, TRBP and AGO1

(C) DROSHA and DGCR8

(D) DICER, DROSHA and AGO1

62. With regards to siRNA, the most efficient siRNA duplexes will contain which of the following features?

(A) Will be 21-23 nucleotides in length

(B) Comprise a sense strand and an antisense strand, where the antisense strand harbours 100% complementarity to the mRNA target sequence

(C) Possess a 2-nucleotide overhang at the 3' termini end and AA bases at the 5' end

(D) All of the above

63. During shRNA sequence design, why is it beneficial to incorporate A-T residues towards the 5' region of the antisense sequence of the target mRNA, and G-C residues towards the 5' region of the sense sequence of the target mRNA?

(A) A-T residues confer more thermodynamic instability than G-C residues. The use of A-T residues at the 5' region of the antisense strand and G-C residues at the 5' region of the sense strand promote loading of the antisense strand at the 5' end into the RISC

(B) A-T residues confer more thermodynamic instability than G-C residues. The use of A-T residues at the 5' region of the antisense strand and G-C residues at the 5' region of the sense strand promote loading of the sense strand at the 3' end into the RISC G-C residues confer more thermodynamic instability than A-T residues.

(C) The use of A-T residues at the 5' region of the antisense strand and G-C residues at the 5' region of the sense strand promote loading of the sense strand at the 5' end into the RISC G-C residues confer more thermodynamic instability than A-T residues.

(D) The use of A-T residues at the 5' region of the antisense strand and G-C residues at the 5' region of the sense strand promote loading of the antisense strand at the 3' end into the RISC

64. Why can it be beneficial to drive expression of the shRNA sequence from an expression vector containing an RNA pol III promoter?

(A) The use of an RNA pol III promoter does not require the addition of an RNA pol

III terminator sequence (TTTTT) in the shRNA design, and so allows for a shorter shRNA sequence for ligation into the expression vecto.:

(B) If an RNA pol III terminator sequence (AAAAA) was incorporated downstream of the antisense strand, the RNA pol III promoter in the expression vector generates an shRNA with a "AA" 3' overhang characteristic of pre-miRNAs; thus negating the need for any pri-shRNA to pre-shRNA processing by Drosha

(C) If an RNA pol III terminator sequence (TTTT) was incorporated downstream of the antisense strand, the RNA pol III promoter in the expression vector generates an shRNA with a "UU" 3' overhang characteristic of pre-miRNAs; thus negating the need for any pri-shRNA to pre-shRNA processing by Drosha

(D) The use of an RNA pol III promoter does not require the incorporation of a stem loop into the shRNA design since the RNA pol III terminator sequence (TTTTT) generates fully-processed mature shRNA duplexes

65. Which of the following statements best describe why i it is beneficial to include a single restriction site in the shRNA expression vector?

(A) A restriction digest can be performed on the purified recombinant plasmid stocks to excise the shRNA sequence out of the flanking vector sequences prior to introduction into the cell cytoplasm and integration into the host genome

(B) Once the recombinant plasmid has been introduced into the cell cytoplasm, the restriction site will be targeted by the endogenous cell machinery to linearize the plasmid prior to integration into the host genome

(C) Once bacteria are transformed with the recombinant plasmid the restriction site is acted upon by the bacterial cells to release linearized plasmid into the growth medium for purification

(D) A restriction digest can be performed on the purified recombinant plasmid working stock to linearize the plasmid prior to introduction into the cell cytoplasm and integration into the host genome

66. Which of the systems below would typically necessitate the use of a viral vector and packaging cell line system for shRNA delivery?

(A) To silence a gene of interest that is known to be expressed at low abundance in a cell line with a fast division rate

(B) To silence a gene of interest that is known to be expressed at high abundance in primary cells

(C) To silence a gene of interest that is known to be expressed at high abundance in a cell line with a slow division rate

(D) To silence a gene of interest that is known to be expressed at low abundance in a cell line with a slow division rate

67. Which of the below hypothetical endpoint RT-PCR results for transient validation of a shRNA demonstrate that it is unsuitable for use downstream in stable cell line generation?

(A) 'Target mRNA-shRNA' cell line shows no amplicon. 'Wild type' cell line shows an amplicon. 'Vector only-shRNA' cell line shows no amplicon. 'Transfection reagent only' cell line shows an amplicon

(B) 'Target mRNA-shRNA' cell line shows no amplicon. 'Wild type' cell line shows an amplicon. 'Vector only-shRNA' cell line shows an amplicon. 'Transfection reagent only' cell line shows an amplicon

(C) 'Target mRNA-shRNA' cell line shows no amplicon. 'Wild type' cell line shows no amplicon. 'Vector only-shRNA' cell line shows no amplicon. 'Transfection reagent only' cell line shows an amplicon.

(D) 'Target rRNA-shRNA' cell line shows no amplicon. 'Wild type' cell line shows no amplicon. 'Vector only-shRNA' cell line shows no amplicon. 'Transfection reagent only' cell line shows an amplicon.

68. The function of Interference RNA to prevent a gene from functioning is also known as
(A) Reverse transcription
(B) Epigenetics
(C) PTGS
(D) RNA splicing

69. Hypercondensation of heterochromatin can result in:
(A) Transcriptional Silencing
(B) mRNA editing
(C) Ubiquitination
(D) Methylation

70. Alteration of the properties of one transcription factor by another is an example of:
(A) Transcriptional silencing
(B) mRNA Editing (C) Ubiquitination
(D) Quenching

71. Targeting of a protein for degradation is accomplished by:
(A) Transcriptional silencing
(B) mRNA Editing
(C) Ubiquitination
(D) Methylation

72. Gene silencing that is heritable is an example of:
(A) Basal Factors (B) Imprinting
(C) Iron response element
(D) Proteasome

73. Transcription factors act as ....
(A) Cis-acting regulatory elements
(B) Trans-acting regulatory elements
(C) Promoters
(D) Enhancers

74. SWI-SNF proteins are involved in:
(A) Basal transcription
(B) RNA interference
(C) Enhancer inactivation
(D) Chromatin remodeling

75. The process of ubiquitination is involved in ....:
(A) Formation of homeodimers
(B) Protein degradation
(C) RNA interference
(D) Genomic imprinting
(E) Chromatin remodeling

76. The siRNAs are involved in which of the following?
(A) Genomic imprinting
(B) Chromatin remodeling
(C) Targeting of specific mrnas
(D) Ubiquitination

77. Which of the following methods of regulation is not present in prokaryotes?
(A) DNA-Binding Proteins
(B) Attenuation
(C) Repressor proteins
(D) Chromatin remodeling

78. The delayed ripening tomato was created by a biotechnology that .... a gene.
(A) Altered (B) Silenced
(C) Replaced (D) Relocated

79. Part of DNA that is transcriptionally repressed is usually having their cytosine residues modified by
(A) Phosphorylation (B) Methylation
(C) Oxidation (D) Acetylation

80. Which process can be used for gene Silencing
(A) Phosphorylation (B) Hydrolysis
(C) Oxidation (D) Hypoacetylation

81. Which of the following gene linked to adventitious root emergence and development, is altered to improve crop productivity of transgenic plants by RNAi technology?
(A) *OsPIN1* (B) *OsGA20ox2*
(C) *HDAC* (D) *HDT702*

82. Which of the following gene has an important role in auxin-dependent adventitious root emergence and tillering?
(A) *OsGA20ox2* (B) *HDAC*
(C) *OsPIN1* (D) *HDT702*

83. Which of the following trait is controlled by suppressing the expression of *OsGLU1* gene encoding a membrane-bound endo-1, 4-b-D-glucanase gene in rice (Oryza sativa) through RNAi?
(A) Leaf colour
(B) Plant height
(C) Protein content
(D) ATP production

84. Which of the following gene is suppressed to develop a dwarf phenotype in rice (*Oryza sativa*) through RNAi?
(A) *OsGA20ox2* (B) *HDAC*
(C) *OsPIN1* (D) *OsGLU1*

85. Which of the following technology is used to suppress the expression of the OsGA20ox2 gene in rice (O. sativa L.), which encodes the regulatory enzyme GA 20-oxidase for the syntheses of biologically active gibberellins (GAs)?
(A) Antisense RNA (B) Mythylation
(C) RNA interference (D) Footprinting

86. Find the correct match between genes and their controlling trait which are suppressed in Arabidopsis by RNAi.

S. No.	Gene	S. No.	Controlling trait
I	*HDA703*	p	Regulation of plant growth and development in transgenic rice under drought stress
Ii	*HDT702*	q	Auxin signaling and seedling growth under abiotic stress conditions
Iii	*TIR1*	r	The production of narrowed leaves and stems
Iv	*RACK1*	s	Rice peduncle elongation and fertility

(A) i – p, ii – q, iii – r, iv – s
(B) i – s, ii – r, iii – q, iv – p
(C) i – r, ii – s, iii – p, iv – q
(D) i – s, ii – p, iii – q, iv – r

87. What is VIGS?
(A) Virus infecting genome sequence
(B) Virion induced gene silencing
(C) Virus induced gene silencing
(D) Viroid induced gene silencing

88. Food allergy is hypersensitive response to normally harmless food components that are mediated by mainly antibodies ....
(A) IgA (B) IgB
(C) IgE (D) IgD

89. Which immunodominant allergen protein is silenced transgenic soybean seeds through RNAi?
(A) Gly m Bd 30 K
(B) Mal d 1
(C) Lyc e 1.01
(D) Ara h 1

90. In which plant the expression of allergen Mal d 1 is inhibited by RNAi technology which displays IgE antibodies cross-reactivity to the birch pollen allergen Bet v 1?
(A) Transgenic soybean
(B) Transgenic apple
(C) Transgenic banana
(D) Transgenic papaya

91. Find the correct match between allergen and plant in which these allergens were suppressed by RNAi.

S. No.	Allergen	S. No.	Plant
i	Birch pollen allergen Bet v 1	p	*Arachis hypogaea*
ii	Lipid transfer protein Lyc e 1	q	*Glycine max*
iii	Trypsin inhibitor Ara h 2	r	*Malus domestica*
iv	Gly m Bd 30 K (P34)	s	*Solanum lycopersicum*

(A) i – p, ii – q, iii – r, iv – s
(B) i – s, ii – r, iii – q, iv – p
(C) i – r, ii – s, iii – p, iv – q
(D) i – s, ii – p, iii – q, iv – r

92. Which of the following anther-specific gene is down-regulated in tobacco (*N. tabacum cv. Samsun*) to display male sterility by RNAi?
(A) *OsGEN-L*
(B) *TA29*
(C) *DCW11*
(D) *AOX1a*

93. Which of the following trait in transgenic rice down-regulated by silencing of genes OsGEN-L and DCW11through RNAi?
(A) Delayed ripening
(B) Cytoplasmic male sterility
(C) Flower color and scent
(D) Anthocyanin and flavonoid biosynthesis

94. What is called to place the amplified PCR DNA products on a solid surface?
(A) DNA sequencing
(B) Microarray
(C) SAGE
(D) DNA Recombination

95. Which technique is similar to DNA microarray?
(A) Photolithography
(B) SAGE
(C) Reverse western
(D) Reverse northern

96. Which of the following technique can be used to determine the expression of gene at a growth stage of organism?
(A) DNA sequencing (B) Microarray
(C) ELISA (D) TILLING

97. Which of the following step is not involved in microarray technique?
(A) Isolation of total complement of mRNA from cells
(B) Converted to cDNA using reverse transcriptase
(C) Production of protein regarding to gene
(D) Fluorescently labeled deoxynucleotides

98. Which type of probe is used to hybridize cDNA in microarray?
(A) Fluorescent cDNAs
(B) Radiolabelled DNA
(C) Fluorescent mRNA
(D) Fluorescent ssDNA

99. Which type of colour spots are not showed in microarray after probe hybridization?
(A) Red spots (B) Green spots
(C) Yellow spots (D) Transparent spots

100. Which of the following result is incorrect obtained during microarray?
(A) Colorless spots show cDNAs that represent non-coding regions of mRNA
(B) Fluoresce green represent mRNAs more abundant at the single-cell stage
(C) Fluoresce red represent sequences more abundant later in development
(D) mRNAs are equally abundant at both stages of development fluoresce yellow

101. What is Photolithography?
(A) A technique to determine the sequence of amino acids in a particular protein
(B) A computer is programmed with the oligonucleotide sequences synthesized on a solid surface
(C) An alternative strategy is to synthesize DNA directly on the solid surface
(D) An array of antibodies immobilized as individual spots on a solid surface

102. What is called the technology in which proteins can be immobilized on a solid surface and used to help define the presence or absence of other proteins in a sample?
(A) Protein Chip Technique
(B) SAGE
(C) DNA Microarray
(D) Photolithography

103. What is protein chip technology?
(A) A technique to determine the sequence of amino acids in a particular protein
(B) A computer is programmed with the oligonucleotide sequences synthesized on a solid surface
(C) An alternative strategy is to synthesize DNA directly on the solid surface
(D) An array of antibodies immobilized as individual spots on a solid surface

104. Who was first developed DNA microarray at the Stanford University in1995?
(A) Patrick Brown
(B) Roger Kornberg (2006)
(C) Phil Sharp (1999)
(D) Thomas Tuschl and Greg Hannon (2000)

105. Agriculturally important traits such as yield, quality and some forms of disease resistance are known as .... that controlled by many genes
(A) Quantitative traits
(B) Polygenic traits
(C) Multifactorial traits
(D) All of the above

106. The regions within genomes that contain genes associated with a particular quantitative trait are known as
(A) QTLs (B) MAS
(C) Genes (D) Dots

107. What is the main use of DNA markers in agricultural research?
(A) Used for development of hybrids through interspecies crosses
(B) To develop RIL and NIL population
(C) To Construct linkage maps for diverse crop species
(D) To help in crossing over

108. Why does not linkage maps use?
(A) For identifying chromosomal regions for genes controlling simple traits
(B) For identifying chromosomal regions for genes controlling quantitative traits
(C) To develop RIL and NIL population
(D) For QTL analysis

109. The process of constructing linkage maps and to identify genomic regions associated with traits known as ....
(A) QTL mapping
(B) Marker assisted selection
(C) Chromosome walking
(D) Gene tagging

110. DNA markers that are tightly linked to agronomically important genes may be used as molecular tools for marker-assisted selection (MAS) in plant breeding.
(A) Gene tapping
(B) Marker linkage tendency
(C) QTL mapping
(D) Gene tagging

111. .... involves using the presence/absence of a marker as a substitute for or to assist in phenotypic selection.
(A) Marker linkage tendency
(B) Marker assisted selection
(C) QTL mapping
(D) Gene tagging

112. The use of DNA markers in plant and animal breeding developed as a new realm in agriculture for crop improvement called ....
(A) Molecular breeding
(B) Molecular Biology
(C) Molecular pharming
(D) Molecular tagging

113. All genetic markers occupy specific genomic positions within chromosomes like genes called
(A) Mark (B) Label
(C) Loci (D) Sign

114. Which of the following is not a major type of genetic markers?
(A) Morphological markers
(B) Biochemical markers
(C) Allele markers
(D) DNA markers

115. Morphological markers are based on ....
(A) Phenotypic traits
(B) Allelic variants of enzymes
(C) Sites of variation in DNA
(D) Sites of variation in RNA

116. Biochemical markers include allelic variants of enzymes called ....
(A) Ribozymes (B) Abzymes
(C) Isozymes (D) Molecular tags

117. Molecular markers reveal ....
(A) Phenotypic traits
(B) Allelic variants of enzymes
(C) Sites of variation in DNA
(D) Sites of variation in proteins

118. Which of the following is not a morphological marker?
(A) Flower colour
(B) Seed shape
(C) Pigmentation
(D) RFLP

119. Technique is used for analyzing isozyme markers.
(A) Electrophoresis
(B) Trizole method
(C) CTAB method
(D) HPLC

120. What is the major disadvantage of morphological and biochemical markers?
(A) They may be limited in number
(B) They are depends on variation in proteins
(C) They are influenced by environmental factors
(D) They are influenced by the developmental stage of the plant

121. How DNA markers arise?
(A) Substitution mutations
(B) Rearrangements (insertions or deletions)
(C) Errors in replication of tandemly repeated DNA
(D) All of the above

122. Which of the following is not a characteristic of DNA markers?
(A) They are selectively neutral due to presence in non-coding regions of DNA
(B) DNA markers are practically unlimited in number
(C) They are affected by plant developmental stage
(D) They are not affected by environmental factors

123. What are the applications of DNA markers?
(A) To assess the level of genetic diversity within germplasm
(B) Cultivar identity
(C) To Construct linkage maps for diverse crop species
(D) All of the above

124. Which of the following is not a class of DNA markers based on the method of their detection?
(A) Hybridization-based
(B) Isozyme based
(C) DNA sequence-based
(D) PCR-based

125. DNA markers reveal genetic differences that can be visualised by using a technique except
(A) Gel electrophoresis with ethidium bromide staining
(B) Detection with radioactive or colourimetric probes
(C) SDS electrophoresis with comassive brilliant blue staining
(D) Gel electrophoresis with silver staining

126. What are the polymorphic DNA markers?
(A) They show differences between individuals of the same or different species
(B) They do not discriminate between genotypes
(C) They cannot be used to construct linkage map
(D) They cannot be used in QTL mapping

127. DNA markers that do not discriminate between individuals of the same or different species called ....
(A) Polymorphic markers
(B) Selective markers
(C) Scorable markers
(D) Monomorphic markers

128. What are the uses of polymorphic DNA markers?
(A) Markers can discriminate between homozygotes and heterozygotes
(B) They do not discriminate between genotypes
(C) They cannot be used to construct linkage map
(D) They cannot be used in QTL mapping

129. What are called the different forms of a gene or DNA marker with different sized bands on gels?
(A) Loci (B) Tags
(C) Centimorgan (D) Alleles

130. Codominant DNA markers may have .... alleles whereas a dominant marker only has .... alleles.
(A) Two and Many
(B) Many different and two
(C) One and Many
(D) Many different and one

131. A .... construct as a road map of the chromosomes derived from two different parents.
(A) Linkage map
(B) Chromosome bar coding
(C) Parental polymorphism
(D) Gene tagging

132. Which of the following statement is not true about linkage map?
(A) A road map of the chromosomes derived from two different parents
(B) It indicate the position and relative genetic distances between markers along chromosomes
(C) It is used to identify chromosomal locations containing genes and QTLs associated with traits of interest
(D) It can be constructed using monomorphic DNA markers

133. .... is based on the principle that genes and markers segregate via chromosome recombination during meiosis thus allowing their analysis in the progeny.
(A) Linkage mapping
(B) QTL mapping
(C) Gene mapping
(D) Chromosome mapping

134. Genes and markers segregate via chromosome recombination called .... during meiosis in sexual reproduction?
(A) Synapsis
(B) Chromosome jumping
(C) Crossing-over
(D) Chromosome walking

135. A segregating population contains ....
(A) Only recombinant genotypes
(B) Only homozygous parental genotypes
(C) A mixture of parental and recombinant genotypes
(D) A mixture of same genetic frequent genotypes

136. The frequency of recombinant genotypes using genetic markers can be used to .... in segregating population.
(A) Infer the genetic distance between markers
(B) Infer the genetic distance between genotypes
(C) Analyze of QTLs
(D) Separate homozygous and heterozygous lines

137. How many a recombination frequency should be between markers and gene to describe as 'unlinked' and assumed to be located far apart on the same chromosome or on different chromosomes.
(A) 50% (B) 100%
(C) 10% (D) 25%

138. Mapping functions in QTL mapping are used to convert recombination fractions into map units called ....
(A) Centi-Morgans (cM)
(B) Angstrom (A°)
(C) Percentage (%)
(D) Quantum (Q)

139. What are the main steps of linkage map construction from the analysis of many segregating markers?
(A) Production of a mapping population
(B) Identification of polymorphism
(C) Linkage analysis of markers
(D) All of the above

140. What are the requirements for the construction of a linkage map?
i. A segregating plant population
ii. A population derived from asexual reproduction
iii. The selected parents will differ for one or more traits of interest
iv. Polymorphic markers to differentiate parents and hybrids
(A) ii, iii and iv (B) i, ii, and iv
(C) i, iii, and iv (D) All of the above

141. What should be average size of mapping population for the construction of a linkage map?
(A) 50 to 250 individuals
(B) Only 50 individuals
(C) Less than 50 individuals
(D) More than individuals

142. Generally in self-pollinating species, mapping populations originate from parents that are both highly ....
(A) Homozygous (B) Inbred
(C) heterozygous (D) Both (A) and (B)

143. Many cross pollinating plant species are also polyploidy that contain several sets of
(A) Centromere pairs
(B) Nucleosome pairs
(C) Chromosome pairs
(D) Gene pairs

144. Mapping populations used for mapping cross pollinating species may be derived from a cross between
(A) A heterozygous parent and a homozygous parent
(B) A heterozygous parent and a haploid parent
(C) A heterozygous parent and a diploid parent
(D) Both (A) and (B)

145. Which type of mapping population is not used for the construction of a linkage map?
(A) F_1 populations
(B) Backcross populations
(C) Doubled haploid (DH) populations
(D) Recombinant inbred lines (RILs)

146. F_2 populations are derived from ....
(A) Backcrossing after crossing between F_1 hybrids and parent
(B) F_1 hybrids after selfing
(C) Homozygous parent after selfing
(D) Heterozygous parent after selfing

147. .... are derived by crossing the F_1 hybrid to one of the parents.
(A) F_1 populations
(B) Backcross populations
(C) Doubled haploid (DH) populations
(D) Recombinant inbred lines (RILs)

148. Which of the following is the simplest type of mapping population developed for self pollinating species?
(A) F_1 populations
(B) Backcross populations
(C) Doubled haploid (DH) populations
(D) Recombinant inbred lines (RILs)

149. What is the main advantage of backcross populations developed for self pollinating species?
(A) They are easy to construct and require only a short time to produce
(B) They are inbreed lines derived from individual F_2 plants
(C) They consist of a series of homozygous lines
(D) Produced by regenerating plants by the induction of chromosome doubling

150. Inbreeding from individual F_2 plants allows the construction of .... which consist of a series of homozygous lines containing a unique combination of chromosomal segments from the original parents.
(A) F_1 populations
(B) Backcross populations
(C) Doubled haploid (DH) populations
(D) Recombinant inbred lines (RILs)

151. What is the major disadvantage in the production of RIL populations?
(A) Require more time to develop RILs
(B) No important for linkage map construction
(C) Not appropriate for QTLs
(D) Two to three generations are required

152. Doubled haploid (DH) populations produced by regenerating plants by the induction of .... from pollen grains.
(A) Chromosomal mutation
(B) Chromosome dropping
(C) Centromere doubling
(D) Chromosome doubling

153. What are the major advantages of RI and DH populations developed for QTL analysis?
(A) They produce homozygous lines
(B) They produce true-breeding lines
(C) Produced true-breeding lines can be multiplied without genetic change occurring.
(D) All of the above

154. What is the requirement for the construction of a linkage map?
(A) To identify polymorphic DNA markers
(B) LOD values of <3 are typically used to construct linkage maps
(C) To identify parents with same traits
(D) Markers that have a recombination frequency of more than 50%

155. Why is simple to construct linkage map in self pollinated crops than cross pollinated crops.
(A) Cross pollinating species possess higher levels of DNA polymorphism
(B) Cross pollinating species possess low levels of DNA polymorphism
(C) Self pollinating species possess no DNA polymorphism
(D) Self pollinating species possess higher levels of DNA polymorphism

156. Once polymorphic markers have been identified, they must be screened across the entire mapping population including the parents and F1 hybrid This is known as marker .... of the population.

(A) Phenotyping (B) Genotyping
(C) Karyotyping' (D) Linkage

157. In which species is to identify more complicated polymorphic markers?
(A) Haploid
(B) Double haploid
(C) Polyploid
(D) None of the above

158. Which of the following is polyploidy species?
(A) Rice (B) Sugarcane
(C) Maize (D) Pigeonpea

159. Linkage between markers is usually calculated using odds ratios of ....
(A) Linkage versus no linkage
(B) Polymorphic versus Monomorphic markers
(C) Dominant and co-dominant markers
(D) Absence and presence of bands

160. Linkage between markers is usually calculated using odds ratios is more conveniently expressed as the logarithm of the ratio is called ....
(A) 0/1 matrix (B) cM values
(C) LOD score (D) QTL score

161. What is LOD?
(A) Logarithmic odd distribution
(B) Logarithm of objects
(C) Logarithm of ordinary numbers
(D) Logarithm of odds

162. A LOD value of 3 between two markers indicates that
(A) Linkage is 1000 times more than no linkage
(B) Linkage is 1000 times less than no linkage
(C) No linkage is 1000 times more than linkage
(D) No Linkage is 1000 times less than linkage

163. Which of the following commonly used software programs for constructing linkage map and QTL analysis?
(A) Mapmaker/EXP
(B) MapManager QTX
(C) JoinMap
(D) PROSITE

164. Ideally, mapping populations should consist of a minimum of .... for constructing linkage maps.
(A) 150 individuals
(B) 50 individuals
(C) 100 individuals
(D) 250 individuals

165. Which of the following markers has co-dominant nature?
(A) SSR and RFLP
(B) RAPD and AFLP
(C) AFLP and RFLP
(D) SSR and AFLP

166. Which of the following markers has co-dominant nature?
(A) SSR and RFLP
(B) RAPD and AFLP
(C) AFLP and RFLP
(D) SSR and AFLP

167. Which of the following markers amplified multiple loci from a single primer?
(A) SSR and RFLP (B) RAPD and AFLP
(C) AFLP and RFLP (D) SSR and AFLP

168. SSRs are also known as .... markers.
(A) Sequence tagged microsatellite site (STMS)
(B) Random amplified length polymorphism (RAPD)
(C) Cleaved Amplified polymorphisms (CAPs)
(D) Expressed sequence tags (ESTs)

169. Codominant markers indicate differences in .... whereas dominant markers are .... of bands.
(A) Size and shape
(B) Either present or absent and size
(C) Size and either present or absent
(D) Polymorphic and Monomorphic bands

170. Distance on a linkage map is not directly related to the physical distance of DNA between genetic markers but depends on the .... of the plant species.
(A) Number of genes
(B) Genome size
(C) VNTRs
(D) Number of chromosomes

171. There are recombination 'hot spots' and 'cold spots,' which are chromosomal regions in which recombination occurs ...., respectively
(A) More crossing over or less crossing over
(B) Less frequently or more frequently
(C) More distance or less distance of linkage
(D) More frequently or less frequently

172. .... is based on the principle of detecting an association between phenotype and the genotype of markers.
(A) Bulk segregate analysis
(B) Pure line selection
(C) QTL analysis
(D) Gene transformation

173. What is calculated to obtain for differences between mean trait values indicate linkage between marker and QTL?
(A) *F-value*
(B) *Chi square value*
(C) *P-value*
(D) *Z-value*

174. What should be significant *P-value* for a closely-linked marker to a QTL that will usually be inherited together in the progeny?
(A) P > 0.05 (B) P >0.5
(C) P < 0.05 (D) P <0.5

175. When a marker is loosely-linked or unlinked to a QTL. Unlinked markers located .... to the QTL are randomly inherited with the QTL.
(A) Far apart or on different chromosomes
(B) Near to different chromosomes
(C) On different chromosomes
(D) Not located on different chromosomes

176. QTLs are detected using different methods except ....
(A) Single-marker analysis
(B) Simple interval mapping
(C) Multiplex marker analysis
(D) Composite interval mapping

177. Which is the simplest method for detecting QTLs associated with single markers?
(A) Single-point analysis
(B) Simple interval mapping
(C) Multiplex marker analysis
(D) Composite interval mapping

178. Single-marker analysis is the simplest method for detecting QTLs associated with single markers that is include which statistical methods?
(A) *t*-tests
(B) Analysis of variance (ANOVA)
(C) Linear regression
(D) All of the above

179. Which QTL method is used to make linkage maps and analyses intervals between adjacent pairs of linked markers along chromosomes simultaneously?
(A) Single-point analysis
(B) Simple interval mapping
(C) Multiplex marker analysis
(D) Composite interval mapping

180. Which QTL method is most popular for mapping QTLs.
(A) Single-point analysis
(B) Simple interval mapping
(C) Multiplex marker analysis
(D) Composite interval mapping

181. Which QTL method combines interval mapping with linear regression and includes additional genetic markers in the statistical model in addition to an adjacent pair of linked markers for interval mapping.
(A) Single-point analysis
(B) Composite interval mapping
(C) Multiplex marker analysis
(D) Simple interval mapping

182. Which of the following software is not used to detect QTLs by performing Composite interval mapping (CIM)?
(A) QTL Cartographer
(B) MapManager QTX
(C) GenState
(D) PLABQTL

183. Common markers that are highly polymorphic in mapping populations are called .... markers like SSRs or RFLPs.
(A) Anchor or core
(B) Adapter
(C) Linker
(D) Oligo

184. Which of the following map can indicate which markers are located in a specific region containing a QTL, and thus be used to identify more tightly linked markers?
(A) Genome map (B) DNA map
(C) A consensus map (D) Gene map

185. Conserved marker order is referred to as ....
(A) Contig (B) Synteny
(C) Anchor (D) Core

186. Which of the following population type is ideal to investigate environmental influences on QTLs affecting trait(s) of interest?
(A) Recombinant inbred population
(B) DH population
(C) F_2 population
(D) Both (A) and (B)

187. Which of the following short-cut methods used to identify markers that tag QTLs and require mapping populations?
(A) Bulked segregant analysis (BSA)
(B) Selective genotyping
(C) Both (A) and (B)
(D) Pure line method

188. Which method is used to detect markers located in specific chromosomal regions by forming two pools or 'bulks' of DNA samples from 10–20 individual plants from a segregating population?
(A) Bulked segregant analysis (BSA)
(B) Selective genotyping
(C) Marker assisted selection
(D) Pure line method

189. What is called selective genotyping method to identify markers and tag QTLs?
(A) Distribution extreme analysis
(B) Trait-based marker analysis
(C) Bulked segregant analysis
(D) Both (A) and (B)

190. LOD threshold depend on
(A) Population size
(B) Distance between markers
(C) The length and density of the map
(D) Polymorphism

191. Who explained continuous variation in quantitative traits by the segregation of numerous genes modified by the environment?
(A) Johansen (1909)
(B) Sax (1923)
(C) Lander and Botstein (1989)
(D) Shull (1940)

192. Who discovered quantitative trait loci (QTL) ?
(A) Johansen (1909)
(B) Sax (1923)
(C) Lander and Botstein (1989)
(D) Shull (1940)

193. Who demonstrated for the first time linkage between a qualitative marker (seed color) and a quantitative locus (size of the seed)?
(A) Johansen (1909)
(B) Shull (1940)
(C) Lander and Botstein (1989)
(D) Sax (1923)

194. A LOD score is calculated every .... along the complete chromosome and represented in a graphical way
(A) 5 cM (B) 10 cM
(C) 12 cM (D) 2 cM

195. What is the value of LOD for one interval, 5% confidence?
(A) LOD 0.83 (B) LOD 2.0
(C) LOD 0 (D) LOD 5-10

196. What is the value of LOD for one chromosome 5% confidence?
(A) LOD 0.83 (B) LOD 2.0
(C) LOD 0 (D) LOD 5-10

197. A higher LOD score is required to keep the 5% confidence along
(A) High population size
(B) Interval map
(C) More distance between markers
(D) Complete map

198. All of the genes of an organism are known as ....
(A) Chromosome (B) Nucleosome
(C) Genome (D) DNA

199. Which of the following crop genome has been fully sequenced?
(A) *Arabidopsis thaliana*
(B) Rice
(C) Wheat
(D) Mustard

200. What is genetic marker?
(A) A allele that allows for the tracing of a specific region of DNA
(B) A polypeptide segment that responsible for phenotype
(C) A short ribonuceotide sequence that catalyze RNA transcription
(D) Pigment, plant height and seed size are genetic markers

201. What are the roles of molecular markers?
i. Analysis of associations between economically important traits and genes or QTLs
ii. Facilitate the introgression of desirable genes or QTLs through MAS
iii. To detect genetic diversity among individuals of a species
iv. Genetic mapping
(A) Only iii
(B) i and ii
(C) i, ii and iii
(D) i, ii, iii and iv

202. Which of the following is a class of genetic marker?
(A) Morphological traits
(B) Biochemical markers
(C) Biomolecular markers
(D) DNA markers

203. Morphological markers are based on ....
(A) Visually assessable traits
(B) Agronomic traits
(C) Gene and gene products
(D) Both (A) and (B)

204. Biochemical markers are based on ....
(A) Visually assessable traits
(B) Agronomic traits
(C) Gene products
(D) DNA and Gene

205. Molecular markers are based on ....
(A) Visually assessable traits
(B) DNA and Gene
(C) Agronomic traits
(D) Gene products

206. Which of the following is morphological marker?
(A) Pigments in flower
(B) Sequence of a gene controls pigment production
(C) Sequence of amino acids present in pigment producing gene product
(D) Sequence of mRNA derived from gene controls pigment synthesis

207. Which of the following marker group is classified based on mode of gene action?
(A) Dominant and co-dominant markers
(B) hybridization-based or PCRbased markers
(C) Maternal inheritance and paternal inheritance
(D) Isozymes and phenotypic traits

208. Which of the following marker group is classified based on method of analysis?
(A) Dominant and co-dominant markers
(B) hybridization-based or PCRbased markers
(C) Maternal inheritance and paternal inheritance
(D) Isozymes and phenotypic traits

209. Which of the following is hybridization-based molecular marker?
(A) RFLP
(B) SSR
(C) AFLP
(D) RAPD

210. Which of the following is non PCR-based molecular marker?
(A) AFLP (B) SSR
(C) RFLP (D) RAPD

211. .... markers were first used in 1975 to identify DNA sequence polymorphisms for genetic mapping of a temperature-sensitive mutation of adeno-virus serotypes.
(A) AFLP (B) SSR
(C) RAPD (D) RFLP

212. Which of the following method include restriction enzyme digestion?
(A) RFLP technique
(B) SSR technique
(C) RAPD technique
(D) EST technique

213. Which of the following correct sequence of RFLP technique?

i. Autoradiography
ii. Separation of the restriction fragments in agarose gel
iii. Digestion of the DNA with one or more restriction enzymes
iv. Detection of individual fragments by nucleic acid hybridization with a labeled probes
v. Transfer of separated fragments from agarose gel to a filter by Southern blotting

(A) i, ii, iii, iv, and v (B) iii, ii, v, iv and i
(C) ii, iii, iv, v and i (D) iii, iv, i, v and ii

214. Southern blotting technique used to detect restriction fragments in ....

(A) RFLP technique (B) SSR technique
(C) RAPD technique (D) AFLP technique

215. Restriction enzymes are ....

(A) Exonucleases (B) Ribozymes
(C) Abzymes (D) Endonucleases

216. Molecular markers are classified on the basis of ....

(A) Mode of transmission
(B) Mode of gene action
(C) Method of analysis
(D) All of the above

217. Some of the steps involved in gene cloning are given below

i. Insertion of isolated gene to the vector
ii. Introduction of recombinant vector to the host
iii. Isolation of desired gene
iv. Expression of recombinant gene in host
v. Extraction of recombinant gene product

The correct sequence of steps involved is

(A) iii, i, iv, ii, v (B) iii, i, ii, iv, v
(C) i, ii, iii, iv, v (D) ii, i, iii, iv, v

218. EcoR1 cleaves DNA at

(A) 5'G↓AATTC3'
3'CTTAA↓G5'
(B) 5'GTT↓AAC3'
3'CAA↑TTG5'
(C) 5'C↓AATTG3'
3'GTTAA↑C5'
(D) 5'GGGCC↓T3'
3'CCCGG↑A5'

219. DNA finger printing was first developed by

(A) David Suzuki (B) Khorana
(C) Alec Jaffrey (D) Gilbert

220. A technique called southern blotting is used in

(A) Monoclonal antibody production
(B) In vitro culture
(C) Genetic finger printing
(D) Polymerase chain reaction

221. Restriction enzymes found in only in ....

(A) All prokaryotes
(B) Eukaryotes
(C) Viruses
(D) Bacterial cells

222. The greatest resolution is obtained by using .... enzymes due to many sites in the genome.

(A) Four-cutter (B) Six-cutter
(C) Two cutter (D) Five cutter

223. After digestion by restriction enzymes, the DNA is present as a mixture of

(A) Linear double-stranded molecules of various lengths
(B) Linear single and double-stranded molecules of various lengths
(C) Linear and circular double-stranded molecules of various lengths
(D) Linear double-stranded molecules of same lengths

224. Restriction fragments are saperated by .... in RFLP technique.

(A) Agarose gel electrophoresis
(B) Polyacrylamide gels
(C) Only (A)
(D) Both (A) and (B)

225. Southern blotting is discovered by .... in 1975.

(A) E. Southern (B) F. Sanger
(C) Kary Mullis (D) Ingo Potrykus

226. Which of the following used to hybrize DNA fragments on in RFLP?

(A) Adapter (B) Probe
(C) Linker (D) Ladder

227. The probes used for hybridization are preferably single locus and mostly species-specific probes of .... in size in RFLP.
(A) About 50 to 300 bp
(B) About 5 to 30 bp
(C) About 500 to 3000 bp
(D) About 10 to 50 bp

228. Probes used to identify specific DNA fragments by hybridization are of .... types.
(A) Genomic DNA clones
(B) cDNA clones
(C) mRNA
(D) Both (A) and (B)

229. Which of the following technique is used to detect and visualize hybridized probes with DNA in RFLP?
(A) Autoradiography
(B) Spectroscopy
(C) Ultracentrifugation
(D) Electrophoresis

230. The major strength of RFLP markers are
(A) High reproducibility
(B) Codominant inheritance
(C) Provide locus specific markers
(D) All of the above

231. Which of the following markers is not require prior sequence information?
(A) SSR (B) EST
(C) CAPs (D) RFLP

232. What are the limitations of RFLP?
i. It requires the presence of high quantity and quality of DNA
ii. It depends on the development of specific probe libraries for the species
iii. The level of polymorphism is low, and few loci are detected per assay.
iv. It usually requires radioactively labeled probes
(A) Only iii (B) i and ii
(C) i, ii, iii and iv (D) i, ii and iii

233. Who discovered Polymerase chain reaction (PCR)?
(A) H.G. Khorana (B) Richard Palmiter
(C) Kary Mullis (D) Robert Rushmer

234. Kary Banks Mullis was born in 1944?
(A) Lenoir (North Carolina)
(B) Columbia (North Carolina)
(C) California (Berkeley)
(D) Emeryville (California)

235. Which technique does allow making millions of copies of DNA in a short period of time?
(A) Sequencing
(B) PCR
(C) Electrophoresis
(D) Blotting

236. In which year Karry Mullis was awarded by Nobel Prize in chemistry for his contributions to the development of DNA-based chemical methods?
(A) 2000 (B) 1983
(C) 2004 (D) 1993

237. Who shared the Nobel Prize with Karry Mullis in 1993 for his contributions to the development of DNA-based chemical methods?
(A) Michael Smith
(B) H.G. Khorana
(C) Frederick Sanger
(D) Maxam Gilbert

238. In which year did Karry Mullis discovered PCR?
(A) 1980 (B) 1983
(C) 1953 (D) 2000

239. Who discovered ribozymes?
(A) Ingo Potrykus (B) F. Sanger
(C) T. Cech (D) Stull and Szoka

240. What are called enzymatic RNA molecules that cleave the target RNA?
(A) Abzymes (B) Ribozymes
(C) Enzymes (D) Lysozymes

241. What arc ribozymes?
(A) A sense RNA molecules with enzymatic activity that cleaves RNA Strands
(B) An antisense RNA molecules with enzymatic activity stops transcription
(C) An complementerystrand of mRNA with enzymatic activity cleaves protein synthesis
(D) It is DNA based therapeutic to suppress genes

242. Which of the following bonds are cleaved in ribozyme technology
(A) Hydrogen bonds between bases
(B) Glycosidic bonds between base and sugar
(C) Covalent bonds between neighbouring sugars
(D) Phosphodiester linkage between sugar and phosphate

243. How functions ribozymes?
(A) Inactivate target RNA molecules with antisense sequence-specific hybridization
(B) Stop transcription adding complementary RNA strand to mRNA
(C) Block translation breaking glycosidic linkage between ribose sugar and bases
(D) Stop transcription and translation adding complementary RNA strand to DNA

244. Ribozymes selectively bind to target mRNAs to hydrolyze mRNA that may be used for
(A) Stop transcription to knock out specific gene
(B) Block transcription and translation to overexpress specific gene
(C) Targeted suppression of specific gene by blocking translation
(D) Creat mutation in sequence code for amino acids to suppress specific gene

245. Which of the following group of RNA used in ribozyme technology?
(A) hnRNA and snRNA
(B) Hammerhead RNA and Hairpin RNA
(C) miRNA and mnRNA
(D) mRNA, tRNA and rRNA

246. What are hammerhead ribozymes?
(A) Cleave RNA at the nucleotide sequence by breaking phosphodiester bonds
(B) Contain 25 nucleotide sequence C–U–G as their cleavage site
(C) These molecules are biologically stable in vivo
(D) Used in RNAi technology to supreeass specific gene

247. Which statements are true about hairpin ribozymes?
i. Utilize 25 nucleotide sequence C–U–G as their cleavage site
ii. They easily targeted for degradation by RNases
iii. They are susceptible to RNases
iv. These are used to knockout specific gene
(A) i and ii (B) Only i
(C) i, ii and iv (D) i, ii, iii and iv

248. What is the application of ribozymes?
(A) Can be used for knockout gene therapy
(B) Used to inhibit transcription to increase fruit ripening period in tomato
(C) Target normaly expressed oncogenes for overexpression
(D) To reduce fertility restoration and form male sterility in musturd

249. What is limitation of the use of ribozymes for gene therapy?
(A) They degrade only DNA strads forms for mRNA
(B) They are degraded by RNases
(C) They creat mutations in mRNA sequences
(D) They permanentaly occur in genome

250. Ribozyme is .... used to suppress specific genes.
(A) Posttranslation approach
(B) Posttranscriptional approach
(C) Both Posttranscriptional and Posttranslation approach
(D) DNA-based approach

251. Who developed the determination of DNA sequences (DNA sequencing method)?
(A) James D. Watson and Francis Crick (1953)
(B) Measelson and Stahal (1958)
(C) Rosalind Franklin and Maurice Wilkins (1962)
(D) Alan Maxam and Walter Gilbert (1977)

252. Who developed the determination of dideoxynucleotides DNA sequenced?
(A) James D. Watson and Francis Crick (1953)
(B) Frederick Sanger (1977)
(C) Alan Maxam and Walter Gilbert (1977)
(D) Rosalind Franklin and Maurice Wilkins (1962)

253. What is the general principle of both Sanger and Maxam-Gilbert sequencing in which the reaction producing each set is base-specific?
(A) To reduce the DNA to four sets of labeled fragments
(B) To increase the DNA to four sets of labeled fragments
(C) To remain constant the DNA to four sets of labeled fragments
(D) All of the above

254. Maxam and Gilbert and Sanger's DNA sequencing does not require an improved understanding of which?
(A) Nucleotide chemistry
(B) DNA metabolism
(C) Electrophoretic
(D) Protein metabolism

255. Electrophoresis method separate the segments of DNA based on ....
(A) Size of DNA (B) Charge of DNA
(C) pH of DNA (D) Hydrogen bonds in DNA

256. What is called the digramattic representation of banding pattern of isozymes?
(A) Karyogram (B) Idiogram
(C) Zymogram (D) Karyogram

257. Isozymes are
(A) Morphological traits
(B) Biochemical markers
(C) Biomolecular markers
(D) DNA markers

258. Which of the following is biochemical marker?
(A) Pigments in flower
(B) Sequence of a gene controls pigment production
(C) Sequence of amino acids present in pigment producing gene product
(D) Sequence of mRNA derived from gene controls pigment synthesis

259. Which of the following is molecular marker?
(A) Pigments in flower
(B) Sequence of a gene controls pigment production
(C) Sequence of amino acids present in pigment producing gene product
(D) Sequence of mRNA derived from gene controls pigment synthesis

260. VNTR is ....
(A) Variation numeral train repeat
(B) Variable number tandem repeat
(C) Vary number trailer repeat
(D) Variable number tandem rotation

261. A PCR reaction that continues for 30 cycles will produce approximately how many PCR products from a single template DNA molecule?
(A) 64 (B) 128000
(C) Approximately 1 million
(D) Approximately 1 billion

262. In an SDS-PAGE
(A) Proteins are denatured by the SDS
(B) Proteins have the same charge-to-mass ratio
(C) Smaller proteins migrate more rapidly through the gel
(D) All of the above

263. Proteins can be visualized directly in gels by
(A) Staining them with the dye
(B) Using electron microscope only
(C) Measuring their molecular weight
(D) None of these

264. In SDS-PAGE, the protein sample is first
(A) Treated with a reducing agent and then with anionic detergent followed by fractionation by electrophoresis
(B) Fractionated by electrophoresis then treated with an oxidizing agent followed by anionic detergent.
(C) Treated with a oxidizing agent and then with anionic detergent followed by fractionation by electrophoresis
(D) None of the above

265. Electrophoresis of histones and myoglobin under non-denaturing conditions (pH = 7.0) results in
(A) Both proteins migrate to the anode
(B) Histones migrate to the anode and myoglobin migrates to the cathode
(C) Histones migrate to the cathode and myoglobin migrates to the anode
(D) Both proteins migrate to the cathode

266. In isoelectric focusing, proteins are separated on the basis of their
- (A) Relative content of positively charged residue only
- (B) Relative content of negatively charged residue only
- (C) Size
- (D) Relative content of positively and negatively charged residue

267. In a gel filtration column
- (A) Smaller proteins enter the beads more readily
- (B) Large proteins elute first
- (C) Both (A) and (B)
- (D) Large proteins enter the beads more readily

268. In a native PAGE, proteins are separated on the basis of
- (A) Net negative charge
- (B) Net charge and size
- (C) Net positive charges size
- (D) Net positive charge

269. The subunit molecular weight as well as the number of subunits in the quaternary structure can be determined by
- (A) SDS-PAGE electrophoresis
- (B) Gel filtration chromatography
- (C) Combining information from (a)and (b)
- (D) Isoelectric focusing

270. Proteins are separated in an SDS-PAGE experiment on the basis of their
- (A) Positively charged side chains
- (B) Molecular weight
- (C) Negatively charged side chains
- (D) Different isoelectric points

271. In agarose gel electrophoresis, DNA is moved towards the
- (A) Cathode (B) Anode
- (C) Dna doesn't move (D) None of above

272. In SDS electrophoresis, proteins are separated on basis of
- (A) Charge
- (B) Mass
- (C) Both (A) and (B)
- (D) Structure

273. If proteins are separated according to their electrophoretic mobility then type of electrophoresis is
- (A) SDS page
- (B) Free flow electrophoresis
- (C) Electro focusing
- (D) Affinity electrophoresis

274. A technique which separates charged particles using electric field is
- (A) Hydrolysis
- (B) Electrophoresis
- (C) Protein synthesis
- (D) Protein denaturing

275. The purity of an enzyme at various stages of purification is best measured by:
- (A) Total protein
- (B) Total enzyme activity
- (C) Specific activity of the enzyme
- (D) Percent recovery of protein

276. Which would be best to separate a protein that binds strongly to its substrate?
- (A) Gel filtration
- (B) Affinity chromatography
- (C) Cation exchange
- (D) Anion exchange

277. How many hydrogen bonds can a water molecule potentially take in liquid form?
- (A) one (B) two
- (C) three (D) four

278. How does the hydrophobic effect influence the structures of large molecules?
- (A) Nonpolar molecules are not easily solubilized in water and aggregate
- (B) Polar groups are oriented on the surface, interacting with the water
- (C) Nonpolar molecules can mask the polar characteristics of the hydrophilic molecules
- (D) Both (A) and (B)

279. The most important buffering system for maintaining proper blood pH is:
- (A) the charges on the amino acids
- (B) the bicarbonate buffer system of CO_2, carbonic acid, and bicarbonate
- (C) phosphate groups of serum phosphoproteins
- (D) none of the above

280. In the binding of oxygen to myoglobin, the relationship between the concentration of oxygen and the fraction of binding sites occupied can best be described as:
(A) hyperbolic
(B) linear with a negative slope
(C) linear with a positive slope
(D) sigmoidal

281. Two amino acids of the standard 20 contain sulfur atoms. They are:
(A) cysteine and serine
(B) cysteine and threonine
(C) methionine and cysteine
(D) methionine and serine

282. An octapeptide composed of four repeating glycylalanyl units has:
(A) One free amino group on an alanyl residue
(B) One free amino group on an alanyl residue and one free carboxyl group on a glycyl residue
(C) One free amino group on a glycyl residue and one free carboxyl group on an alanyl residue
(D) Two free amino and two free carboxyl groups

283. In a mixture of the five proteins listed below, which should elute second in size-exclusion (gel filtration) chromatography?

(A)	Cytochrome C	Mr = 13,000
(B)	Immunoglobulin G	Mr = 145,000
(C)	Ribonuclease A	Mr = 13,700
(D)	RNA polymerase	Mr = 450,000

284. Karry Mullis did not get which award in his life?
(A) Robert Koch Prize
(B) Nobel Prize in Chemistry
(C) Japan Prize
(D) Food prize

285. Experiments on denaturation and renaturation after the reduction and reoxidation of the –S-S- bonds in the enzyme ribonuclease (RNase) have shown that:
(A) Folding of denatured RNase into the native, active conformation requires the input of energy in the form of heat
(B) Native ribonuclease does not have a unique secondary and tertiary structure
(C) The completely unfolded enzyme, with all –S-S- bonds broken, is still enzymatically active
(D) The primary sequence of RNase is sufficient to determine its specific secondary and tertiary structure

286. What are the three basic steps of conventional PCR?
(A) Denature, anneal, & strand displacement
(B) Denature, anneal & extension
(C) Strand displacement, synthesis & release
(D) Reverse-transcription, anneal & extend

287. Which of the following is not a stage of PCR
(A) Decay (B) Plateau
(C) Stochastic / lag (D) Exponential

288. A PCR efficiency of '2' means
(A) 100% efficiency / initial target copies are doubled by the end of the reaction
(B) 95% efficiency / each target copy is doubled every cycle
(C) 100% efficiency / each target copy is doubled every cycle
(D) 95% efficiency / initial target copies are doubled by the end of the reaction

289. RNA is copied into complementary DNA (cDNA) by:
(A) *Taq* DNA polymerase
(B) RNA polymerase II
(C) Reverse transcriptase
(D) Uracil-N-Glycosylase

290. The reverse transcriptase reaction can be primed by
(A) Target sequence specific primers
(B) Random hexamers
(C) Oligo dT primers
(D) All of the above

291. The cycle threshold is:
(A) The total number of cycles performed during a real-time PCR reaction
(B) The cycle that a sample crosses a certain point during a real-time PCR reaction
(C) The cycle number that a sample enters the plateau phase of PCR
(D) None of the above

292. Which of the following statements is false?
(A) PCR inhibitors can lead to false negative results
(B) PCR examines a large proportion of the tissue leading to false positive results
(C) Pathogen diversity at primer sites may lead to false negative results
(D) Contamination may lead to false positive results

293. Which of the following is an advantage of nested PCR (nPCR)?
(A) Provides a quantitative assessment of initial starting copy number
(B) Second round PCR products can be a source of laboratory contamination
(C) Is less time consuming than single round conventional PCR
(D) Typically has high sensitivity and specificity

294. Which of the following is not an advantage of quantitative PCR (qPCR)
(A) Reliable indicator of viable infection
(B) No post-PCR handling of products
(C) Highly sensitive, specific and repeatable
(D) Can obtain quantitative results

295. Which is not a property of real-time PCR assays?
(A) Incorporate dyes that bind double-stranded DNA
(B) Incorporate an internal hydrolysis probe
(C) Be performed at single temperature with no specialized instrumentation required
(D) Be interpreted as a plus / minus result or as a quantitative result

296. A plasmid encoding a target sequence of interest will be used as the quantitative PCR standard. Where should you work with the plasmid?
(A) The PCR reagent clean area
(B) A special area designated for high risk templates
(C) The area where all sample templates are prepared
(D) None of the above

297. Choose the statement that correctly finishes the sentence: "A PCR reaction that contains only one copy of the target sequence (1 copy /reaction) ....
(A) Is typically amplified in a highly repeatable manner"
(B) May amplify but is detection is not likely to be highly repeatable"
(C) Can be precisely and accurately quantified using quantitative PCR"
(D) All of the above

298. Which of the following is not a method for stabilizing RNA
(A) 95% ethanol
(B) Liquid nitrogen
(C) RNAlater
(D) EDTA

299. Difference between stability between DNA and RNA
(A) DNA is more stable than RNA
(B) RNA is more stable than DNA
(C) Both are stable for equal time
(D) DNA degrade fast than RNA

300. Samples of known concentration/copy number used to construct a standard curve are called:
(A) Controls
(B) Standards
(C) Exogenous normalizing variables
(D) Endogenous normalizing variables

Answer-Key

Q	Ans	Q	Ans	Q	Ans	Q	Ans	Q	Ans	Q	Ans	Q	Ans	Q	Ans
1.	(C)	41.	(C)	81.	(A)	121.	(D)	161.	(D)	201.	(D)	241.	(C)	281.	(C)
2.	(A)	42.	(D)	82.	(C)	122.	(C)	162.	(A)	202.	(C)	242.	(D)	282.	(C)
3.	(C)	43.	(B)	83.	(B)	123.	(D)	163.	(D)	203.	(D)	243.	(A)	283.	(B)
4.	(D)	44.	(C)	84.	(D)	124.	(B)	164.	(B)	204.	(C)	244.	(C)	284.	(D)
5.	(B)	45.	(D)	85.	(C)	125.	(C)	165.	(A)	205.	(B)	245.	(B)	285.	(D)
6.	(C)	46.	(D)	86.	(B)	126.	(A)	166.	(B)	206.	(A)	246.	(A)	286.	(B)
7.	(A)	47.	(B)	87.	(C)	127.	(D)	167.	(B)	207.	(A)	247.	(D)	287.	(A)
8.	(B)	48.	(A)	88.	(C)	128.	(A)	168.	(A)	208.	(B)	248.	(A)	288.	(C)
9.	(C)	49.	(C)	89.	(A)	129.	(D)	169.	(C)	209.	(A)	249.	(B)	289.	(C)
10.	(A)	50.	(D)	90.	(B)	130.	(B)	170.	(B)	210.	(C)	250.	(B)	290.	(D)
11.	(D)	51.	(C)	91.	(C)	131.	(A)	171.	(D)	211.	(D)	251.	(D)	291.	(B)
12.	(A)	52.	(C)	92.	(B)	132.	(D)	172.	(C)	212.	(A)	252.	(B)	292.	(B)
13.	(B)	53.	(D)	93.	(B)	133.	(B)	173.	(C)	213.	(B)	253.	(A)	293.	(D)
14.	(A)	54.	(C)	94.	(B)	134.	(C)	174.	(C)	214.	(A)	254.	(D)	294.	(A)
15.	(D)	55.	(B)	95.	(A)	135.	(C)	175.	(A)	215.	(D)	255.	(A)	295.	(C)
16.	(A)	56.	(A)	96.	(B)	136.	(A)	176.	(C)	216.	(D)	256.	(C)	296.	(B)
17.	(B)	57.	(A)	97.	(C)	137.	(A)	177.	(A)	217.	(B)	257.	(C)	297.	(B)
18.	(D)	58.	(B)	98.	(A)	138.	(A)	178.	(D)	218.	(A)	258.	(C)	298.	(A)
19.	(C)	59.	(D)	99.	(D)	139.	(D)	179.	(B)	219.	(C)	259.	(B)	299.	(A)
20.	(D)	60.	(A)	100.	(A)	140.	(C)	180.	(D)	220.	(C)	260.	(B)	300.	(B)
21.	(B)	61.	(A)	101.	(C)	141.	(A)	181.	(B)	221.	(D)	261.	(D)		
22.	(A)	62.	(D)	102.	(A)	142.	(D)	182.	(C)	222.	(A)	262.	(A)		
23.	(C)	63.	(A)	103.	(D)	143.	(C)	183.	(A)	223.	(A)	263.	(A)		
24.	(A)	64.	(C)	104.	(A)	144.	(D)	184.	(C)	224.	(D)	264.	(A)		
25.	(B)	65.	(D)	105.	(D)	145.	(A)	185.	(D)	225.	(A)	265.	(C)		
26.	(A)	66.	(B)	106.	(A)	146.	(B)	186.	(C)	226.	(B)	266.	(D)		
27.	(C)	67.	(A)	107.	(C)	147.	(B)	187.	(A)	227.	(C)	267.	(C)		
28.	(B)	68.	(C)	108.	(C)	148.	(B)	188.	(D)	228.	(D)	268.	(B)		
29.	(C)	69.	(A)	109.	(A)	149.	(A)	189.	(C)	229.	(A)	269.	(C)		
30.	(A)	70.	(D)	110.	(D)	150.	(D)	190.	(A)	230.	(D)	270.	(B)		
31.	(B)	71.	(C)	111.	(B)	151.	(A)	191.	(A)	231.	(D)	271.	(A)		
32.	(D)	72.	(B)	112.	(A)	152.	(D)	192.	(B)	232.	(C)	272.	(B)		
33.	(C)	73.	(B)	113.	(C)	153.	(D)	193.	(D)	233.	(C)	273.	(A)		
34.	(B)	74.	(D)	114.	(C)	154.	(A)	194.	(D)	234.	(A)	274.	(B)		
35.	(A)	75.	(B)	115.	(A)	155.	(A)	195.	(A)	235.	(B)	275.	(C)		
36.	(D)	76.	(C)	116.	(C)	156.	(B)	196.	(B)	236.	(D)	276.	(B)		
37.	(C)	77.	(D)	117.	(C)	157.	(C)	197.	(D)	237.	(A)	277.	(D)		
38.	(D)	78.	(B)	118.	(D)	158.	(B)	198.	(C)	238.	(B)	278.	(D)		
39.	(C)	79.	(B)	119.	(A)	159.	(A)	199.	(A)	239.	(C)	279.	(B)		
40.	(A)	80.	(D)	120.	(B)	160.	(C)	200.	(A)	240.	(B)	280.	(A)		

10

Biotechnology in Crop Improvement

Plant breeders and farmers applied different convetional methods to alter the genetic background of the crops and selected the best performing plants and seeds for the next season. Plants have been seleted on the basis of phenotypic characteristics such as early maturity, higher yields, resistant to pest and disease, larger seed size and other quality characters. The breeder uasually uses the natural variation that already exists within the species. Plant breeders used various breeding methods (selection, hybridization, introduction, bulk segregant analysis, mass selection etc.) to transfer desirable characteristics from one plant to another plant and developed new varieties of crops for hundreds of years. Thus, new crop improvement technologies should be developed and utilized.

In agricultural of India, dwarf varieties of wheat and rice have been developed since 1970s to ensure food security and self-sufficiency using plant breeding. Although due to increase population day by day, various techniques of modern biology, especially biotechnology, have been delevoped which provide many advantages compared to conventional plant breeding methods. These techniques have been used to improve yield potential, nutritional quality, decreasing maturity time, delay ripening, production of secondary metabolites, geneome sequencing, increase resistant to abiotic and biotic facors.

Applications of biotechnology for crop improvement

Tolerant to Submergence in rice varieties

Flood-tolerant and high-yielding new varieties of rice developed through identification of a single gene Sub1A responsible for submergence tolerance. Sub1A plant undergoes dormancy during submergence which allows it to conserve energy until flood water recedes. These plants developed through precision breeding, known as marker-assisted selection, are not genetically modified.

Rice with Iron

Iron deficiency is a major nutritional problem especially in women and children of developing countries like India. Therefore, biofortified rice with high iron content may reduce malnutritional due to iron deficiency. Ferritin protein is a iron molecule which produced by genetic engineering by transferring its gene from soybean into rice. After polishing, rice seeds provide more iron content to human being.

Artificial Cell

Craig Venter first developed synthetic cell, controlling human-made genetic instructions. The synthetic genome transplanted into an empty Mycoplasma and transfer it into a new species. This cell normally survives, reproduces itself according to an artificial DNA.

Bt cotton

Government of India approved the cultivation of genetically modified Bt-cotton in 2002. Bt-cotton delepoed through introducing bt gene from *Bacillus thueriengiensis* soil bacterium to improve resistance against different types of pest-insects like cotton bollworm, pink bollworm, spotted bollworm and tobacco caterpillar, developed by Monsanto. Bollgaurd I containts Cry1Ac and Bollgard II contains two Bacillus thuringiensis genes Cry1Ac and Cry2Ab which both control certain Lepidopteran larvae (*H. armigera* and *H. punctigera*) when they feed on it.

Decoding of bread wheat genome

In 2012, Brenchley and associates published the draft genome of bread wheat. However, genome sequence of wheat is not yet complete since these efforts have generated information only on 61% of the entire wheat genome.

Cultivation of Bt Brinjal in Bangladesh

In January 2014, Bangladesh (the first south Asian country) grows the genetically modified food crop Bt brinjal for limited commercial cultivation. However, the productivity of brinjal is severely affected by brinjal fruit and shoot borer (FSB, a lepidopteran pest) and yield losses 50 to 70 percent. Bangladesh government approved four new varieties of brinjal 'BARI Bt Brinjal-1 (Uttara), BARI Bt Brinjal-2 (Kajla), BARI Bt Brinjal-3 (Nayontar) and BARI Bt Brinjal-4.

Golden rice awarded for patent humanity

On April 21st, 2015,Golden Rice Project was awarded to save million of lives around the world. This award was shared by Ingo Potrykus and Peter Beyer along with Dr. Adrian Dubock (former scientist in Syngenta company). Golden rice developed by introducing two genes through genetic engineering for the synthesis of beta-carotene (provitamin A).

Development of waxy corn using CRISPR-Cas9 technology

Recently, Company DuPont developed waxy corn which contains starch composed of brached amylopectin residues. This developed by employing the genome editing technique CRISPR-Cas9 to knockout yhe endogenous waxy gene Wx1 which encodes for the endosperm' granule-bound starch synthase (GBBSI) that is responsible for amylase synthesis.

Key notes

- Biotechnology is usually defined as any technique that uses living organisms, or substances from those organisms, to make or modify a product, to improve plants or animals, or to develop microorganisms for specific uses.
- The key areas in which biotecimology has application to agriculture include:
  1. Multiplication of selected/elite clones from hybrid or specific parental lines and rapid production of disease-free propagates through conventional micropropagation or somatic embryogenesis
  2. Development of plant varieties, in conjunction with whole plant breeding, through genetic engineering of plant species
  3. Application of agricultural microbiology to produce microorganisms to crop cultivation
  4. Development of techniques, based on the use of monoclonal antibodies and nucleic acid probes, for the diagnosis of plant pests and diseases and the detection of foreign chemicals in food
  5. Application of genetic mapping techniques, based on the use of molecular markers, as an aid to conventional plant breeding programs

- Totipotency is ability of cells are autonomic and capable of regenerating to give a complete new plant allows a genetic change at the cellular level to become an established traits of a whole plant.
- Tissue culture is particularly important for all crop species since it reduces the labor associated with line maintenance and germplasm conservation.
- Meristem culture is used for the development of disease-free plant material.
- Anther or pollen culture facilitates in early achievement of homozygosity and is particularly useful to breed for recessive traits.

Biotechnological techniques for crop improvement

Technology	Application
Meristem and bud culture	Micropropagation for commercial purposes, genetic conservation, and exchange of material
Zygotic embryo culture	Interspecific crosses
Anther and microspore culture	Haploid plant production
Cell and tissue culture	In vitro selection, somaclonal variation, somatic embryogenesis, artificial seeds
Chromosome engineering	2n gametes for interspecific crosses
Protoplast culture	Fusion for somatic hybridization
Genetic engineering	Gene transfer
Monoclonal antibodies	Diagnosis of plant diseases

- *In vitro* culture of zygotic embryos (embryo rescue) has enabled to overcome barriers to a number of interspecific crosses from zygotic failure.
- Several interspecific hybrids have been obtained in this way between *Lycopersicunm* species, cucurbits and legumes.
- *In vitro* culture of anthers allows the regeneration of large numbers of haploid plants from which by chromosome doubling via colchicine treatment or via further *in vitro* culture, diploid plants, homozygous at all loci, can be obtained.
- Direct culture of microspores has proved to be efficient in *Brassica* species.
- Brcrssica, new germplasm derived from somatic hybridization and characterized by cytoplasmic male sterility and atrazine resistance has been released.
- Several crops, for which transformation and regeneration systems are available, can be genetically engineered.
- All the transgenic plants that have been released by genetic engineering using Agrobacterium tumefaciens as the vector.

Species/cultivar originating from in vitro and in vivo induced haploids

Species/cultivar	Source	Characteristics
Brassica napus	Microspore culture	Agronomic
Nicotiana tabacum CTDII 38 CTDI-I 40 Hl, LI	 Maternal haploid Maternal haploid Anther culture	 verticillium wilt verticillium wilt nicotine levels

Species/cultivar	**Source**	**Characteristics**
Zea mays - AC 4115	Anther culture	Agronomic
Hordeurn vulgaris Mingo, Rodeo, Craig, Etienne, Winthrop, TBR 579-5, TBC 555-1, Gwylan	Maternal haploid	Agronomic
Oryza sativa - Texmont	Anther culture	earliness, yield
Triticum aestivum - Florin	Anther culture	Agronomic

Germplasm, released or in field trials, originating from gene transfer using either somatic hybridization or genetic engineering

Species	Sources	Characteristics
Brassica napus	Somatic hybridization	Cytoplasmic male sterility, atrazine resistance
	Gene transfer	Glyphosate resistance
Nicoticana tabacum	Gene transfer	Alfalfa mosaic virus resistance
Solanum tuberosum	Gene transfer	Virus resistance
		Phosphinothricin
Lycopersicon esculentum	Gene transfer	Tobacco mosaic virus resistance
		Insect-Bacillus thuringiensis
		Fruit ripening
Linunz sp.	Gene transfer	Glyphosate resistance
Aleclicago sativa	Gene transfer	Glyphosate resistance
		Alfalfa mosaic virus resistance
Glycine max	Gene transfer	Glyphosate tolerance

Applications of RFLP marker in molecular breeding

Traits	**Technique**
Simply inherited traits	
Resistance to TMV in tomato	RFLP markers linked to Tm-2 gene on chromosome 9
Resistance to dwarf mosaic virus in maize	One major gene identified on chromosome 6
Resistance to Heterodera schactii in sugar beet	One major gene identified by using additional lines
Resistance to Globodera rostochiensis in potato	One major gene on chromosome 9
Resistance to small animal predators in Trifolium repens	Analyses of Li locus controlling the presence of limarase, which functions as a defense mechanism
Quantitative traits	
Resistance to insects due to 2- trideranone in tomato	RFLPs on three different linkage groups were correlated to this character
Soluble solids content in tomato	4 QTL increase the soluble solid concentration

Traits	Technique
Fruit pH and fruit weight in tomato	6 QTL control fruit weight and 5 QTL control fruit pH
Water use efficiency in tomato	3 RFLP markers were associated with this character

- The use of molecular markers is based on naturally occurring polymorphism whose magnitude has only recently been appreciated.
- Understand the nature of naturally occurring polymorphism is a basis for designing strategies to exploit it for applied genetic practices.
- The concept of DNA-based markers has revolutionized our ability to follow chromosome segments, including minute regions, and has led to new opportunities such as map-based cloning and directed plant breeding.
- Meristem culture used to produce virus-free plants.
- In 1994, genes for resistance to Taiwan tomato leaf curl virus, the local geminivirus strain, were successfully incorporated into three AVRDC improved tomato lines from *Lycopersicon chiiense*, using this technique.

Biotech crop highlights in 2015

- 2015 is the 20th year of commercialization of biotech crops.
- The area was increased from 1.7 million hectares in 1996 to 179.7 million hectares in 2015.
- In 2015, year-to-year growth of biotech crops was higher in developing countries at 0.9 million hectares (1%) than in industrial countries which were reduced by 3% (2.7 million hectares).
- Increase in soybean plantings in Brazil and Argentina as well as cotton plantings in Pakistan, Myanmar, and Sudan was recorded.
- Biotech crops were grown commercially in all six continents of the world.
- Of the 28 countries planting biotech crops in 2015, 19 countries planted 50,000 hectares or more to biotech crops.
- These mega-countries include the USA, Brazil, Argentina, India, Canada, China, Paraguay, Pakistan, South Africa, Uruguay, Bolivia, Philippines, Australia, Burkina Faso, Myanmar, Mexico, Spain, Colombia, and Sudan.

Global Area of Biotech Crops in 2014 and 2015 (million hectares)

Rank	Country	2015	Biotech crop
1	USA*	70.9	Maize, soybean, cotton, canola, sugar beet, alfalfa, papaya, squash, potato
2	Brazil*	44.2	Soybean, maize, cotton
3	Argentina*	24.5	Soybean, maize, cotton
4	India*	11.6	Cotton
5	Canada*	11.0	Canola, maize, soybean, sugar beet
6	China*	3.7	Cotton, papaya, poplar
7	Paraguay*	3.6	Soybean, maize, cotton
8	Pakistan*	2.9	Cotton

Rank	Country	2015	Biotech crop
9	South Africa*	2.3	Maize, soybean, cotton
10	Uruguay*	1.4	Soybean, maize
11	Bolivia*	1.1	Soybean
12	Philippines*	0.7	Maize
13	Australia*	0.7	Cotton, canola
14	Burkina Faso*	0.4	Cotton
15	Myanmar*	0.3	Cotton
16	Mexico*	0.1	Cotton, soybean
17	Spain*	0.1	Maize
18	Colombia*	0.1	Cotton, maize
19	Sudan*	0.1	Cotton
20	Honduras	<0.1	Maize
21	Chile	<0.1	Maize, soybean, canola
22	Portugal	<0.1	Maize
23	Vietnam	<0.1	Maize
24	Czech Republic	<0.1	Maize
25	Slovakia	<0.1	Maize
26	Costa Rica	<0.1	Cotton, soybean
27	Bangladesh	<0.1	Brinjal/eggplant
28	Romania	<0.1	Maize
	Total	**179.7**	

* 19 biotech mega-countries growing 50,000 hectares, or more, of biotech crops
** Rounded off to the nearest hundred thousand

- The total area is 364 million hectares, of which 49% or 179.7 million hectares were biotech.
- Globally, four principal as biotech crops soybean, cotton, maize, and canola adopted in 2015.

Crop	Biotech cultivated area in 2015 (mha)	Total cultivated area in 2015 (mha)
Soybean	92.1	111
Cotton	24	32
Maize	53.6	185
Canola	8.5	36

- The first genetically modified (GM) maize research trials have been planted on October 5, 2016 in Tanzania's Dodoma region, a semi-arid area in the central part of the country. The confined field trial aims to demonstrate the effectiveness and safety of a drought tolerant GM maize hybrid developed by the Water Efficient Maize for Africa (WEMA) project.
- Scientists from the Chinese Academy of Sciences used gene silencing technology to control *Verticillium dahliae*, a major fungal pathogen of cotton. Results showed that cotton plants infected with *V. dahliae* boost production of two microRNAs and export both to the fungal hyphae for specific silencing. Two *V. dahliae* genes were targeted by the two microRNAs that are vital for fungal virulence. *V. dahliae* strains expressing either of the genes showed resistance to the respective microRNA exhibited drastically enhanced virulence in cotton plants.

- Nabeela Abid, from the Forman Christian College in Pakistan, together with a team of researchers developed transgenic wheat expressing the *Aspergillus japonicus* phytase gene (*phy*A) in its endosperm. The transgenic lines exhibited 18–99 % increase in phytase activity and as high as 76% reduction of phytic acid in seeds, with the lowest phytic acid content observed in chapatti.
- There was no significant difference in nutrient composition between transgenic and non-transgenic seeds. However, assays for iron and zinc in the dough and bread made from the transgenic lines revealed a significant increase in iron and zinc contents.
- Phytate is a major component of wheat seeds which chelates metal ions, thus, reducing the nutritional value of wheat grains. Transgenic plants expressing heterologous phytase are expected to have enhanced degradation of phytic acid and are hypothesized to have increased mineral nutrients in their seeds.
- Yuanyuan Bu and Bo Sunwe from Northeast Forestry University in Japan, recently discovered that the expression of the Arabidopsis gene, *AtOxR*, is a response to multiple abiotic stresses. The *AtOxR*-overexpressing transgenic *Arabidopsis* had enhanced tolerance to abiotic stresses. Further analysis revealed that overexpression of *AtOxR* gene resulted in vitamin C accumulation in the transgenic plants. The results showed that the *AtOxR* gene responds to multiple abiotic stresses and overexpressing it improves tolerance to abiotic stresses by increased vitamin C content in *Arabidopsis thaliana.*
- Adrienne Forsyth described a new breeding method that integrates a TALEN-mediated induction of double strand breaks (DSBs) with a non-autonomous marker selection to insert a transgene into a selected, active region of the potato (*Solanum tuberosum*) genome. The TALEN was designed to create a DSB in the genome sequence following a constitutive promoter. The donor vector contained the gene of interest cassette and a promoter-less, plant-derived, herbicide resistance gene to act as the non-autonomous marker. Transformed potato events occurred with high frequency, with each of the resulting events consistently expressing the gene of interest.
- Viral vectors, which are engineered replication-defective viruses, recently gained an important role as delivery vehicles for donor DNA templates and designer nucleases, such as zinc-finger nucleases (ZFNs), transcription activator-like effector nucleases (TALENs) and clustered, regularly interspaced, short palindromic repeats-associated Cas9 (CRISPR–Cas9).
- Biosafety is the prevention of large-scale loss of biological integrity, focusing both on ecology and human health.
- The international Cartagena Protocol on Biosafety deals primarily with the agricultural definition. The Cartagena Protocol on Biosafety to the Convention on Biological Diversity is an international agreement on biosafety as a supplement to the Convention on Biological Diversity effective since 2003. The Biosafety Protocol seeks to protect biological diversity from the potential risks posed by genetically modified organisms resulting from modern biotechnology.
- Biosafety in agriculture, chemistry, medicine, exobiology and beyond will likely require application of the precautionary principle, and a new definition focused on the biological nature of the threatened organism rather than the nature of the threat.
- Biosafety level refers to the stringency of biocontainment precautions deemed necessary by the Centers for Disease Control and Prevention (CDC) for laboratory work with infectious materials.

- Bioterrorism is terrorism involving the intentional release or dissemination of biological agents. These agents are bacteria, viruses, or toxins, and may be in a naturally occurring or a human-modified form. For the use of this method in warfare, see biological warfare.
- A biosafety level is a set of biocontainment precautions required to isolate dangerous biological agents in an enclosed laboratory facility. The levels of containment range from the lowest biosafety level 1 (BSL-1) to the highest at level 4 (BSL-4). In the United States, the Centers for Disease Control and Prevention (CDC) have specified these levels. In the European Union, the same biosafety levels are defined in a directive. Facilities with these designations are also sometimes given as P1 through P4 (for Pathogen or Protection level), as in the term "P3 laboratory".
- Environment Protection Act (EPA, 1986) provide for the protection and improvement of environment
- Six Committees notified under Rules 1989 for genetically modified organisms as following.
- **Recombinant DNA Advisory Committee (RDAC):** The functions are of an advisory nature and involve review of developments in biotechnology at national and international levels and recommend suitable and appropriate safety regulations for India in recombinant research, use and applications from time to time.
- **Review Committee on Genetic Manipulation (RCGM)** established under the Department of Biotechnology, ministry of Science and technology is to monitor the safety related aspects in respect of on-going research projects and activities (including small scale field trials) and bring out manuals and guidelines specifying procedure for regulatory process with respect to activities involving genetically engineered organisms in research, use and applications including industry with a view to ensure environmental safety.
- **Genetic Engineering Appraisal Committee (GEAC)** established under MoEFCC is the apex body to accord notified under Rules 1989. For approval of activities involving large scale use of hazardous microorganisms and recombinants in research and industrial production from the environmental angle. The GEAC is also responsible for approval of proposals relating to release of genetically engineered organisms and products into the environment including experimental field trials (Biosafety Research Level trial-I and II known as BRL-I and BRL-II).
- **State Biotechnology Coordination Committee** (SBCC's) have a major role in monitoring. It also has powers to inspect, investigate and take punitive action in case or violations of statutory provisions.
- **District Level Committees (DLCs)** have a major role in monitoring the safety regulations in installations engaged in the use of genetically modified organisms/hazardous microorganisms and its applications in the environment.
- **Institutional Biosafety Committee (IBSC)** is established under the institution engaged in GMO research to oversee such research and to interface with the RCGM in regulating it.

MAJOR MILESTONES IN INDIAN BIOTECHNOLOGY INDUSTRY

1978 : India's first biotech firm, Biocon, was setup

1981 : Centre for Cellular and Molecular Biology setup in Hyderabad

1984 : Institute for Microbial Technology, Chandigarh was setup

1986 : Department of Biotechnology (DBT) was formed

1987 : National Institute of Immunology was setup by DBT

1989 : Bangalore Genei commenced operations

1991 : National Centre for Biological Sciences pursued R&D in molecular biology

1994 : Syngene, India's first Contract Research Organisation (CRO), started R&D services

1997 : Centre for Biological Technology (CBT) was established to focus on bioinformatics and genomics

1998 : Monsanto Research established an R&D centre for plant genomics

1998 : DBT approved MahycoMonsanto to grow Bt cotton

2007 : National Biotechnology Development Strategy launched

2009 : National Biotechnology Regulatory Authority Bill 2008 introduced in the parliament and Biotechnology Industry Partnership Programme was launched

2011 : Government approved setting up the Biotechnology Industry Research Assistance Council (BIRAC)

2012 : India ranked 12th in the world in biotech and 3 rd largest in AsiaPacific, after Japan and Korea

2013 : India became the biggest producer of Hepatitis B vaccine recombinant

2015 : DBT allocated USD0.735 million to University of Agricultural Sciences to support a national multiinstitutional project.

Multiple Choice Questions

1. Microorganism used for the production of acetone-butanol.
   (A) *Aspergilus niger*
   (B) *Clostridium acetobutylicum*
   (C) *Lactobacillus lacti*
   (D) *Bacillus amyloliquifacience*
2. Microorganism used for the production of citric acid.
   (A) *Aspergilus niger*
   (B) *Clostridium acetobutylicum*
   (C) *Lactobacillus lacti*
   (D) *Bacillus amyloliquifacience*
3. Microorganism used for the production of penicillin.
   (A) *Aspergilus niger*
   (B) *Clostridium acetobutylicum*
   (C) *Penicillium notatum*
   (D) *Bacillus amyloliquifacience*
4. Which is resistance gene against Asian soybean rust in *Cajanus cajan*?
   (A) *Xa*21 (B) *Wx*1
   (C) *GBSS*I (D) *CcRpp*1
5. What is the rate of spontaneous mutation per gene per generation?
   (A) 10^{-10} to 10^{-20} (B) 10^{-4} to 10^{-7}
   (C) 10^{-2} to 10^{-6} (D) 10 to 50
6. Mutations create the variations that are important for ....
   (A) Recombination (B) Reproduction
   (C) Crossing (D) Evolution
7. Which type of replication is generally performed in living organism?
   (A) Conservative (B) Dispersive
   (C) Semi-conservative (D) All of these
8. Why the high fidelity of replication is required?
   (A) To ensure the transmission of genes without changes
   (B) To control phenotypes over generation
   (C) To develop new traits in organism
   (D) To reduce the mutations
9. Mutation is the main source to create ....
   (A) Similarities (B) Variations
   (C) Uniformity (D) Homogeny
10. What is gene?
    (A) The segment of RNA
    (B) Codon
    (C) The segment of DNA
    (D) Short polypeptide chain
11. Which is responsible to produce phenotypic effects through replication and transcription?
    (A) Chromosome (B) Gene
    (C) Codon (D) Factors
12. Which of the following is natural process of gene transfer?
    (A) Cloning
    (B) Asexual reproduction
    (C) Apomixis
    (D) Sexual reproduction
13. A gene is responsible for high protein content in maize with low yield and susceptible to diseases?
    (A) *SalTol* (B) *PSTOL1*
    (C) *Opaque2* (D) *Sr31*
14. In which year *Agrobacterium* is first time used to gene transfer in plant genome?
    (A) 1973 (B) 1983
    (C) 2003 (D) 1963
15. A tolerant gene is a good resource in the group of rice variety Kasalath called *aus*-type originated in India?
    (A) *pi 54* (B) SUB1A
    (C) *SalTol* (D) *PSTOL1*
16. Genetically modified organism promise for controlling diseases, manufacturing synthetic drugs and biofuel production?
    (A) Mushrooms (B) Bionic bacteria
    (C) Spirulina (D) Lactic acid bacteria
17. What are the characteristics of Opaque2 maize variety?
    (A) Chalky appearance and high protein content
    (B) High amylose content and black powder in seeds
    (C) Woody stems and high nitrogen fixing roots
    (D) High photosynthetic rate and improved oil content

18. Government of India approved cultivation of genetically modified Bt-cotton in ....
(A) 2012 (B) 2002
(C) 2006 (D) 2010

19. What is characteristic of genetically modified bionic bacteria?
(A) Carry unnatural amino acids
(B) Produce high protein content
(C) It contains longest genome in prokaryotes
(D) Used in nitrogen fixation

20. Which of the following gene control tannin biosynthesis in sorghum that was cloned using molecular techniques?
(A) Sr33 (B) Tan1
(C) PSTOL1 (D) Ugg99

21. Which of the following fungal cause stem rust in wheat?
(A) *Puccinia graminis f sp. tritici*
(B) *Ustilago nuda*
(C) *Puccinia graminis f sp. striformis*
(D) *Pithium aphenidermatum*

22. How much loss was recorded by *Puccinia graminis f sp. tritici* cause stem rust in wheat?
(A) 60-70% (B) 10-20%
(C) 40-60% (D) 80-90%

23. What is called improved maize variety with much higher level of tryptophan and lysine?
(A) QTL (B) QPM
(C) PSTOL (D) Flavr Savr

24. A biotechnological approach is used to develop genetically modified bionic bacteria for controlling diseases, manufacturing synthetic drugs and biofuel production?
(A) RNA interference
(B) Protoplast culture
(C) Recombinant DNA technology
(D) Antisense RNA technology

25. Which of the following gene enable to control stem rust in wheat?
(A) *Opaque2* (B) *Sr31*
(C) *Cry* (D) *Nif*

26. Which of the following amino acids lack in maize that requires to produce balanced protein?
(A) Methionine and tyrosine
(B) Arginine and lysine
(C) Tryptophan and lysine
(D) Phenylalanine and Tryptophan

27. Bt cotton currently cultivated in an area of ...., out of total cotton area of .... in India by 2010.
(A) 11 million hectares and 19 million hectares
(B) 53 million hectares and 70 million hectares
(C) 11 million hectares and 33 million hectares
(D) 9.4 million hectares and 11 million hectares

28. Which of the following disease is controlled by *Sr31, Sr33, Sr35* and *Ugg99* genes in wheat?
(A) Stem rust (B) Karnal bunt
(C) Powdery mildew (D) Loos smut

29. A gene reduces the reactive oxidative species (ROS) activity that responsible for postharvest physiological deterioration in transgenic cassava?
(A) Mitochondrial alternative oxidase (*AOX1A*)
(B) *Pi54* (C) Cry1AC
(D) Phosphorus starvation tolerant 1 (*PSTOL1*)

30. A phenolics compound is synthesized by controlling gene *Tan1* in sorghum?
(A) Tarpene (B) Ranin
(C) Tannin (D) Saponine

31. A gene absent for phosphorus deficiency tolerance in rice variety Nipponbare?
(A) *pi 54* (B) *shat 1*
(C) *sr 33* (D) *PSTOL 1*

32. A chronic disorder in people results by lacking of tryptophan and lysine in maize?
(A) Pellagra (B) Leishmaniasis
(C) Dwarfism (D) Kala Azar

33. A gene *Sr35* for controlling stem rust is identified in which diploid wheat variety?
(A) *Triticum durum*
(B) *Triticum aestivum*
(C) *Triticum monococcum*
(D) *Agelopsis tauschii*

34. Which of the following biotechnological approach is used to transfer QPM traits to popular maize hybrid Vivek9?
(A) Marker Assisted Selection (MAS)
(B) Antisense RNA technology
(C) Protoplast culture
(D) Molecular pharming

35. Which of the following transgenic rice line contain blast resistant gene?
(A) IR 64 (B) TP-Pi54
(C) Kranti (D) IR 8

36. Marker Assisted Selection (MAS) is used to transfer QPM traits to popular maize hybrid Vivek9 at .... ?
(A) Borlogue institute for Maize and Wheat, Jabalpur
(B) CYMMIT, Mexico
(C) Vivekanand Parvatiya Krishi Anusandhan Sansthan, Almora
(D) Natonal Bureau of Plant Genetic Resources, New Delhi

37. A maize variety containing gene *Opaque2* was developed in .... with high tryptophan and lysine content through MAS.
(A) 1973 (B) 1983
(C) 2003 (D) 1963

38. Which of the following crop genome has not been sequenced till now?
(A) Wheat (B) Lentil
(C) Pigeonpea (D) Rice

39. In cassava, postharvest physiological deterioration is due to an oxidative process that is triggered by ....
(A) High temperature and high salinity
(B) Ugg99 gene
(C) ROS accumulation
(D) Chemical herbicides

40. A gene *Sr33* is identified for controlling stem rust in which diploid wheat variety?
(A) *Triticum durum*
(B) *Triticum aestivum*
(C) *Triticum monococcum*
(D) *Agelopsis tauschii*

41. Which of the following gene against stem rust absent in tetrapliod (pasta) and hexaploid (bread) wheat?
(A) *Sr33* (B) *Ugg99*
(B) *Sr35* (C) *Pi 54*

42. Which of the following enzyme is encoded by the phosphate metabolizing gene deploying phosphite metabolizing gene in tobacco plants?
(A) Tryptophan-2-oxygenase
(B) Phosphite-specific oxidoreductase
(C) PHB synthase
(D) Penicillin acylase

43. .... was created by coating different catalysts on each side of a playing-card sized silicon wafer for more efficient photosynthesis than natural leaf.
(A) Artificial immature seeds
(B) Artificial leaf
(C) Artificial immature fruit
(D) Artificial chlorophyll

44. Which of the following transgenic wheat is developed by transferring gene *Sr35* for controlling stem rust?
(A) Hexaploid (B) Diploid
(C) Tetraploid (D) Triploid

45. Which of the following organism is used to deploying a phosphite metabolizing gene in plants?
(A) Bacterial strain *Pseudomonas stutzeri*
(B) Grass-weed *Brachypodium distachyon*
(C) Tobacco *Nicotiana tobaccum*
(D) Mushroom

46. Which of the following crop is used as biofuel and a source of industrial starch?
(A) Arabidopsis
(B) Cassava
(C) Jatropa
(D) Coconut

47. Artificial uses a stable electrode material known as ....
(A) Indium tin oxide
(B) Volatile organic compounds
(C) Asbestos
(D) Aluminum rod

48. Which of the following crop is highly produced during green revolution?
(A) Rice (B) Maize
(C) Wheat (D) Sorghum

49. What is the sheer size of wheat genome measured that is 6 times the size of human genome?
(A) 50 billion base pairs
(B) 17 billion base pairs
(C) 25 billion base pairs
(D) 6 billion base pairs

50. Which of the following tissue culture technique was used to sequence of dihapliod Valencia sweet orange genome by whole genome shotgun sequencing?
(A) Meristem tip culture
(B) Anther culture
(C) Embryo culture
(D) Protoplast culture

51. Which of the following fungal pathogen cause serious disease blast in rice at global level?
(A) *Ustilago nuda*
(B) *Pithium aphenidermatum*
(C) *Magnaporthe oryzae*
(D) *Agelopsis tauschii*

52. Plants acquire phosphorus in the term of .... both as primary and secondary orthophosphate ion.
(A) Inorganic phosphate (Pi)
(B) Phosphoric acid (H_3PO_4)
(C) Phosphate ion (PO_4^{-3})
(D) Hydrogen phosphate (HPO_4^-)

53. Which of the following gene confer the disease blast in rice?
(A) *Sr33* (B) *Tan1*
(C) *Pi-54* (D) *Ugg99*

54. Which of the following method is used to improve popular hybrid variety HHB-67 of Bajra for resistance against downy mildew due to lost disease resistance?
(A) Mutation
(B) MAS
(C) Bulk segregation method
(D) TILLING

55. Which of the following rice line is developed as blast resistance line through marker assisted selection developed at IARI, New Delhi?
(A) Pusa basmati 1 (B) IR64
(C) Kranti (D) IR36

56. Which of the following genes is *Pup*1-specific protein kinase gene responsible for phosphorus deficiency tolerance in rice variety Kasalath?
(A) Blast resistance gene (*pi 54)*
(B) Submergence-tolerance gene (*SUB1A*)
(C) Phosphorus uptake 1 gene (*Pup 1*)
(D) Phosphorus starvation tolerance gene (*PSTOL 1*)

57. Which of the following major terpene is produed by a wild tomato species *S. habrochaites* in its chloroplast against several insect pests?
(A) Ferulic acid
(B) 7-epizingiberene
(C) Quercetin
(D) Azadirehtin

58. Which of the following is soil bacterium that contain Bt gene for controlling insects in crops by producing insecticidal crystal proteins?
(A) *Agrobactrium Tumefecience*
(B) *Bacillus subtilis*
(C) *Bacillus thuringiensis*
(D) *Agrobacterium rhizogenes*

59. What is the medium of larval gut of insect that damaged Bt proteins for insect control in crops?
(A) Acidic (B) Alkaline
(C) Basic (D) Neutral

60. Which of the following are major insects in brinjal for which transgenic Bt brinjal was developed?
(A) Shoot borer and pod borer
(B) Fruit borer and white fly
(C) Fruit borer and Shoot borer
(D) Leaf miner and ball worm

61. Which crystal protein is produced by Bt gene for insect control in crop?
(A) Cry1Ac (B) Interferon
(C) Humulin (D) Interleukin

62. Which of the following is widely used as energy currency in all metabolic processes in plant cell?
(A) FAD (B) ATP
(C) NADP (D) GTP

63. What is the function of silicon wafer in artificial leaf created by coating different catalysts on each side of a playing-card sized silicon wafer?
(A) Produces crystal protein Cry1AC
(B) Control insect attacking on crop
(C) Absorbs sunlight and passes charged electrons
(D) Protect from high temperature and humidity

64. A dual fertilization and weed control system by generating transgenic plants that can use .... as a sole phosphorus source.
(A) Phosphate (B) Phosphoric acid
(C) Phospholipid (D) Phosphite

65. *Bacillaus thuringiensis* (Bt) synthesizes insecticidal protein that are highly effective against the .... of insect pests.
(A) Larval stages (B) Adult stages
(C) Pupa stages (D) Nymph staged

66. For which insect in cotton has not been developed transgenic Bt cotton?
(A) Pink Bollworm
(B) Spotted bollworm
(C) Fruit sucking fly
(D) *Helicoverpa armigera*

67. How many percentage of total edible oil production constitute cotton oil?
(A) 20% (B) 37.5%
(C) 13.7% (D) 1.5%

68. Which of the following enzymatic pathway is significantly upregulated by GalUR gene involved in vitamin C metabolism in sweet orange fruit?
(A) Galacturonase pathway
(B) α-amylase pathway
(C) Aminoacylac pathway
(D) Glucoamylase pathway

69. Which of the following popular hybrid variety of Bajra has been again released by improving through by MAS for resistance against downy mildew?
(A) HHB-67
(B) HHB-67 improved
(C) HHB-76
(D) HHB-76 improved

70. Which of the following cell organelle involved in terpene pathway for enhancing tolerance against several insect pests in a wild tomato species S. habrochaites?
(A) Mitochondria (B) Golgy body
(C) ER (D) Chloroplast

71. What is not a fundamental advantage of MAS compared to conventional phenotypic selection?
(A) MAS can be applied at seedling stage
(B) Single plant can be selected with high reliability
(C) MAS is based on DNA markers that is more reliable
(D) DNA markers are influenced by environmental factors

72. Hybrid variety HHB-67 of Bajra is resistant to downy mildew disease and ....
(A) Terminal drought tolerance
(B) Heat tolerant
(C) Cold tolerant
(D) Salinity tolerant

73. Which of the following recently isolated gene is responsible for shattering in mutant rice?
(A) *tan 1* (B) *sr 33*
(C) *shat 1* (D) *pi 54*

74. Which of the following method is used for artificial selection of non-shattering rice with isolating shattering responsible gene *shat 1*.
(A) Hybridization (B) MAS
(C) Backcrossing (D) Mutation

75. Which of the following element is the component of ATP used as energy currency in plan?
(A) Boron (B) Molybdenum
(C) Phosphorus (D) Calcium

76. Which of the following gene to develop phosphorus efficiency has been identified from traditional rice varierty Kasalath from India?
(A) *PSTOL 1* (B) *sr 33*
(C) *shat 1* (D) *pi 54*

77. Which of the following quantitative trait loci was identified and sequenced for phosphorus deficiency tolerant in rice variety Kasalath?
(A) *pi 54* (B) SUB1A
(C) *Pup 1* (D) *PSTOL 1*

78. Which of the following pathway is modified for enhancing tolerance against several insect pests in cultivated tomatoes?
(A) Terpene pathway
(B) Beta-galacturonase pathway
(C) Aromatic amino acid pathway
(D) Bt response pathway

79. What is the total estimated size of sweet orange that was sequenced by whole genome shotgun sequencing method?
(A) 367 Mb (B) 412 Mb
(C) 16,000 Mb (D) 130 Mb

80. One dalton is equal to
(A) 1 mol (B) 1 g
(C) 1 g/mol (D) 1 kg/mol

81. What is the average molecular weight of three subsequent bases in dsDNA?
(A) 1000 dalton (B) 5000 dalton
(C) 110 dalton (D) 2000 dalton

82. What is the average molecular weight of three subsequent bases in ssRNA?
(A) 1000 dalton (B) 5000 dalton
(C) 110 dalton (D) 2000 dalton

83. What is the molecular weight of one amino acid in a polypeptide chain?
(A) 1000 dalton (B) 5000 dalton
(C) 110 dalton (D) 2000 dalton

84. If 300 amino acid protein weight is 33 kg/mol. The stored information at the DNA level has a mass
(A) 9900 kg/mol (B) 600 kg/mol
(C) 999 kg/mol (D) 900 kg/mol

85. Which of the following amino acids are coded by only one codon?
(A) Lysine and Cystein
(B) Methionine and Tryptophan
(C) Phenylalanine and Arginine
(D) Histidine and Glycine

86. Which genome-editing technology was used to develop a waxy corn which contains starch exclusively composed of the branched polysaccharide amylopectin by DuPont company?
(A) RNA interference
(B) CRISPR-Cas9
(C) Gene gun
(D) Gene chip

87. A method of breeding most commonly used when the desired variation is required to be induced in a vegetatively propagated crop is
(A) Polyploidy
(B) Pedigree method of breeding
(C) Mutation breeding
(D) Backcross method of breeding

88. The plant 'Brassica napus' has been modified to produce ....
(A) Vitamin A
(B) Lysine rich seeds
(C) Hirudin
(D) Insulin

89. A promising rapeseed gene *BnGRF*2 that enhances seed oil yield through regulating cell number and ....
(A) Respiration
(B) Vernalization
(C) Photoperiod
(D) Photosynthesis

90. Crossing between different parents is known as ....
(A) Selfing (B) Hybridization
(C) isogamy (D) Heterogamy

91. Genetically most pure seed ....
(A) Commercial seed
(B) Registered seed
(C) Inproved seed
(D) Nucleus seed

92. Inbreeding depression is observed in ....
(A) Often self pollinated
(B) Self pollinated
(C) Cross pollinated
(D) None of the above

93. Clonal selection is practiced in
(A) Vegetative propagated crop
(B) Cross pollinated
(C) Self pollinated crop
(D) None of all

94. A mutation in one of the stem cell producing gene CLAVATA3 leads to ....
(A) Early ripening in tomato
(B) Giant tomato fruit
(C) Enhances seed oil yield in mustard
(D) Starch quality improvement in rice

95. .... is an ancient method of crop improvement
(A) Hybridization (B) Introduction
(C) Selection (D) Mutagenesis

96. Floral organ development is controlled by overlapping expression of 'A' class, 'B' class and 'C' class genes in different whorls. In an *Arabidopsis* mutant, the flowers had sepals, sepals, carpels and carpels in the four whorls. Mutation in which one of the following is the cause for the mutant phenotype?
(A) 'A' class gene alone
(B) 'B' class gene alone
(C) 'A' and 'B' class genes
(D) 'C' class gene alone

97. Which of the following is natural process of gene transfer?
(A) Cloning
(B) Asexual reproduction
(C) Apomixis
(D) Sexual reproduction

98. Endochitinase can be used as an
(A) Antibacterial (B) Antiviral
(C) Antifungal (D) Antinematodal

99. The herbicide, dichlorophenyldimethylurea, is an inhibitor of ....
(A) Shikimate pathway for biosynthesis of aromatic amino acids
(B) Electron transport from P700 to ferredoxin
(C) Branched chain amino acid pathway
(D) Electron transport from P680 to P700

100. FlavrSavr is a variety of
(A) Tomato (B) Potato
(C) Brinjal (D) Cotton

101. Which of the following is an herbicide resistant gene?
(A) Bar (B) Hpt
(C) Chitinase (D) Oxidase

102. Which of the following is a green tissue specific promoter?
(A) Pea vicilin
(B) Arabidopsis small subunit of RUBISCO
(C) 35s CaMV
(D) Bean phytahemaglutinin

103. In *Agrobacterium* mediated plant transformation which one of the following is used to check the *Agrobacterium* growth
(A) PPT
(B) Kanamycin
(C) Timentin
(D) Pencillin

104. The gene transferred from *E.coli* to tobacco to increase the level of manitol is
(A) Manitol dehydrogenase
(B) Manitol carboxylase
(C) Manitoloxygenase
(D) Manitol hydrolase

105. .... is a science that began only in the last century. It is concerned with the control and minimization of risks from biological technologies.
(A) Biosafety (B) Biology
(C) Biotechnology (D) Bioethics

106. What is the basic foundation of biosafety while assuring the progress of biotechnology?
(A) To protect human health
(B) To protect animal health
(C) To save the environment
(D) All of the above

107. Which of the following U.S. agency examine plants and plant products and regulates the development and field testing of genetically engineered plants and microorganisms?
(A) CFIA
(B) CTNBio
(C) USDA
(D) CONABia

108. What is meaning of APHIS that provides procedures for obtaining a permit or for providing notification of the intent to field test prior to importation, interstate movement, or release into the United States?
(A) Animal and Plant Health Inspection Service
(B) Agriculture patent for health information service
(C) Agency for Plant Health Inspection Service
(D) Agency for patent for health information service.

109. The USDA regulates the development and field testing of genetically engineered plants, microorganisms, and certain other organisms under the authority of ....
(A) Federal Plant Pest Act
(B) Plant Quarantine Act
(C) Patent Act
(D) Both (A) and (B)

110. U.S. agency Animal and Plant Health Inspection Service (APHIS) regulations grant permission based on the genes that ....
....
(A) Naturally transferred into plants against pests
(B) Control plants from pests by the transgenes
(C) Control plants from abiotic factors by the transgenes
(D) Control plants from diseases by the transgenes

111. What is meaning of FFDCA
(A) Federal Foodstuff, Drug, and Committee Act
(B) Federal for Fundamental Drug Commission Act
(C) Fundamental Foodstuff, Drug, and Cosmetic Act
(D) Federal Food, Drug, and Cosmetic Act

112. Which of the following naturally soil bacterium is considered a biological pesticide?
(A) *Bacillus thuringiensis*
(B) *Bacillus subtilis*
(C) *Phythium aphenidermatum*
(D) *Pseudomonas citri*

113. Physical containment relates to which of the following?
(A) Containment equipment
(B) Laboratory design
(C) Special procedures
(D) All of the above

114. Which of the following a transgenic crop variety had to be withdrawn from cultivation deaths due to allergy?
(A) Maize (B) Soybean
(C) Tomato (D) *Brassica napus*

115. Which of the following term is not used as synonymous of plant introduction?
(A) Field testing
(B) Deliberate release
(C) Contained use
(D) Deliberate introduction

116. The biosafety guidelines are developed to insure an adequate level of protection during which of the following activities of modern biotechnology?
(A) Safe transfer (B) Handling
(C) Use of GMOs (D) All of the above

117. In case of which of the following products, an imperial approach based on data obtained through experimentation about the source of potential hazards has proved satisfactory?
(A) Pesticides (B) Drugs
(C) Vaccines (D) All of the above

118. The α-amylase inhibitor gene was transferred to modify which of the following trait?
(A) Insect resistance
(B) Amino acid balance
(C) Food digestibility
(D) Fungal resistance

119. The α-amylase inhibitor from *Phaseolus vulgaris* (Razma) caused allergy when expressed in which of the following crop by Australian scientists?
(A) Soybean (B) Tomato
(C) Peas (D) Maize

120. Which of the following is not correct about *cry* protein?
(A) Cry proteins are rapidly degraded by the stomach juices of vertebrates
(B) They could have harmful affect on non target insect species
(C) Cry proteins leak into soil from transgenic maize roots where they persist for over three weeks
(D) Cry protein present in the soil become associated with soil clay/humus, and become prone to microbial degradation

121. The first recombinant microorganism (Tn5-containing strain of Pseudomonas fluorescence) was released in which of the following country?
(A) USA (B) Canada
(C) Netherland (D) Germany

122. Which of the following noninfectious illness is associated with production of steroids?
(A) Increased body weight
(B) Feminization of males
(C) Hypertension
(D) All of the above

123. Which of the following noninfectious illness is associated with endotoxin production?
(A) Flu symptoms (B) Dermatitis
(C) Asthma (D) Cholera

124. Non infectious illnesses like allergic asthma and dermatitis are associated with which of the following production processes?
(A) Brewing (B) Single cell protein
(C) Antibiotics (D) Drugs

125. Crop yield can be stabilized by introducing which of the following genes in transgenic varieties?
(A) Genes for improved nutritional quality
(B) Genes for better tolerance/resistance to biotic and/or abiotic stresses
(C) Genes for herbicide resistance
(D) All of the above

126. Strategy of temporal and tissue-specific expression for preventing transgene transfer from transgenic varieties of crops is based on which of the following techniques?
(A) Tapetum-specific promoters allow expression only when the suppressor is present
(B) Terminator technology
(C) Recoverable block of function
(D) Inducible promoters allow expression only when the inducer is present

127. Seed ferility can be reduced by using which of the following techniques?
(A) Terminator technology
(B) Transgenic technology
(C) Apomixis
(D) Parthenogenesis

128. For minimizing pollen-mediated gene-flow from transgenic varieties of *Solanum tuberosum* how many meters of isolation distance is recommended?
(A) 10 (B) 20
(C) 40 (D) 30

129. A careful evaluation of toxin encoding genes for toxic action on non-target insects is warranted since it seems to be a function of which of the following?
(A) Target disease (B) Crop species
(C) Target insect pest
(D) The toxin encoding gene

130. Which of the following is not correct about bromoxynil?
(A) It is a non-biodegradable herbicide
(B) It causes birth defects in laboratory animals
(C) It is toxic to fish and may cause cancer in humans
(D) Bromoxynil is absorbed through the skin

131. Which of the following is not correct about bromoxynil?
(A) Transgene isolation from a food crop guarantees the food safety of the transgene product
(B) Acute toxic effect depends on the concentration of the compound in question
(C) Psoralen is known to be toxic carcinogenic and mutagenic
(D) Any outstanding insect resistance whose biochemical basis is not known should be considered as the prime candidate for toxicological evaluations

132. Which of the following toxic effects is not associated with proteinase inhibitors?
(A) Pancrease enlargement
(B) Adenoma
(C) Hyperplasia
(D) Weight loss

133. Which of the following is known to be a food allergen?
(A) Wheat germ agglutinin
(B) Soybean Kunitz trypsin inhibitor
(C) Early ripened tomatos
(D) Peanut lecitins

134. The rate of gene flow is not affected by which of the following factors?
(A) Mode of seed dispersal
(B) Composition of gene
(C) Mating system
(D) Mode of pollination

135. Who must contact the Department of Biological Safety prior to beginning work on their project?
(A) Principal Investigators who work with recombinant DNA
(B) Principal Investigators who work with infectious agents
(C) Principal Investigators who work with potential biohazards
(D) All of these Principal Investigators

136. Which committee(s) must approve research involving on human tissue from the hospital operating room, viral vectors, and mouse cell lines from ATCC?
(A) IRB, IBC and IACUC
(B) IRB and IBC
(C) IBC
(D) IRB and IACUC

137. Which committee(s) must approve research involving on mice that will not be exposed to any recombinant DNA, infectious agents, or potential biohazards?
(A) IRB and IACUC (B) IBC and IACUC
(C) IACUC (D) IBC

138. Which type of disposable glove provides the greatest tear and puncture resistance?
(A) Nitrile (B) Latex
(C) Rubber (D) Plastic

139. What type of lotion can be used with latex gloves to prevent degradation of latex?
(A) Petroleum Based Lotions
(B) Non-Petroleum Based Lotions
(C) Moisturizing Hand Lotions
(D) Detergent based lotions

140. A cabinet that protects both the user and the material inside the cabinet and that the exhaust is HEPA filtered is ....
(A) Biological safety cabinet
(B) Fume hood
(C) Growth chamber
(D) Incubation chamber

141. PPE is ....
(A) Personal Protective Equipment
(B) Public Protective Equipment
(C) Possibly Protective Equipment
(D) Probably Protective Environment

142. Surgical masks and N-95 respirators provide the same level of protection against ....
(A) Infectious aerosols (B) Fog
(C) Acidic rain (D) Pond water

143. .... -based hand sanitizers are just as effective as hand-washing for removal of a broad-spectrum of infectious organisms.
(A) Toluene (B) Alcohol
(C) Acetone (D) Benzene

144. Which of the following procedures could generate aerosols?
(A) Sonicating tissue culture cells
(B) Inoculating animals
(C) Operating a flow cytometer
(D) All of the above

145. Which of the following items are not considered sharps?
(A) Needles
(B) Scalpels
(C) Forceps and spatula
(D) Microscope slides and cover slips

147. Transgenic plant containment involves preventing accidental release into the environment of
(A) Seeds (B) Pollen
(C) Whole Plants (D) All of the above

148. .... can not be utilized as the sole form of decontamination in a biological safety cabinet.
(A) UV light
(B) Surface sterilization
(C) Flame sterilization
(D) None of the above

149. Which of the following statements about UV light is true?
(A) The Department of Biological Safety highly recommends the use of UV light in BSCs.
(B) UV light sources in the lab are no more dangerous than being outside on a sunny day.
(C) Organisms exposed to UV light have no mechanisms to repair the damage.
(D) UV light can reflect off cabinet surfaces and out of the BSC if the sash is not fully closed.

150. Which of the following practices should not be utilized when working in a biological safety cabinet?
(A) Disinfect the work surface of the BSC before and after work
(B) Disinfect all items which go into and come out of the BSC
(C) Do store any items in the BSC
(D) Allow the BSC to operate before work begins and after work ceases

151. Ethanol is not a preferred disinfectant because:
(A) It cannot be used in a BSC
(B) It has a long contact time and a fast evaporation rate
(C) It does not have broad spectrum disinfectant capability
(D) All of the above

152. In which year, DBT brought out separate guidelines for carrying out research in transgenic plants including the guidelines for toxicity and allergenicity of transgenic seeds, plants and plant parts.
(A) 2000 (B) 1998 (C) 1992 (D) 2004

153. Which of the following committee has been set up to monitor over a period of time and the impact of transgenic plants on the environment by the RCGM?
(A) Gene Technology Community Consultative Committee
(B) Gene Technology Technical Advisory Committee
(C) Monitoring cum Evaluation Committee
(D) Gene Technology Ethics Committee

154. Which of the following practices are allowed in the laboratory?
(A) Eating and Drinking
(B) Applying Cosmetics
(C) Handling Contact Lenses
(D) Hand washing

155. Which of the following statements about Personal Protective Equipment (PPE) is not correct?
(A) PPE should be worn and stored only inside the laboratory
(B) PPE should be kept in personal room outside laboratory
(C) PPE should be chosen based upon the work being completed
(D) Employees utilizing PPE should be properly trained

156. How should biological materials that need to be transported from the lab to another location be handled?
(A) Wear a lab coat and transport materials in your pocket
(B) Wear gloves and carry the material in your hands
(C) Seal materials in a leak-proof, shatter-resistant secondary container
(D) Cells in cell culture flasks and dishes are fine for transport

157. When autoclaving biohazardous waste it is not important to
(A) Overfill autoclave bags
(B) Record autoclave use in the user log
(C) Place autoclave bag in a secondary container
(D) Add water to the autoclave bag

158. An ideal container for autoclave bags containing biohazardous waste would not
(A) Be hard-sided
(B) Have a foot operated lid
(C) Be easily decontaminated
(D) Use only one time

159. Which of the following materials may be autoclaved?
(A) Infected animal carcasses
(B) Sharps
(C) Biological infectious waste
(D) Explants

160. Open flames are not allowed in a biological safety cabinet because:
(A) They disrupt the normal airflow patterns designed to protect user and product
(B) They can damage HEPA filters
(C) They can cause a fire in the BSC
(D) All of the above

161. A good alternative to using a Bunsen burner in the BSC is a:
(A) Bacticinerator
(B) Wrist-operated Touch-O-Matic
(C) Blow Torch
(D) Both (A) and (B)

162. In order to avoid wearing gloves outside the laboratory you should:
(A) Never leave the lab
(B) Not work with infectious agents
(C) Use a disinfected secondary container for transport
(D) Have someone else transport your materials

163. Biological Safety Cabinets must be certified ....
(A) Daily (B) Monthly
(C) Yearly (D) Once

164. Which of the following is not correct about lectins?
(A) They are polysaccharides
(B) They are involved in defense mechanism of plant
(C) Several lectins have been identified as food allergen
(D) α-amylase/trypsin inhibitor family proteins as lectins are potential allergen

165. According to biosafety guidelines of India physical containment envisages to limit the spread of dangerous microorganisms by which of the following?
(A) Laboratory design and facilities
(B) Laboratory practice
(C) Safety equipments
(D) All of the above

166. Transgenic plants could serve as vectors for transgene dissemination through which of the following?
(A) Seed
(B) DNA released from debris of transgenic plants
(C) Pollen (D) All of the above

167. Which of the following is not correct?
(A) In European Union, the maximum permissible limit for GMO content in conventional food products and food ingredients is 1% at the marketing stage
(B) All food that contain more than 1% GMOs related to the different ingredients have to be labeled
(C) Volunteers are not likely to pose problems in crops like rapeseed and mustard
(D) In case of organic farming, utilization of presence of GMOs is excluded

168. Which of the function is not related to the Review Committee on Genetic Manipulation (RCGM) under the Department of Biotechnology?
(A) To bring out manuals of guidelines specifying producers for regulatory process on GMOs in research, use and applications including industry with a view to ensure environmental safety
(B) To not authorize imports of GMOs/transgenes for research purposes
(C) To lay down producers for restriction or prohibition, production, sale, import & use of GMOs both for research and applications
(D) To review all ongoing r-DNA projects involving high risk category and controlled field experiments

169. Which of the following committee functions as a body under the Ministry of Environment and Forests and is responsible for approval of activities involving large scale use of hazardous microorganisms and recombinant products in research and industrial production from the environment angle?
(A) Genetic Engineering Approval Committee (GEAC)
(B) Review Committee on Genetic Manipulation (RCGM)
(C) Institutional Biosafety Committee (IBSC)
(D) The Recombinant DNA Advisory Committee (RDAC)

170. Which of the following committee is considered to be smallest authoritative unit to monitor the safety regulations in installations engaged in the use of GMOs in research and applications?
(A) Review Committee on Genetic Manipulation
(B) State Biotechnology Coordination Committee
(C) District Level Committee
(D) Institutional Biosafety Committee

171. Which of the following is made as separate section on transgenic plant varieties for environment and biosafety before their commercial release as per the regulations on guidelines of the EPA in 2002?
(A) Seed Policy
(B) Drug policy
(C) Cosmetics policy
(D) Public policy

172. What is EPA?
(A) Environment Programme Act
(B) Exclusive Programme and Protection Agency
(C) European Protection Agency
(D) Environmental Protection Act

173. The Patent Act was originally issued to protect
(A) Any art, machine or material that is new and useful
(B) Any revised research material which previously has been used
(C) Only for human drugs of cancer and hepatitis B
(D) Only genetically modified foods

174. In 1988, Harvard University received the first patent for a transgenic animal, a mouse with a human gene for cancer. it has great usefulness in the development of anticarcinogenic drugs.
(A) A chimeric mouse
(B) A carcinogenic mouse
(C) An oncomouse
(D) A nude mouse

175. Which of the following book is written by a famous anthropologist Carleton Stevens Coon in 1962?
(A) Origin of species
(B) Origin of Races
(C) Big bang theory
(D) Secret of life

176. A gene from a soil-borne bacterium engineered into corn to provide resistance to
(A) A devastating diseases and insect species
(B) A devastating diseases
(C) A devastating virus
(D) A devastating insect species

177. What is called the exchange of genetic information among individuals, populations, or species, with the preservation of that genetic information in the generations?
(A) Gene bank (B) Gene flow
(C) Gene myth (D) Gene escape

178. Loss of genetic diversity and variability caused by either natural or man-made processes.
(A) Genetic diversity
(B) Genetic pollution
(C) Genetic erosion
(D) Genetic vulnerability

179. Condition of low genetic diversity that might influence an organism to susceptibility to pests.
(A) Genetic diversity
(B) Genetic pollution
(C) Genetic erosion
(D) Genetic vulnerability

180. The correction of a genetic deficiency in a cell, tissue, or organ by the addition of a normal, functioning copy of a gene and its insertion into the genome.
(A) Genetic engineering
(B) Gene therapy
(C) Gene silencing
(D) Gene escape

181. Genetically modified crop variety has a high content of □-carotene is a vitamin A precursor and developed by Ingo Potrykus and Peter Beyer at the Swiss Federal Institute of Technology and University of Freiburg, Germany.
(A) Golden Rice
(B) Bt cotton
(C) Dwarf wheat
(D) Quality protein maize

182. The process whereby a compound is decomposed by natural biological activity and a substance is one in which a microbe can reduce it to CO_2 and H_2O by mineralization.
(A) Bioremediation
(B) Biosafety
(C) Biodegradation
(D) Biodiversity

183. .... deals with the evaluation, control, and minimization of risks from biotechnology and concerns with risks to human and animal health and to the environment.

(A) Bioremediation
(B) Biosafety
(C) Biodegradation
(D) Biodiversity

184. .... is the term used to describe the offensive employment of biological substances or toxins with the objective of causing harm to an individual or a group of individuals.

(A) Bioremediation
(B) Biosafety
(C) Biodegradation
(D) Bioterrorism

185. .... can be defined as the deliberate and cruel use of biological or chemical agents as weapons against agriculture, and more specifically, the food supply.

(A) Bioremediation (B) Biosafety
(C) Agroterrorism (D) Biodegradation

186. .... seeks to study the moral vision, decisions, and politics of human behavior in relation to biological phenomena or events.

(A) Bioethics (B) Biosafety
(C) Biodegradation (D) Bioterrorism

187. What is the aim of biosafety?

(A) Develop a list of facilities and experts who are familiar with the pathogens.
(B) To study, monitor, and control the potential impacts of biotechnology
(C) Patenting any art, machine or material that is new and useful
(D) Improve the understanding of host–pathogen interaction at a molecular level.

188. Which of the following trait is not improved in the plant species through transgenic technology?

(A) Yield improvement
(B) Insect and disease resistance
(C) Herbicide tolerance
(D) Nutritional quality

189. The transgenic Bt corn contains a protein that failed an *in vitro* digestion test indicating

(A) Cancerous disease in human
(B) Less nutritional quality
(C) Yield loss
(D) The potential of becoming an allergen

190. Gene escape or gene flow from transgenic crops is an often-discussed problem related to biotechnology in the development of superweeds and loss of natural diversity is an important topic. This phenomenon cannot occur in following one way that

(A) When the transgenic plant itself becomes a weed
(B) When the transgene is transferred by crossing from an adapted variety to a wild relative
(C) When the transgene is sexually transmitted to other species and organisms
(D) When the transgene is asexually transmitted to other species and organisms

191. Match the correct approximate genome size with crop

i.	Rice	p.	2400Mbp
ii.	Wheat	q.	135 Mb
iii.	Maize	r.	430Mb
iv.	Arabidopsis	s.	17 Gb

(A) i-p, ii-q, iii-r, iv-s
(B) i-r, ii-s, iii-p, iv-q
(C) i-s, ii-r, iii-p, iv-q
(D) i-r, ii-q, iii-s, iv-p

192. .... estimated that incompatibility occurs in more than 3,000 species among 20 families of the flowering plants.

(A) D. De Winton
(B) M.B. Hughes
(C) W.H. Gabelman
(D) E.M. East

193. .... provides the best documented example of long time effects of selection.

(A) Sugarcane (B) Sweet sorghum
(C) Sugarbeet (D) Sugar maple

194. .... deteriorates drastically upon inbreeding.

(A) Carrot (B) Okara
(C) Brinjal (D) Broccoli

195. .... in Ladino clover is governed by two dominant alleles, one produces cyanogenic glycoside, while the other releases the enzyme to catalyze its release from gloycoside.
(A) Oxalic acid (B) IAA
(C) HCN (D) NAA

196. .... are examples of protandrous species.
(A) Avocados (B) Walnuts
(C) Carrots (D) None of these

197. Origin of Species by means of Natural Selection' was given by ....
(A) Lamark (B) Linnaeus
(C) Charles Darwin (D) Erasmus

198. "Experiments in Plant Hybridization" a paper of .... was first published in the year ....
(A) Morgan, 1905
(B) Linnaeus, 1787
(C) Mendal, 1866
(D) Hugo de Veries, 1896

199. "If a gene is represented in an infinitely large random-mating population by adaptively neutral alleles, A and a, in the ration of qA: (1-q)2 aa = 1" is known as ....
(A) Stanford-Suneson law
(B) Weibe-Thomas law
(C) Briggs-Harlan law
(D) Hard-Weinberg law

200. 2n - 1 is a ....
(A) Monosomic (B) Transonic
(C) Tetrasomic (D) Alloploid

201. 2n denotes ....
(A) Number of homozygote produces by selfing a F1
(B) Number of types of gametes produced by F1
(C) Number of types of heterozygote produced by selling F1
(D) None of the above

202. A cell or an organism having one chromosome pair in addition to the normal somatic compliment of the species is ....
(A) Monosonic (B) Nullisomic
(C) Tetrasomic (D) Trisomic

203. A character dermined by many genes and does not show discrete variation is known as ....
(A) Qualitative character
(B) Oligogenic character
(C) Quantitative character
(D) Multiple-allelic character

204. A complex of communities of plants and animals of a region existing under identical climatic conditions is called as
(A) Biome (B) Ecosystem
(C) Ecological complex
(D) Environment

205. A condition in which pollination and fertilization take place in an unopened flower is termed as
(A) Homogamy (B) Apogamy
(C) Polygamy (D) Cleistogamy

206. A criss-cross inheritance is shown by
(A) Sex-influenced character
(B) Polygamy
(C) Sex-linked character
(D) Sex-quantitative character

207. A cybrid is
(A) A sex hybrid
(B) A hybrid formed by cell fusion
(C) A plant produced by cell culture
(D) Synonym to hybrid

208. A dry, one chambered and one seeded fruit developing from a superior ovary with the hard and woody pericarp is
(A) Achene (B) Caryopsis
(C) Composite fruit (D) None of the above

209. A generative nucleus in the male gametophyte divides .... to form two male gametes.
(A) Mitotically
(B) Meiotically
(C) Mitotically and Meiotically
(D) Amitotically

210. A homozygous line developed by self-fertilization in a cross-pollinated crop is termed as
(A) Pureline (B) Inbred
(C) Clone (D) Mutant

211. A hybrid is always
(A) Heterozygous
(B) Homozygous
(C) Both of the above
(D) Often homozygous

212. A large unit of natural vegetation in an area under identical condition (climatic) is called
(A) Plant formation (B) Association
(C) Community (D) Climax

213. A maintainer line is
(A) Male fertile and can restore fertility
(B) Male sterile but can restore fertility
(C) Male sterile but cannot restore fertility
(D) Male fertile but cannot restore fertility

214. A male child would be born to parents if
(A) The father is healthier than the mother
(B) The mother eats well during pregnancy
(C) The genetic composition of the child includes XY combination of chromosomes
(D) The genetic composition of the child includes XX combination of chromosomes

215. A male gametophyte in angiosperms contains
(A) One gametes only two male gametes only
(B) Two male gamete only
(C) Two male gametes and one tube nucleusnucleus
(D) One male gamete and one tube nucleus

216. A mendelian population means
(A) Self-pollinated population
(B) Inbred population
(C) Random-mating population
(D) Hybrid population

217. A method of breeding most commonly used when the desired variation is required to be induced in a vegetatively propagated crop is
(A) Polyploidy
(B) Pedigree method of breeding
(C) Mutation breeding
(D) Backcross method of breeding

218. A plant bearing both male and female flowers is said to be
(A) Dioecious (B) Monoecious
(C) Polygamous (D) None of above

219. A plant breeder is interested to control pollination to
(A) Prevent cross-pollination
(B) Control pollination
(C) None of these
(D) Both of these

220. A plant cell when put into a hypotonic solution, it unlike animal cell does not brust because
(A) It has no semi-permeable membrane
(B) It has inelastic cell wall
(C) All both (A) & (B)
(D) None of the above

221. Which is the right option for the tallest and the smallest Gymnosperm plant ?
(A) *Eucalyptus* and *Zamiapygmea*
(B) *Wolffia globosa* and *Eucalyptus*
(C) *Sequoia sempervirens* and *Zamia pygmea*
(D) *Sequoia sempervirens* and *Wolffia globosa*

222. Which one is odd for species diversity?
(A) α-diversity (B) β-diversity
(C) γ-diversity (D) λ-diversity

223. How many biosphere reserves are present in India?
(A) 41 (B) 34
(C) 14 (D) 43

224. Which micro organismis responsible for synthesis of antibiotics?
(A) Bacteria (B) Virus
(C) Fungus (D) Algae

225. Which scientist has classified species diversity?
(A) Thoeprestus (B) Lineus
(C) Whittaker (D) Treshaw

226. Which group is meant for Endemic species of birds?
(A) Nilgiri pipit, Rofous babbler, Lesser-Florican
(B) Lesser-Florican, Nilgiri wood pigeon, Malabar parakeet
(C) Malabar parakeet, Niligiri pipit, Rofousbabbler
(D) All the above

227. How many Indian plant species are used to extract essential oils and scents?
(A) 50 (B) 500
(C) 50,000 (D) 5000

228. Which is the correct option?
(A) There is chance in Natural selection in evolutionprocess due to alpha biodiversity
(B) There is chance in Natural selection in process of evolution due to genetic diversity
(C) There is chance in Natural selection in process of evolution due to Ecosystem biodiversity
(D) There is chance in Natural selection in process of due to biocommunitydiversity

229. Which is the state plant of Gujarat?
(A) Polyalthia (B) Prosopis
(C) Ficus (D) Neem

230. Which can be used for cryopreservation at 196°C temprature?
(A) Liquid N_2 (B) Free N_2
(C) Liquid CO_2 (D) Solid N_2

231. How manyprotected areas are present in India?
(A) 89 (B) 581
(C) 492 (D) 34

232. What can be done by the number of species in habitat?
(A) Measurement of species diversity in habitat
(B) Measurement of gene diversity ofspecies in habitat
(C) Measurement ofinteraction of biocommunities
(D) Measurement offunction of Ecosystem

233. Which is the right sequence for Ecological diversity?
(A) Biomes → habitat → Ecosystem → population
(B) Biomes→Ecosystem→habitat→niches
(C) Biomes → Ecosystem → population → niches
(D) Biomes→habitat→niches→Ecosystem

234. Where the knowledge of biodiversity is applicable?
(A) To study, classification of animals and plants
(B) To study, Ecosystem
(C) To study, Biogeological region
(D) All the given

235. Which information is correct for our country?
(A) India is one of the twelve mega biodiversity countries of the world.
(B) India comprises 2.4% biodiversity of world.
(C) India contains more than 7 % plant species of world.
(D) India stands at 7th rank in traditional crop varieties.

236. Which statement is correct for buffer zone of biosphere reserve?
(A) It is legally protected
(B) Authority takes cooperation with localpeople
(C) Environmental Educationis facilitated bythis zone
(D) Hunting and felling of trees occur in this zone

237. What is ecosystem diversity?
(A) Similarity ofspecies diversity in Ecosystem
(B) Variation in species diversity in Ecosystem
(C) Mutation in species diversityin Ecosystem
(D) Homozygosityin species diversity in Ecosystem

238. How the biodiversity can be conserved?
(A) By maintenance of different species
(B) Bymaintenance of genes of different species.
(C) Bymaintenance of different Ecosystem
(D) All of the given

239. Which regions are included in Biodiversity Hot-spot?
(A) Sanctuary (B) National park
(C) Only Hotspot (D) All the given

240. Which organisation is active for conservation of biodiversity at worldlevel?
(A) WWF (B) WCU
(C) Both (A) and (B) (D) EE

241. Which statements are true?

1. At present there are 14 biosphere reserves in India
2. Biospheres includes 3 zones
3. Bandipur, Periyar, Kaziranga and Hazirabaughare biosphere reserve
4. UNO has formulated the concept of biosphere reserve
5. Core Zone is the inner most region of biosphere reserve

(A) 1, 2, 4 (B) 3, 5, 2
(C) 1, 3, 4, 5 (D) 1, 2, 5

242. Which one is correct for individual of the same species?

(A) Population (B) Biotic Community
(C) Ecosystem (D) All of the above

243. Which is the example of ex-situ conservation?

(A) National park (B) Sanctuary
(C) Biosphere reserve (D) Zoo

244. Which is the main cause of extinction ofwild life?

(A) Destruction of habitat
(B) Hunting for flesh
(C) Pollution of medicine and water
(D) All of the above

245. At which place animals and plants are most protected?

(A) Botanicalgardens (B) National Park
(C) Zoos (D) Sanctuary

246. MAB means ....

(A) Man and biosphere programme
(B) Mammal and biological programme
(C) Mammal and biosphere programme
(D) Men and biological programme

247. Who publish Red-list?

(A) WWF (B) IUCN
(C) MAB (D) IBWL

248. IUCN means ....

(A) International unionfor conservation of nature and natural resources
(B) Indian unionfor conservation of nature and natural resources
(C) International unionfor conservation of nature and nutrients resources
(D) Indian Union chemicalnomenclature

249. What is important of gene diversity?

(A) Maintenance of species
(B) Speciation
(C) Research of genetic code
(D) Maintenance and research of spices

250. What is called the biodiversity of habitat?

(A) α-diversity (B) β-diversity
(C) γ-diversity (D) Biosphere

251. Which is the modern concept of conservation?

(A) Biosphere reserve (B) Sanctuary
(C) National park (D) Protected forest

252. Match the following

p.	Biosphere reserve	i.	Eastern Himalayan
q.	Hot-spot	ii.	Amazon
r.	Biodiversity of Gujarat	iii.	Great Nicobar
s.	Greatest biodiversity	iv.	Marine National Park

(A) r - i, p – ii, s - iii, q - iv
(B) r - i, q – ii, p - iii, s - iv
(C) p - i, r – ii, q - iii, s - iv
(D) q - i, s – ii, p - iii, r – iv

253. Match the following

p.	National Bureau of Plant Genetic Resource	i.	Conservation of biodiversity
q.	World Wild Life Fund	ii.	Conservation of gene of domesticated animal
r.	National Bureau of Fish Genetic Resource	iii.	Conservation of gene of crop
s.	National Bureau of Animal Genetic Resource	iv.	Conservation of gene of ecologically valuable species

(A) r - i, p – ii, q - iii, s - iv
(B) r - i, q – ii, p - iii, s - iv
(C) r - i, p – ii, s – iii, q - iv
(D) q - i, s – ii, p - iii, r – iv

254. Match the following

p. diversity	i. Richness of different species in a habitat
q. diversity	ii. Richness ofdifferent species along with a gradiant from one habitat to another habitat within the community
r. diversity	iii. Richness of different species in different habitat

(a) (p - i), (r – ii), (q - iii)
(b) (p - i), (q – ii), (r - iii)
(c) (q - i), (p – ii), (e - iii)
(d) (r - i), (q – ii), (p - iii)

255. A plant with 2n = 20 will have how many linkage group
(A) 20 (B) 10
(C) 40 (D) 5

256. A plant with genotype Gg Ww will produce gametes of which of the following types?
(A) GW, gw, gW, Gw
(B) GG, WW, gg, ww
(C) Gw, GG, GW
(D) All of them

257. The solubility of solid in a liquid with the increase in temperature.
(A) Increases (B) Decreases
(C) Both (A) & (B) (D) Constant

258. A restorer gene is present in
(A) Nucleus (B) Cytoplasm
(C) Mitochondria (D) Chloroplast

259. A state of meiotic cell division when homologous chromosomes separate and move to different poles is known as ....
(A) Metaphase I (B) Anaphase I
(C) Telophase I (D) Anaphase II

260. With four bases A, U, G and C the possible number of triplet cordons are 64. If there is no uracil, the number of codons will be ....
(A) 3 (B) 9
(C) 27 (D) None of above

261. Acclimatisation is
(A) A process of adjustment by crop plants to environment
(B) Related with the clay
(C) Removal of female parts of a flower
(D) Changing of climatical condition

262. Adjustment of individual to their environment by mean of special structures or of function of spoken of
(A) Mutation (B) Variation
(C) Heredity (D) Adaptation

263. After the cell has undergone mitosis, DNA amount in daughter cells will be
(A) Equal to the parent cell
(B) Half of the parent cell
(C) Double of the parent cell
(D) None of the above

264. Albinism in plants is associated with
(A) Epistatis
(B) Recessive lethal
(C) Dominant lethal
(D) Chromosome duplication

265. Alternative forms of a gene is known a
(A) Mutation (B) Gamete
(C) Allele (D) Genotype

265. An anticodonis a sequence of three nitrogenous bases found on
(A) DNA (B) mRNA
(C) rRNA (D) tRNA

266. An egg or sperm cell in plants is called
(A) Gene (B) Gamete
(C) Zygote (D) None of them

268. An embryo sac of angiosperms generally contains
(A) Two nuclei (B) Three nuclei
(C) Four nuclei (D) Eight nuclei

269. An endospermic cell of a seed (2n = 20) will have
(A) 30 Chromosomes (B) 20 Chromosomes
(C) 10 Chromosomes (D) 40 Chromosomes

270. An individual having different alleles for any gene pair and producting two kinds of gametes is known as
(A) Hemizygous (B) Homozygous
(C) Heterozygous (D) Heterogenous

271. An organism having the gametic chromosome numbers is called as
(A) Genome (B) Hybrid
(C) Gamete (D) Haploid

272. An organism with one extra chromosome (2n + 1), is instead of the normal diploid (2n) is known as
(A) Monoploid (B) Monosomic
(C) Trisomic (D) Tetrasonic

273. In a .... protocol, bacteria with engineered abilities to detoxify pollutants are intentionally released in an area.
(A) Microcosm Establishment
(B) Mibridization
(C) Bioremediation
(D) Rhizosecretion

274. The controversy regarding the use of Bt corn is that it
(A) Is potentially harmful to monarch butterflies
(B) Is a potential allergen to humans
(C) Both (A) and (B)
(D) Can contaminate groundwater

275. Starch content of potatoes can be increased by using a bacterial gene, known as
(A) Sucrose phosphate synthase gene
(B) ADP glucose pyrophosphorylase gene
(C) Polygalactouranase gene
(D) None of the above

276. Hybridoma technology is used in the production of
(A) mRNA (B) Monoclonal serum
(C) Monoclonal antibodies
(D) Polyclonal antibodies

277. Barnase/barstar gene obtained from...........
(A) *Bacillus amyloliquefaciens*
(B) *Lactobacillales*
(C) *Streptococcus agalactiae*
(D) *Erwiniarubrifaciens*

278. Transgenic crop approved on 1 June 2013.
(A) Bt corn (B) Bt cotton
(C) Bt brinjal (D) Both (A) and (B)

279. StarLink is transgenic variety of .... and released in 1998.
(A) Cotton (B) Tomoto
(C) Alfalfa (D) Corn

280. Golden rice developed by ....
(A) Thomas Cech
(B) Ingo Potrykus
(C) Monsanto
(D) Biocon

281. Squash developed for virus resistant through ....
(A) Agrobacterium tumefaciens
(B) Gene gun
(C) Electroporation
(D) Lipofection

282. Which of the following is alkaloid?
(A) Terpene (B) Morphine
(C) Cyanidin (D) Delphinidin

283. Which involved in meat tenderization?
(A) Ficin (B) Amylase
(C) Papain (D) Lactase

284. Which Haploid rice variety developed through double haploid?
(A) IR36 (B) IR8
(C) IR64 (D) Âbel

285. Which is used as biofertilizers?
(A) Rizosphere (B) Rhizobium
(C) Bacteria (D) Fungi

286. Which is third generation diesel plant?
(A) Alfalfa (B) Mucor
(C) Jatropha (D) Eucalyptus

287. Gene npt produced by
(A) Viral transposon
(B) SINES
(C) LINES
(D) Bacterial transposon

288. Technique used for detection of 3D structure of proteins.
(A) NMR (B) TEM
(C) SAGE (D) SEM

289. DNA replication takes place .... of cell cycle.
(A) G1 phase (B) G2 phase
(C) S phase (D) M phase

290. What is zinc finger domain?
(A) Protein motif bind to RNA
(B) Transcription factor bind to DNA
(C) Protein motif bind to RNA
(D) Protein motif bind to DNA

291. Herbicide resistant gene group
(A) nptII, barnase and bar
(B) bar, als and chs
(C) TA29, nptII, lux
(D) Chs, baenase, bar

292. Gene cryI used to improve resistance for ....
(A) Coleoptera
(B) Diptera
(C) Lepidoptera
(D) Nematodes

293. Guha and Maheshwari developed haploid plants in ....
(A) Arabidopsis (B) Datura
(C) Eucalyptus (D) Banana

294. LEA is related to ....
(A) Protection of plants from insects
(B) Protection of plants under water stress
(C) Protection of plants from weeds
(D) Protection of plants from cold

295. Fungus resistance developed by ....
(A) Chitinase and Glucanase
(B) Chitinase
(C) EPSP
(D) Tn5

296. Quinolone used as ....
(A) Replication inhibitor
(B) Transcription inhibitor
(C) Translation inhibitor
(D) PCR inhibitor

297. Find the incorrect match
(A) Endosperm – food storage part of seed
(B) α-amylase – insect resistant
(C) Aleuron – seed protein
(D) Pectinase – cell wall degrading enzyme

298. Artificial seeds are synthesisez using ....
(A) Sodium Chloride
(B) Calccium Carbonate
(C) Sodium alginate
(D) Colchicine

299. Polyethylene glycol is
(A) Fusogenic chemical
(B) Electro fusion stimulant
(C) Callus stimulant
(D) Differentiation stimulant

300. The first hormone artificially produced by bacteria is
(A) Insulin (B) Auxin
(C) Adrenalin (D) Gibberaline

Answer-Key

Q	Ans	Q	Ans	Q	Ans	Q	Ans	Q	Ans	Q	Ans	Q	Ans	Q	Ans
1.	(B)	41.	(B)	81.	(D)	121.	(C)	161.	(D)	201.	(B)	241.	(D)	281.	(A)
2.	(A)	42.	(B)	82.	(A)	122.	(D)	162.	(C)	202.	(C)	242.	(A)	282.	(B)
3.	(C)	43.	(B)	83.	(C)	123.	(A)	163.	(C)	203.	(A)	243.	(D)	283.	(C)
4.	(D)	44.	(A)	84.	(B)	124.	(B)	164.	(A)	204.	(A)	244.	(A)	284.	(D)
5.	(B)	45.	(A)	85.	(B)	125.	(B)	165.	(D)	205.	(D)	245.	(B)	285.	(B)
6.	(D)	46.	(B)	86.	(B)	126.	(D)	166.	(D)	206.	(C)	246.	(A)	286.	(C)
7.	(C)	47.	(A)	87.	(A)	127.	(A)	167.	(C)	207.	(B)	247.	(B)	287.	(D)
8.	(A)	48.	(C)	88.	(C)	128.	(B)	168.	(B)	208.	(C)	248.	(A)	288.	(A)
9.	(B)	49.	(B)	89.	(D)	129.	(D)	169.	(A)	209.	(A)	249.	(B)	289.	(C)
10.	(C)	50.	(B)	90.	(B)	130.	(A)	170.	(C)	210.	(B)	250.	(A)	290.	(D)
11.	(B)	51.	(C)	91.	(D)	131.	(A)	171.	(A)	211.	(A)	251.	(A)	291.	(B)
12.	(D)	52.	(A)	92.	(C)	132.	(D)	172.	(D)	212.	(A)	252.	(D)	292.	(C)
13.	(C)	53.	(C)	93.	(A)	133.	(C)	173.	(A)	213.	(D)	253.	(D)	293.	(B)
14.	(B)	54.	(B)	94.	(B)	134.	(B)	174.	(C)	214.	(C)	254.	(B)	294.	(B)
15.	(B)	55.	(A)	95.	(C)	135.	(D)	175.	(B)	215.	(C)	255.	(B)	295.	(A)
16.	(B)	56.	(D)	96.	(B)	136.	(B)	176.	(D)	216.	(C)	256.	(A)	296.	(B)
17.	(A)	57.	(B)	97.	(D)	137.	(C)	177.	(B)	217.	(A)	257.	(A)	297.	(B)
18.	(B)	58.	(C)	98.	(C)	138.	(A)	178.	(C)	218.	(B)	258.	(A)	298.	(C)
19.	(A)	59.	(B)	99.	(D)	139.	(B)	179.	(D)	219.	(D)	259.	(B)	299.	(B)
20.	(B)	60.	(C)	100.	(A)	140.	(A)	180.	(B)	220.	(B)	260.	(C)	300.	(A)
21.	(A)	61.	(A)	101.	(A)	141.	(A)	181.	(A)	221.	(C)	261.	(A)		
22.	(C)	62.	(B)	102.	(B)	142.	(A)	182.	(C)	222.	(D)	262.	(D)		
23.	(B)	63.	(C)	103.	(B)	143.	(B)	183.	(B)	223.	(D)	263.	(A)		
24.	(C)	64.	(D)	104.	(A)	144.	(D)	184.	(D)	224.	(C)	264.	(B)		
25.	(B)	65.	(A)	105.	(A)	145.	(D)	185.	(C)	225.	(C)	265.	(C)		
26.	(C)	66.	(C)	106.	(D)	146.	(C)	186.	(A)	226.	(A)	266.	(B)		
27.	(D)	67.	(C)	107.	(C)	147.	(D)	187.	(B)	227.	(B)	267.	(B)		
28.	(A)	68.	(A)	108.	(A)	148.	(D)	188.	(A)	228.	(B)	268.	(B)		
29.	(A)	69.	(B)	109.	(D)	149.	(D)	189.	(D)	229.	(D)	269.	(A)		
30.	(C)	70.	(D)	110.	(B)	150.	(C)	190.	(C)	230.	(A)	270.	(C)		
31.	(D)	71.	(D)	111.	(D)	151.	(D)	191.	(B)	231.	(B)	271.	(A)		
32.	(A)	72.	(A)	112.	(A)	152.	(B)	192.	(D)	232.	(A)	272.	(C)		
33.	(C)	73.	(C)	113.	(D)	153.	(C)	193.	(C)	233.	(B)	273.	(C)		
34.	(A)	74.	(D)	114.	(B)	154.	(D)	194.	(A)	234.	(D)	274.	(C)		
35.	(B)	75.	(C)	115.	(C)	155.	(B)	195.	(C)	235.	(A)	275.	(B)		
36.	(C)	76.	(A)	116.	(D)	156.	(C)	196.	(B)	236.	(C)	276.	(C)		
37.	(D)	77.	(C)	117.	(D)	157.	(A)	197.	(C)	237.	(B)	277.	(A)		
38.	(B)	78.	(A)	118.	(A)	158.	(D)	198.	(C)	238.	(D)	278.	(B)		
39.	(C)	79.	(A)	119.	(C)	159.	(C)	199.	(D)	239.	(A)	279.	(D)		
40.	(D)	80.	(C)	120.	(D)	160.	(D)	200.	(A)	240.	(C)	280.	(B)		

Memory Based Previous Year Exam Papers

ASRB-NET I – 2013

1. Which of the following pairs are anomers?
(A) Ribose and Ribulose
(B) L-Glucose and D-Glucose
(C) α-D-glucose and α -D-mannose
(D) α-D-glucose and β-D-glucose

2. Hybridoma technology is used in the production of
(A) mRNA
(B) Monoclonal serum
(C) Monoclonal antibodies
(D) Polyclonal antibodies

3. For mitochondrial structure and function, mt-DNA specifies
(A) 10% (B) 20%
(C) 30% (D) 40%

4. Which of the following reagent is useful for visualizing DNA?
(A) Urea
(B) EDTA
(C) Isopropanol
(D) Ethidium bromide

5. Homologous pairs line up along equatorial plane during
(A) Anaphase II (B) Metaphase I
(C) Telophase (D) Interphase

6. Between mitotic divisions, the cell is in
(A) G0 (B) G1
(C) G2 (D) S phase

7. Most plant cells are surrounded by a rigid cell wall made primarily of
(A) Monosaccharides (B) Pectin
(C) Chitins (D) Polysaccharides

8. The organelle that functions in the breakdown of cells and tissues is the
(A) Episomes (B) Mitochondrion
(C) Centrosomes (D) Lysosomes

9. Endoplasmic reticulam
(A) Found only in animals
(B) A system of membrane bound channels
(C) Is called rough if mitochondria is attached to it
(D) A site of ATP production

10. "9+2" describes the basic structure of which of the following one
(A) chromosome
(B) Basal body
(C) Chloroplast
(D) Flagellum

11. The function of nucleus includes
(A) Cellular respiration
(B) Synthesis of proteins
(C) Housing the hereditary information
(D) Synthesis of carbohydrates

12. Bacculoviruses
(A) RNA viruses
(B) DNA viruses
(C) Both (A) and (B)
(D) None of the above

13. Secondary metabolites like antibiotics are not essential for
(A) Log phase growth
(B) Exponential growth
(C) Both (A) and (B)
(D) None of these

14. The restriction enzymes useful to the molecular biologist belongs to
(A) Type I (B) Type II
(C) Type III (D) Type IV

15. Which of the following vector is suitable for DNA sequencing
(A) EMBL (B) PBR 322
(C) M 13 (D) Lambda

16. During which process, the free end of the chromosome moves from Hfr donor in to the F cell across an intracellular bridge called piles
(A) Transformation
(B) Transduction
(C) Conjugation
(D) All of these

17. Francis Jacob and Monod Proposed
(A) Lock and Key Hypothesis
(B) Gene for Protein Synthesis
(C) Operon Synthesis
(D) Gene for protein synthesis

18. State which of the following is not a desired characteristic of a vector?
(A) Unique restriction site
(B) Large Size
(C) Gene that confer antibiotic resistance
(D) Autonomous replication

19. Which of the following is a carrier hosts for new genes?
(A) EcoRI (B) Bacteriophages
(C) Nucleus
(D) Transplanted Organs

20. Which of the following chemical is used for preparation of competent cell?
(A) HCl (B) NaCl
(C) $CaCl_2$ (D) Glycine

21. Endochitinase can be used as an
(A) Antibacterial (B) Antiviral
(C) Antinematodal (D) Antifungal

22. The direct DNA uptake by the protoplast is stimulated by the chemical
(A) Cellulose (B) Sucrose
(C) PEG (D) Mannose

23. Which of the following is an herbicide resistant gene?
(A) Bar (B) Hpt
(C) Chitinase (D) Oxidase

24. Restriction enzymes that cuts straight across the DNA produces
(A) Nick
(B) Sticky ends
(C) Blunt ends
(D) None of these

25. Lambda has short single stranded 5' projection of
(A) 8 nucleotide length
(B) 10 nucleotide length
(C) 12 nucleotide length
(D) 14 nucleotide length

26. In *Agrobacterium* mediated plant transformation which one of the following is used to check the *Agrobacterium* growth
(A) PPT (B) Kanamycin
(C) Timentin (D) Pencillin

27. The T-DNA region of all the Ti and Ri plasmids are flanked by direct repeat sequence
(A) 15bp (B) 20 bp
(C) 25 bp (D) 35 bp

28. During *Agrobacterium* infections the plant cell begins to synthesise an Arginine derivative called as
(A) Hygromycin (B) Opines
(C) Acetobenzylpurine
(D) Acetosyringone

29. During *Agrobacterium* transformation, the T-DNA is coated by protein, which is the product of gene
(A) Vir B (B) Vir C
(C) Vir D (D) Vir E

30. Most commonly used strategy to develop transgenic plant against any viral disease is to engineer a
(A) Antisense RNA for viral genome
(B) Coat or capsid protein gene
(C) Direct DNA transfer to viral gene
(D) Liposome Mediated transfer of viral RNA

31. The *vir* region is organised into six operons viz., vir A, vir B, vir C, vir D, vir E, vir G out of these which four operons are required for virulence
(A) A,B,C,D (B) B,C,D,E
(C) C,D,E,G (D) A,B,D,G

32. Replication of DNA by rolling circle mechanism give rise to a
(A) Blunt end DNA (B) Theta structure
(C) Concatamer
(D) Single strand DNA with nick

33. The gene transferred from *E.coli* to tobacco to increase the level of manitol is
(A) Manitol dehydrogenase
(B) Manitol carboxylase
(C) Manitoloxygenase
(D) Manitol hydrolase

34. Relaxed plasmid
(A) Have only single copy per cell
(B) Cannot be purified
(C) Does not exist in bacteria
(D) Have multiple copies per cell

35. Which of the following is a green tissue specific promoter?
(A) Pea vicilin
(B) Arabidopsis small subunit of RUBISCO
(C) 35s CaMV
(D) Bean phytahemaglutinin

36. In vitro packaging is required for which one of the following vector
(A) BAC (B) YAC
(C) PAC (D) None of these

37. Cosmids have the Cos site of
(A) Lambda (B) Plasmid
(C) ML3 (D) YAC

38. *Thiobacillus thioxidans* grows on
(A) Irron (B) Sulphur
(A) Aluminium (B) Pyrites

39. An example of commercially important primary metabolites is
(A) Toxin (B) Alkaloid
(C) Amine (D) Alcohol

40. Secondary metabolites like antibiotics are not essential for
(A) Log phase growth
(B) Exponential growth
(C) Both a and b
(D) None of these

41. Pure line theory was given by
(A) Bateson (B) Johannson
(C) Mendel (D) Morgan

42. Which of the following enzyme is used for making cDNA
(A) DNA ligase (B) Taq polymerase
(C) Restriction enzyme
(D) Reverse transcriptase

43. The chemical used as an inducer of the vir operons is
(A) Kanamycin (B) Hygromycin
(C) Pencillin (D) Acetosyringone

44. "9+2" describes the basic structure of which of the following one
(A) Chloroplast (B) Flagellum
(C) Mitochondria (D) Lysosome

45. The organelle that functions in the breakdown of cells and tissues is the
(A) Centrosomes (B) Centriole
(C) Lysosome (D) Liposome

46. An example for naturally occurring Auxin
(A) IAA (B) ABA
(C) BAP (D) 2, 4-D

47. A general method to confirm expression of transgene at translation level is
(A) western blotting
(B) northern blotting
(C) eastern blotting
(D) southern blotting

48. The browning of culture medium induced by some explants can be prevented by the addition of
(A) GA (B) 2-4-D
(C) IBA
(D) PVP (polyvinyl pyrollidone)

49. Which of the following is not used as a cryoprotectant
(A) Glycerol (B) Dmso
(C) Polyclar (D) PEG

50. Anther culture provide a method for the production of
(A) Polyploidy
(B) Heterozygous line
(C) Apomictic line
(D) Homozygous line

51. The first stable product of photosynthesis
(A) Glucose (B) DHAP
(C) RUBP (D) 3PGA

52. Oxidation of water during photosynthesis involves
(A) PSI (B) PS II
(C) Both
(D) None of the above

53. Which one of the following chemical causes frame shift mutation
(A) 2-amino purine (B) Hydroxylamine
(C) 2 bromo uracil (D) Acridine orange

54. The restriction enzyme useful to the molecular biologist belongs to
(A) Type I (B) Type II
(C) Type III (D) Both (A) and (B)

55. A good molecular marker must be
(A) Polymorphic (B) Heritable
(C) Both (A) and (B) (D) None

56. Shortgun approach is used in
(A) gene mapping
(B) DNA sequencing
(C) gene transformation
(D) genomic library

57. The length of Okazaki fragments in *E. coli* is about
(A) 100-200 nucleotides
(B) 500-1000 nucleotides
(C) 1000-2000 nucleotides
(B) 10000-100000 nucleotides

58. The sigma factor of RNA polymerase in *E. coli* involved
(A) RNA synthesis
(B) promoter binding
(C) transcription initiation
(D) binding with template DNA

59. The Casparian stripes is present in
(A) Epidermis (B) Endodermis
(A) Cortex (B) Root Hair

60. FlavrSavr is a variety of
(A) Potato (B) Tomato
(C) Pomato (D) Avocado

61. During *Agrobacterium* infections the plant cell begins to synthesize an Arginine derivative called as
(A) Opines (B) Nopaline
(C) Lux (D) Isopantenyl

62. Enzyme used for making cDNA from RNA.
(A) Transcriptase
(B) DNA polymerase
(C) Reverse transcriptase
(D) Replicase

63. Reagent used for visualizing DNA?
(A) CTAB (B) EDTA
(C) EtBr (D) Isopropanol

64. Enzyme used in PCR (Polymerase Chain Reaction)
(A) Taq polymerase (B) Liagse
(C) Lipase (D) Amylase

65. PCR invented by
(A) Kary Mullis (B) Mendel
(C) Watson and crick (D) Aleck Jeffery

66. In PCR, the double stranded DNA is generally denatured by
(A) Pressure (B) Humidity
(C) Temperature (D) Salinity

67. DBM paper is used for
(A) Southern Blotting
(B) Northern Blotting
(C) Western Blotting
(D) Dot blot

68. Number of base pair units in a single turn of DNA is
(A) 12 (B) 11
(C) 9.4 (D) 10

69. Which amino acids can stabilize protein structure by forming covalent cross –links between polypeptide chains?
(A) Nitrogen (B) Sulphur
(C) Phosphorus (D) Sodium

70. Acid fast staining used for which bacteria?
(A) Streptococcus (B) Penicillium
(C) Acetobacteria (D) Mycobacterium

71. Inhibitory action of Penicillin is
(A) Cell membrane formation
(B) Cell wall synthesis
(C) ATP synthesis in mitochondria
(D) Chlorophyll synthesis in chloroplast

72. DNA staining is done by
(A) Feulgen staining
(B) Coomassie brilliant blue
(C) Crystal violet
(D) jassen green

73. The most widely used chemical for protoplast fusion, as fusogens, is
(A) PEG (B) EDTA
(C) ETBR (D) DMSO

74. In order to obtained virus free plants through tissue culture the best method is
(A) Protoplast culture
(B) Meristem culture
(C) Anthur culture
(D) Embryo culture

75. Endochitinase can be used as an
(A) Antibacterial
(B) Antiviral
(C) Antifungal
(D) Antiseptic

Answer-Key

Q	Ans	Q	Ans	Q	Ans	Q	Ans	Q	Ans	Q	Ans	Q	Ans	Q	Ans
1.	(D)	11.	(C)	21.	(D	31.	(D)	41.	(B)	51.	(D	61.	(A)	71.	(B)
2.	(C)	12.	(B)	22.	(C	32.	(B)	42.	(D)	52.	(B	62.	(C)	72.	(A)
3.	(A)	13.	(C)	23.	(A	33.	(A)	43.	(D)	53.	(D	63.	(C)	73.	(A)
4.	(D)	14.	(B)	24.	(C	34.	(D)	44.	(B)	54.	(B	64.	(A)	74.	(B)
5.	(B)	15.	(C)	25.	(C	35.	(B)	45.	(C)	55.	(C	65.	(A)	75.	(C)
6.	(A)	16.	(C)	26.	(B	36.	(C)	46.	(A)	56.	(D	66.	(C)		
7.	(D)	17.	(C)	27.	(C	37.	(A)	47.	(A)	57.	(C	67.	(B)		
8.	(D)	18.	(B)	28.	(B	38.	(D)	48.	(D)	58.	(B	68.	(D)		
9.	(B)	19.	(B)	29.	(D	39.	(D)	49.	(D)	59.	(B	69.	(B)		
10.	(D)	20.	(C)	30.	(B	40.	(C)	50.	(D)	60.	(B	70.	(D		

ASRB-NET II – 2013

1. Method of gene transfer through virus into bacteria.
(A) Conjugationg (B) Lipofection
(C) Transduction (D) Electroporation

2. Who discovered PCR?
(A) Karry Mullis (B) Kornberg
(C) Maxum-Gilbert (D) Paul Berg

3. Enzyme used to cut DNA at specific site.
(A) Exonuclease (B) Endonuclease
(C) Restriction endonuclease
(D) Ligase

4. Enzyme used to join okazaki fragments
(A) Exonuclease (B) Endonuclease
(C) Restriction endonuclease
(D) Ligase

5. Enzyme used to add nucleotides at 5' end of DNA.
(A) Terminal transferase
(B) T4 nucleotide kinase
(C) Restriction endonuclease
(D) Alkaline phosphatase

6. Enzyme used to energy for rection.
(A) Terminal transferase
(B) DNA ligase
(C) Restriction endonuclease
(D) Alkaline phosphatase

7. Enzyme used to prevent circulation of linear DNA.
(A) Terminal transferase
(B) DNA ligase
(C) Restriction endonuclease
(D) Alkaline phosphatase

8. Enzyme used to remove ssDNA sequence.
(A) T4 nucleotide kinase
(B) Alkaline phosphatase
(C) S1 nuclease
(D) Ligase

9. Enzyme used to add nucleotides at 3' end of DNA.
(A) T4 nucleotide kinase
(B) Alkaline phosphatase
(C) S1 nuclease
(D) Ligase

10. Enzyme used to form short RNA sequence at DNA template.
(A) Primase (B) Ligase
(C) Helicase (D) Polymerase

11. Which property is not present in klenow fragments?
(A) 3'→5' exonuclease
(B) 5'→3' polymerase
(C) 5'→3' exonuclease
(D) 5'→3' endonuclease

12. Enzyme used to produce blunt ended DNA fragments.
(A) *Eco*RII (B) *Bam*I
(C) *Sau*I (D) *Eco*RV

13. Virus purification method
(A) Ultracentrifugation
(B) Electrophoresis
(C) PCR
(D) Microarray

14. CTAB method is used for ....
(A) RNA isolation
(B) DNA isolation
(C) Protein isolation
(D) Lectin isolation

15. What is the molecular weight of polypeptide chain containing 10 amino acids? (MW of an amino acid = 110).
(A) 1100 (B) 1000
(C) 938 (D) 990

16. What is used of blue-white screening method?
(A) Test for the DNA insert in plasmid DNA
(B) Test of virus contamination
(C) Test of transposable elements in genome
(D) Test of chromosomal abbreviations

17. cDNA formed from ....
(A) tRNA (B) rRNA
(C) snRNA (D) mRNA

18. Enzyme used for tRNA synthesis
(A) RNA polymerase I
(B) RNA polymerase II
(C) RNA polymerase III
(D) DNA polymerase I

19. Enzyme used for mRNA synthesis
(A) RNA polymerase I
(B) RNA polymerase II
(C) RNA polymerase III
(D) DNA polymerase I

20. Enzyme used for adding amino acids to tRNA during translation.
(A) Aminoacylsynthetase
(B) Translocase (C) Peptidase
(D) RNA polymerase

21. Which technique used to check the expression of gene?
(A) RT-PCR (B) Microarray
(C) Transcriptome analysis
(D) All of the above

22. Best method for plasmid transformation.
(A) Filter Reclamation
(B) Electroporation
(C) $Cacl_2$ method
(D) Gene gun

23. Which marker require sequence information for synthesis
(A) AFLP (B) ISSR
(C) SSR (D) Microarray

24. CaMV has .... genome.
(A) Linear ssDNA (B) Linear dsDNA
(C) Circular ssDNA (D) Circular dsDNA

25. CryI gene related to
(A) Coleopteran (B) Lepidoptera
(C) Nematodes (D) Diptera

26. Which is insect resistant gene?
(A) CryIA (B) CryAB
(C) Cry (D) F

27. Enzyme is not produced in plants.
(A) Papain (B) Pectinase
(C) Ficin (D) Amylase

28. Suitable temperature for slow growth culture.
(A) 10-15°C (B) 4-9°C
(C) 0°C (D) 121°C

29. Which is used for cryopreservation?
(A) Liquid carbondioxide
(B) Solid hyrdogen
(C) Liquid carbondioxide
(D) Liquid nitrogen

30. Which vector has shortest size?
(A) M13 (B) Lamda
(C) Plasmid (D) YAC

31. Which require for the synthesis of STS markers?
(A) Sequence of RNA
(B) Taq polymerase
(C) Sequence of DNA
(D) Primer

32. More fine demonstrated group for C-value paradox?
(A) Rice/wheat (B) Rice/Arabidopsis
(C) Rice/popular (D) Rice/pea

33. What is the symbol of turnover of enzyme?
(A) V_{max} (B) ½ V_{max}
(C) K_m (D) K_{cat}

34. Suitable buffer for reactions performed in body?
(A) Lactate (B) Acetate
(C) Carbonate (D) Phosphate

35. Which is used to locate molecular marker in organism genome?
(A) Gene (B) DNA
(C) Genome (D) Chromosome

36. Molecular markers based only on PCR.
(A) RFLP and RAPD
(B) SSR and RFLP
(C) AFLP and RAPD
(D) RAPD and ISSR

37. Collection of clones of plasmids/phages which recombinant DNA.
(A) Genomic library
(B) cDNA library
(C) RNA library
(D) DNA-cDNA library

38. True statement regarding cell cycle.
(A) Largest phase is M-phase
(B) G1 and G0 equilibrium
(C) Spindle fibres form in S-phase
(D) Tubulin protein form in G1 phase

39. What is number of Arabidopsis chromosomes (n)?
(A) 7 (B) 10
(C) 5 (D) 24

40. How many enzymes involved in glycolysis?
(A) 5 (B) 10
(C) 12 (D) 11

41. How many ATP produced in glycolysis?
(A) 5 (B) 3
(C) 2 (D) 10

42. Lamda phage genome.
(A) dsDNA (B) ssDNA
(C) ssRNA
(D) ssDNA before replication and dsDNA after replication

43. Barnase/barstar gene obtained from..........
(A) *Bacillus amyloliquefaciens*
(B) *Lactobacillales*
(C) *Streptococcus agalactiae*
(D) *Erwiniarubrifaciens*

44. What is TA29?
(A) Auxin synthesis promoter
(B) CaMV promoter
(C) Tapetum specific promoter
(D) Seed protein specific promoter

45. Which plasmid has best running in electrophoresis?
(A) Open circular (B) Linear
(C) Nicked circular (D) Supercoil

46. Vector used for gene transfer from eukaryotes to prokaryotes?
(A) Plasmid (B) Shuttle
(C) Phasmid (D) Lamda

47. Cofactor for nitrogenase enzyme.
(A) Mo (B) B
(C) Ni (D) K

48. What is the function of opine?
(A) Infection of bacteria in host
(B) Lysozyme production
(C) Carbon and nitrogen source
(D) Help in transformation of plasmid

49. What is the source of npt gene?
(A) Viral transposon
(B) Bacterial transposon
(C) TA29 promoter
(D) SINES and LINES

50. Function of cefetoxin?
(A) Inhibition of bacterial replication
(B) Inhibition of flagella
(C) Carbon and nitrogen source
(D) Inhibition of bacterial cell wall

51. Which of the following works as a cryoprotectant?
(A) Sucrose (B) Glucose
(C) Trehlose (D) Xylose

52. Which of the following works as a cryoprotectant?
(A) Proline (B) DMSO
(C) Glycerol (D) PEG

53. Substract for *Lac A* gene.
(A) Galactose (B) Lactose
(C) Sucrose (D) Glucose

54. Enzyme codes for *Lac A* gene?
(A) β-galactosidase
(B) Acetyltransferase
(C) Permease
(D) Both a and c

55. Largest DNA insert used by which vector for gene transfer and cloning?
(A) BAC (B) YAC
(C) PAC (D) Plasmid

56. Which of the following is not reporter gene?
(A) *GUS* (B) *GFP*
(C) *CAS* (D) *LacZ*

57. Which works as antibodies?
(A) Reorter gene (B) Immunogen
(C) Lectin (D) Primers

58. Virus detection technique based on antibody-antigen protein reaction?
(A) PCR (B) Microarray
(C) ELISA (D) Sequencing

59. What is called if one mRNA contains more than one functional gene transcripts?
(A) Monocistronic (B) Polycistronic
(C) Polysome (D) Ribosome

60. Functional unit of gene?
(A) Muton (B) Recon
(C) Cistron (D) All of the above

61. Expression technique is based on screening.
(A) PCR screening
(B) Screening by ELISA
(C) Blue –white colony screening
(D) Electrophoresis

62. Rabies vaccine produced by ....
(A) LaSota virus (B) Adenovirus
(C) Parvovirus (D) Bacteria

63. Transgenic crop approved on 1 June 2013.
(A) Bt corn (B) Bt cotton
(C) Btbrinjal (D) Both a and b

64. .... involves in post-translational method or gene silencing?
(A) siRNA (B) mRNA
(C) snRNA (D) hnRNA

65. What is size of siRNA?
(A) 23-25nts (B) 10-20nts
(C) 21-23nts (D) Less than 50nts

66. StarLink is transgenic variety of .... and released in 1998.
(A) Cotton (B) Tomoto
(C) Alfalfa (D) Corn

67. Golden rice developed by ....
(A) Thomas Cech (B) Ingo Potrykus
(C) Monsanto (D) Biocon

68. What is called the Cleaning of pollutants from environment through microbes using plants?
(A) Biofortification (B) Bioengineering
(C) Bioremediation (D) Gene escaping

69. Squash developed for virus resistant through ....
(A) Agrobacterium tumefaciens
(B) Gene gun (C) Electroporation
(D) Lipofection

70. Determination of nucleotide sequence of an organism.
(A) DNA microarray (B) DNA labeling
(C) DNA sequencing (D) Nick translation

71. Which of the following sequence located at upstream site in eukaryotes?
(A) CAAT box
(B) Stop codons
(C) Structural genes
(D) Coding sequencing

72. Which binds to TATA box?
(A) SD sequence
(B) Cozak sequence
(C) Transcription factors
(D) Primer

73. Shine-delgarno sequence in prokaryotic translation binds to ....
(A) 28S rRNA (B) 18S rRNA
(C) 23S rRNA (D) 16S rRNA

74. .... used to replace some of the nucleotides from radiolabelled dsDNA.
(A) Nick translation
(B) DNA Blotting
(C) Yeast two hybrid system
(D) SAGE

75. .... used to detect glycoproteins and specific carbohydrates.
(A) Southern blotting
(B) Agarose electrophoresis
(C) Lectin blotting
(D) PAGE

76. What is the first step of PCR?
(A) Annealing (B) Denaturation
(C) Extension (D) Digestion

77. Which map describes most structural and functional information of genome of an organism?
(A) Linkage map (B) Physical map
(C) Gene map (D) All of the above

78. First developed molecular marker.
(A) RAPD (B) AFLP
(C) SSR (D) RFLP

79. Fluorescent dye used in RT-PCR.
(A) Ethidium bromide
(B) Bromophenol blue
(C) GFP
(D) SYBR green

80. Where located DNA in cell?
(A) Nucleus (B) Mitochondria
(C) Chloroplast (D) All of the above

81. Who is the father of tissue culture?
(A) Haberlandt (B) Cocking
(C) Murashig (D) Skoog

82. Plant part used in artificial seed synthesis?
(A) Meristem tip (B) Somatic embryo
(C) Shoot bud (D) Protoplast

83. Which hormone released under drought stress?
(A) Glycerol (B) Auxin
(C) Gibberellin (D) Proline

84. Which is related to apoptosis?
(A) Capsid (B) Cascades
(C) Callus (D) Homeostasis

85. Alkaloids are ....
(A) Acid (B) Neutral
(C) Chemical
(D) Basic nitrogenous

86. What is ribozyme?
(A) RNA with enzyme activity
(B) Type of proteinase
(C) DNA with enzyme activity
(D) RNA with more than one ribosomes

87. Plant cell does not contain cell organelle ....
(A) Ribosome (B) Nucleolus
(C) Lysosome (D) Mitochondria

88. Responsible gene in Ti pasmid for transformation.
(A) *Iaa* gene (B) *Vir* gene
(C) *Opt* gene (D) *Ori* gene

89. Who discovered QTL?
(A) Michelmore (B) Morgan
(C) Sax (D) Mendel

90. Which is Kornberg enzyme?
(A) DNA pol I (B) RNA pol I
(C) DNA pol II (D) DNA pol III

91. Protein involved in plasmid movement in bacterial cell?
(A) ParM protein (B) Cdk proteins
(C) Vir protein (D) Iaam protein

92. Translation start codon.
(A) UAA (B) UGA
(C) AUG (D) UAG

93. What is open reading frame?
(A) A continuous stretch of codons that do not contain a stop codon
(B) A continuous stretch of codons that also contain a stop codon
(C) A continuous stretch of codons that code for specific DNA
(D) A continuous stretch of coding sequence of DNA which code for a protein

94. What is size of border sequence in T-DNA?
(A) 20bp (B) 25bp
(C) 30bp (D) Less than 50bp

95. What is holoenzyme?
(A) Coenzyme + cofactor
(B) Apoenzyme + cofactor
(C) Metol ion + coenzyme
(D) Cofactor + coenzyme + metal ion

96. Rice variety improved by MAS.
(A) IR64
(B) Golden rice
(C) Quality protein rice
(D) Improved basmati rice

97. Generally used tissue culture mediafor plant propagation.
(A) PDA media
(B) Yeast extarct
(C) MS media
(D) Trypticase soy agar

98. Distance hybridization used for ....
(A) Embryo rescue
(B) Artificial seed
(C) Protoplast culture
(D) Meristem culture

99. Which of the following is alkaloid?
(A) Terpene (B) Morphine
(C) Cyanidin (D) Delphinidin

100. Agrobacterium is ....
(A) Soil born gram positive bacterium
(B) air born gram negative bacterium
(C) Soil born gram negative bacterium
(D) Seed born gram positive bacterium

101. Induced fit theory given by
(A) Emil Fischer (B) DanielKoshland
(C) Wilhelm Kühne (D) Eduard Buchner

102. Which involved in meat tenderization?
(A) Ficin (B) Amylase
(C) Papain (D) Lactase

103. After 42 days, radioactive molecule contains 1.25g weight. What is the starting quantity, if half-life of the molecule is 14 days?
(A) 12 g (B) 2.5g
(C) 5g (D) 10g

104. Which codon present in mRNA against to CAU anticodon of tRNA?
(A) GUA (B) GTA
(C) ATG (D) AUG

105. Enzyme used for conversion of inorganic form.
(A) Peroxidase (B) Isomerase
(C) Rubisco (D) Endonuclease

106. What is called the transformation of a plant part into a glassy, translucent and brittle solid form?
(A) Vitrification (B) Cryopreservation
(C) Solidification (D) Vernalization

107. Hormone not responsible for somatic embryogenesis?
(A) Abscisic acid (B) Auxins
(C) Cytokinins (D) Gibberellins

108. What is charge on DNA?
(A) Positive
(B) Negative
(C) Nutral
(D) None of the above

109. Single cell proteins produced by ....
(A) Yeast (B) Algae
(C) Fungi (D) Mycoplasma

110. Mostly used reporter gene in transformation.
(A) *GFP* (B) *Lux*
(C) *GUS* (D) *Cat*

111. Best method for plasmid transfer
(A) Electroporation (B) Gene gun
(C) PEG treatment (D) Liposome

112. Mostly used energy source in tissue culture.
(A) Glucose (B) Sucrose
(C) Fructose (D) Mannose

113. Mineral as macronutrient is not present in tissue culture.
(A) Fe (B) Ca
(C) Mg (D) S

114. Chromosome doubling used for the development of ....
(A) Double haploid (B) Double diploid
(C) Tetraploid (D) Haplohaploid

115. Which is used to develop homozygous line from heterozygous line in tissue culture?
(A) Double diploid
(B) Single seed method
(C) Double haploid
(A) Haplohaploid

116. Which Haploid rice variety developed through double haploid?
(A) IR36 (B) IR8
(C) IR64 (D) Ábel

117. Which is Heat labile hormone?
(A) Zeatin (B) 2,4-D
(C) Cytokinine (D) Ethylene

118. Framshift mutation caused by ....
(A) UV (B) Acridine
(C) 5-bromouracil (D) EMS

119. More active part of chromatin for transcription?
(A) Heterochromatin (B) Euchromatin
(C) Centromere (D) Both (A) and (B)

120. How herbicide resistant transgenic crop absorb UV radiation?
(A) Due to aliphatic amino acids
(B) Due to sulphur containing amino acids
(C) Due to aromatic amino acids
(D) Due to phosphoinothricin

121. Which amino acid is not present in legumes?
(A) Tryptophan
(B) Phenylalanine
(C) Glycine
(D) Methionine

122. Which is used as biofertilizers?
(A) Rizosphere (B) Rhizobium
(C) Bacteria (D) Fungi

123. Which is third generation diesel plant?
(A) Alfalfa (B) Mucor
(C) Jatropha (D) Eucalyptus

124. Gene npt produced by
(A) Viral transposon
(B) SINES
(C) LINES
(D) Bacterial transposon

125. Technique used for detection of 3D structure of proteins.
(A) NMR (B) TEM
(C) SAGE (D) SEM

126. DNA replication takes place .... of cell cycle.
(A) G1 phase (B) G2 phase
(C) S phase (D) M phase

127. What is zinc finger domain?
(A) Protein motif bind to RNA
(B) Transcription factor bind to DNA
(C) Protein motif bind to RNA
(D) Protein motif bind to DNA

128. Bacteria does not contain ....
(A) Exons (B) Ori
(C) Introns (D) Transposons

129. What is nucleoside?
(A) Sugar + phosphoric acid
(B) Sugar + Base
(C) Sugar + base + phosphoric acid
(D) Base + phosphoric acid

130. Gene cryIkills insect by ....
(A) Producing alkaline juice in mid gut of insect
(B) Producing poisonous compound which blocks excretory system
(C) Producing plasmalemma degrading enzymes
(D) Producing chitin degrading enzymes

Answer-Key

Q	Ans	Q	Ans	Q	Ans	Q	Ans	Q	Ans	Q	Ans	Q	Ans
1.	(C)	21.	(D)	41.	(C)	61.	(C)	81.	(A)	101.	(B)	121.	(D)
2.	(A)	22.	(C)	42.	(A)	62.	(B)	82.	(B)	102.	(C)	122.	(B)
3.	(C)	23.	(C)	43.	(A)	63.	(B)	83.	(D)	103.	(D)	123.	(C)
4.	(D)	24.	(B)	44.	(C)	64.	(A)	84.	(B)	104.	(A)	124.	(D)
5.	(B)	25.	(B)	45.	(B)	65.	(C)	85.	(D)	105.	(C)	125.	(A)
6.	(B)	26.	(A)	46.	(B)	66.	(D)	86.	(A)	106.	(A)	126.	(C)
7.	(B)	27.	(B)	47.	(A)	67.	(B)	87.	(C)	107.	(A)	127.	(D)
8.	(C)	28.	(B)	48.	(C)	68.	(C)	88.	(B)	108.	(B)	128.	(C)
9.	(A)	29.	(D)	49.	(B)	69.	(A)	89.	(C)	109.	(D)	129.	(B)
10.	(A)	30.	(A)	50.	(D)	70.	(C)	90.	(A)	110.	(C)	130.	(A)
11.	(C)	31.	(C)	51.	(C)	71.	(A)	91.	(A)	111.	(D)		
12.	(D)	32.	(B)	52.	(D)	72.	(C)	92.	(C)	112.	(B)		
13.	(A)	33.	(D)	53.	(B)	73.	(D)	93.	(A)	113.	(A)		
14.	(B)	34.	(C)	54.	(B)	74.	(A)	94.	(B)	114.	(A)		
15.	(C)	35.	(B)	55.	(B)	75.	(C)	95.	(B)	115.	(C)		
16.	(A)	36.	(D)	56.	(C)	76.	(B)	96.	(D)	116.	(D)		
17.	(D)	37.	(A)	57.	(B)	77.	(C)	97.	(C)	117.	(A)		
18.	(C)	38.	(B)	58.	(C)	78.	(D)	98.	(C)	118.	(B)		
19.	(B)	39.	(C)	59.	(B)	79.	(D)	99.	(B)	119.	(B)		
20.	(A)	40.	(B)	60.	(C)	80.	(D)	100.	(C)	120.	(C)		

ASRB-NET II — 2014

1. What is the function of enhancer during transcription?
   (A) Helps in binding between sigma factor and promoter
   (B) Decreases the activity of promoter
   (C) Increases the activity of promoter
   (D) Role in early termination of transcription

2. Factors effecting the diffusion through plasmalemma.
   (A) Temperature (B) Pressure
   (C) Size of molecules (D) Both a and c

3. Study of evolutionary relationship between species.
   (A) Morphological genetics
   (B) Natural genetics
   (C) Phylogenetics
   (D) Molecular genetics

4. Blue light produces in plants by ....
   (A) Phytochrome (B) Cryptochrome
   (C) Chromosome (D) Both a and b

5. A tissue culture media contains zeatin as .... And picloram as ....
   (A) Auxin and cytokinine
   (B) Cytokinine and auxin
   (C) Gibberellin and auxin
   (D) Cytokinine and Gibberellin

6. Barnas gene codes enzyme .... for male sterility.
   (A) DNase (B) Protease
   (C) Lipase (D) RNase

7. Where found enzymes for krebs cycle in mitochondria?
   (A) Thylakoids
   (B) Mitochondrial matrix
   (C) Inner membrane
   (D) Perplasmic space

8. Arrange in chronological order following discovery of techniques.
   I. PCR,
   II. DNA sequencing,
   III. Recombinant DNA technology,
   IV. Double helix DNA
   (A) IV, III, II, I (B) I, II, III, IV
   (C) IV, II, III, I (D) II, I, IV, III

9. Sanger discovered ....
   (A) PCR
   (B) DNA sequencing
   (C) Recombinant DNA technology
   (D) Double helix DNA

10. What is karyotype?
    (A) Arrangement of chromosome according to their size
    (B) Diagrammatic representation of chromosome according to their size
    (C) Arrangement of chromosome according to their shape and size
    (D) Diagrammatic representation of chromosome according to their shape and size

11. .... is not a translation inhibitor.
    (A) Chloramphenicol
    (B) Puromycin
    (C) Cyclohexamide
    (D) Oligomycine

12. Isolation of ethidium bromide fro plasmid DNA in clone is done by ....
    (A) n-butanol
    (B) Acetone
    (C) Chromatography
    (D) Lysozyme

13. First patented transgenic organism.
    (A) Mice with mik gene
    (B) Dolly
    (C) *E. coli*
    (D) Rozy cow

14. 3-D structure of proteins stabilized by
    (A) Covalent and non-covalent bonds
    (B) Hydrophobic bonds
    (C) Ionic bonds
    (D) Van der wall bonds

15. Cofactor for pyruvate kinase enzyme.
    (A) Potassium ion (B) Sodium ion
    (C) Zinc ion (D) Iron ion

16. What is holoenzyme?
    (A) Coenzyme + cofactor
    (B) Apoenzyme + cofactor
    (C) Metol ion + coenzyme
    (D) Cofactor + coenzyme + metal ion

17. What are the components of ribosomes?
(A) RNA and DNA
(B) DNA, RNA and protein
(C) RNA and proteins
(D) DNA and proteins

18. What is RNA molecule with enzymatic activity?
(A) Lysozyme (B) Polysome
(C) Zymogene (D) Ribozyme

19. Jute firbres are ....
(A) Primary xylem
(B) Secondary xylem
(C) Primary phloem
(D) Secondary phloem

20. What is called transposable element?
(A) Jumping gene (B) Junk DNA
(C) Selfish DNA (D) Recon

21. Reverse transcription discovered by
(A) Jacob and Monad
(B) T.H. Morgan
(C) Temin and Batimore
(D) Baltimore and Baltimore

22. Diagrammatic representation of chromosome according to their size.
(A) Karyotyes (B) Idiogram
(C) Karyogram (D) Both (B) and (C)

23. Secretory pathway of any produces protein in cell
(A) RER to Golgi vesicles to golgi cisternae to cell membrane
(B) Golgi cisternae to golgi vesicles to RER to cell membrane
(C) Golgi vesicles to golgi cisternae to SER to cell membrane
(D) RER to Golgi cisternae to golgi to SER to cell membrane

24. Which does not require substance for activity?
(A) Beta glucoronidase
(B) Luciferase
(C) Beta galactosidase
(D) GFP

25. Method for producing multiple copy of DNA
(A) Gene tagging
(B) Gene sequencing
(C) Gene amplification
(D) Gene mapping

26. Enzyme used in PCR for genomic DNA amplification
(A) Taq RNA polymerase
(B) Taq DNA polymerase
(C) Reverse transcriptase
(D) Restriction endonuclease

27. Which is auxin?
(A) 2,4-D (B) Picloram
(C) Zeatin (D) Both a and b

28. High crossing over frequency occurs when two gene located ....
(A) Near to centromere on single chromosome
(B) Away from centromere single chromosome
(C) Near on different homologous chromosome
(D) Near to telomere

29. Which is used for cryopreservation for long time storage?
(A) Liquid N_2 (B) Solid CO_2
(C) Liquid O_2 (D) Freon

30. Which are cryoprotectants?
(A) DMSO (B) Proline
(C) Glycerol (D) All of the above

31. Somatic embryo encapsulated in artificial seed by ....
(A) Sodium alginate (B) Agarose
(C) Agar (D) Calcium nitrate

32. Reactive oxygen species
(A) H_2O_2
(B) Singlet oxygen
(C) O_2
(D) Both (A) and (B)

33. From which O_2 released in photosynthesis?
(A) CO_2 (B) H_2O
(C) Sun light (D) Chlorophyll

34. Feedback inhibition involves ....
(A) Competitive inhibition
(B) High K_m
(C) Allosteric enzymes
(D) Low V_{max}

35. What is K_m
(A) Substrate concentration at half the maximum velocity
(B) Maximum velocity at a time
(C) Half of the maximum velocity at a time
(D) Amount of the enzyme at half the maximum velocity

36. ATP produced by one moleculae of glucose in glycolysis.
(A) 4 (B) 10
(C) 12 (D) 8

37. Golden rice developed by
(A) Thomas Cech
(B) InogoPotrykus
(C) Mariani and coworkers
(D) Cocking and Coworkers

38. Glyphosate related to
(A) EPSP (B) Tn5
(C) Cry (D) Cpt gene

39. Which type of amino acids inhibited in herbicide resistant crops?
(A) Aliphatic amino acids
(B) Sulphur containing amoni acids
(C) Proline and glycine
(D) Aromatic amino acids

40. Herbicide resistant gene group
(A) nptII, barnase and bar
(B) bar, als and chs
(C) TA29, nptII, lux
(D) Chs, baenase, bar

41. Gene cryI used to improve resistance for ....
(A) Coleoptera (B) Diptera
(C) Lepidoptera (D) Nematodes

42. nptII gene obtained from ....
(A) Seed storage tissue
(B) Tn5 bacterial transposon
(C) GFP
(D) Barnase

43. .... Is used in microarray.
(A) Carbon chip
(B) Silicon chip
(C) GFP
(D) Sybr green

44. Technique used for detection of probe and target DNA.
(A) Southern blotting (B) Microarray
(C) PCR (D) PAGE

45. Jatropha is used for the production of
(A) Morphine (B) Biodiesel
(C) Antibiotics
(D) Monoclonal antibodies

46. Hairy root culture characteristics of Agrobacterium rhizogenes.
(A) Absent geotropism and lateral branching
(B) Produce crown gall disease with hairy roots
(C) Formation of proliferative main roots without branching at the site of infection
(D) All of the above

47. Similarity between SDS-PAGE and isoelectric point analysis?
(A) Temperature dependent
(B) pH dependent
(C) Ruining proteins from positive to negative in both analysis
(D) Used for protein digestion and sequencing

48. AFLP resolved on
(A) 1% Agarose (B) 4% PAGE
(C) 10% PAGE (D) 2% PAGE

49. What is the use of GISH?
(A) QTL analysis
(B) Cytogenetic study
(C) Genome analysis
(D) Linkage mapping

50. Which is co-dominant marker?
(A) RAPD (B) ISSR
(C) SSR (D) AFLP

51. What is cybrid?
(A) A cell contains nuclear material from one parent and cytoplasm from both parents
(B) A hybrid cell contains cytoplasm from one parent and nucleus from both parents
(C) A hybrid cell contains nucleus and cytoplasm from both parents
(D) A cell contains recombinant exotic nucleus and cytoplasm from both parents

52. Somaclonal variations produced due to ....

(A) Recombination during meiosis and mutation
(B) Only mutation
(C) Mitosis and meiosis
(D) Due to artificial mutations

53. Guha and Maheshwari developed haploid plants in ....

(A) Arabidopsis (B) Datura
(C) Eucalyptus (D) Banana

54. Haploid plant produced by ....

(A) Meristem culture (B) Embryo culture
(C) Anther culture (D) Artificial seed

55. Third generation biodiesel

(A) Arabidopsis (B) Eucalyptus
(C) Jatropha (D) Cotton

56. Electroporation is method of gene ....

(A) Tagging (B) Transfer
(C) Mapping (D) Sequencing

57. What is called the total RNA complements of a cell?

(A) Genome
(B) Proteome
(C) Transcriptome
(D) None of the above

58. What is the use of FISH?

(A) QTL analysis
(B) Cytogenetic study
(C) Genome analysis
(D) Linkage mapping

59. Study of 3-D structure protein is done in

(A) Functional genomics
(B) Structural genomics
(C) Functional transcriptomics
(D) Structural and functional genomics

60. Which is used to detect DNA sequence polymorphism?

(A) Microarray (B) SDS-PAGE
(C) RFLP and AFLP (D) Zymograme

61. Phenolic compounds in plants work as ....

(A) Cryoprotectants (B) Insect resistant
(C) Male sterility (D) Cryopreservation

62. .... removes phosphate from 5' end of DNA.

(A) Terminal transferase
(B) T4 nucleotide kinase
(C) Restriction endonuclease
(D) Alkaline phosphatase

63. What the collection of cDNA cloned bacteria and phages?

(A) Genomic library
(B) cDNA library
(C) Microarray
(D) Operon library

64. Operon of amino acid synthesis is ....

(A) Inducible
(B) Repressible
(C) Both
(D) None of the above

65. Tail PCR amplified .... of gene.

(A) Border sequences
(B) Anchor region
(C) Only cDNA
(D) Coding sequences

66. Which is used in RT-PCR?

(A) Ethidium bromide
(B) Xylene cyanol
(C) Bromophenol
(D) Sybr green

67. The sequence at both sides of coding regions of gene is called ....

(A) Flanked sequences
(B) Transposon sequences
(C) Selfish sequences
(D) Junk sequences

68. LEA is related to ....

(A) Protection of plants from insects
(B) Protection of plants under water stress
(C) Protection of plants from weeds
(D) Protection of plants from cold

69. Which method is used to transfer of transgene in protoplast?

(A) Biolistic gene gun
(B) Liposome
(C) Electroporation
(D) Microinjection

70. Protoplast fusion is done by
(A) Electroporation
(B) High pH and High Ca^{2+}
(C) Conventional hybridization between plants
(D) Antisense RNA technology

71. Fungus resistance developed by ....
(A) Chitinase and Glucanase
(B) Chitinase
(C) EPSP
(D) Tn5

72. In lac operon, structural gene 'A' encodes ....
(A) β-galactosidase (B) Permease
(C) Acetyltransferase (D) Catalase

73. What is not called the copy of mRNA from DNA?
(A) Matured RNA
(B) Primary RNA transcript
(C) Transcript RNA
(D) Pre-RNA

74. DNA sequence used to produce mRNA is called ....
(A) Transcription bubble
(B) Transcriptional unit
(C) Pre-transcript
(D) *OriC*

75. Quinolone used as ....
(A) Replication inhibitor
(B) Transcription inhibitor
(C) Translation inhibitor
(D) PCR inhibitor

76. Enzyme used for the synthesis of mRNA and snRNA.
(A) RNA polymerase I
(B) RNA polymerase III
(C) RNA polymerase II
(D) RNA polymerase IV

77. Mixture of water and alcohol is 55.5 ml. it contains the ratio 15:85 water:alcohol. What is quantity of water in this mixture?
(A) 10ml
(B) Approximate 8.4ml
(C) Approximate 7.5ml
(D) 9.0ml

78. Error rate of replication in *E. coli* is ....
(A) 10^{-10} (B) 10^{-8}
(C) 10^{8} (D) 10^{-10}

79. Formula of deoxyribose
(A) $C_5H_{10}O_5$ (B) $C_6H_{10}O_5$
(C) $C_5H_{10}O_4$ (D) $C_6H_6O_6$

80. Which nucleotide is not labeled in gene sequencing?
(A) Thymine (B) Guanine
(C) Cytosine (D) Adenine

81. A haploid cell contains 12 chromosomes. What is the number of chromosomes in trisomic cell?
(A) 25 (B) 24
(C) 23 (D) 26

82. One cell contains three alleles of a gene. What is the formula according to Hardy–Weinberg?
(A) $p+q+r \neq 1$
(B) $(p+q)^2 = 1$
(C) $p^2 + 2pq + q^2 = 1$
(D) $p + q + r = 1$

83. Initiation of replication in eukaryotes is done by ....
(A) DNA polymerase β
(B) DNA polymerase δ
(C) DNA polymerase α
(D) DNA polymerase γ

84. In which process of translation, high energy phosphate is not used?
(A) When the tRNA charged with N-formylmethionine
(B) When 30S subunit binds to the mRNA template at a purine-rich region
(C) When fMet-tRNA enters the P site
(D) When termination codon moves into the A site

85. What is the function of TFIID in transcription?
(A) Stabilizes TBP and TAF binding
(B) Recruitment of RNA pol II
(C) Recognizes TATA box
(D) Influences selection of the start point for transcription

86. DNA-protein interaction detected by ....

(A) Microarray
(B) DNA fingerprinting
(C) Footprinting
(D) PCR

87. How many enzymes involve in glycolysis?

(A) 10 (B) 11
(C) 9 (D) 12

88. Basic unit of starch?

(A) α-D-glucose and β-D-glucose
(B) α-D-glucose
(C) β-D-glucose
(D) α-D-glucose and β-D-fructose

89. Enzymatic reaction depend on ...., after mixing enzyme and substrate.

(A) Enzyme concentration
(B) Substrate concentration
(C) First enzyme concentrationand then substrateconcentration
(D) Both a and b

90. During fermentation under aerobic condition, how many ATP are produced?

(A) 4 (B) 2
(C) 10 (D) 12

Answer-Key

Q	Ans	Q	Ans	Q	Ans	Q	Ans	Q	Ans	Q	Ans
1.	**(C)**	16.	**(B)**	31.	**(A)**	46.	**(A)**	61.	**(B)**	76.	**(C)**
2.	**(D)**	17.	**(C)**	32.	**(D)**	47.	**(B)**	62.	**(D)**	77.	**(B)**
3.	**(C)**	18.	**(D)**	33.	**(B)**	48.	**(B)**	63.	**(B)**	78.	**(B)**
4.	**(D)**	19.	**(D)**	34.	**(C)**	49.	**(C)**	64.	**(B)**	79.	**(C)**
5.	**(B)**	20.	**(A)**	35.	**(A)**	50.	**(C)**	65.	**(B)**	80.	**(A)**
6.	**(D)**	21.	**(C)**	36.	**(A)**	51.	**(A)**	66.	**(D)**	81.	**(A)**
7.	**(B)**	22.	**(D)**	37.	**(B)**	52.	**(A)**	67.	**(A)**	82.	**(D)**
8.	**(A)**	23.	**(A)**	38.	**(A)**	53.	**(B)**	68.	**(B)**	83.	**(C)**
9.	**(B)**	24.	**(D)**	39.	**(D)**	54.	**(C)**	69.	**(C)**	84.	**(A)**
10.	**(A)**	25.	**(C)**	40.	**(B)**	55.	**(C)**	70.	**(B)**	85.	**(C)**
11.	**(D)**	26.	**(B)**	41.	**(C)**	56.	**(B)**	71.	**(A)**	86.	**(C)**
12.	**(C)**	27.	**(D)**	42.	**(B)**	57.	**(C)**	72.	**(C)**	87.	**(A)**
13.	**(C)**	28.	**(A)**	43.	**(B)**	58.	**(B)**	73.	**(A)**	88.	**(B)**
14.	**(A)**	29.	**(A)**	44.	**(A)**	59.	**(B)**	74.	**(B)**	89.	**(D)**
15.	**(A)**	30.	**(D)**	45.	**(B)**	60.	**(C)**	75.	**(B)**	90.	**(B)**